FUNCTIONAL FOODS IN ENHANCING IMMUNE RESPONSE

FUNCTIONAL FOODS IN ENHANCING IMMUNE RESPONSE

Edited by

Dr. Pravin Udhavrao Ghatge

Delve PUBLISHING

www.delvepublishing.com

Functional Foods in Enhancing Immune Response

Dr. Pravin Udhavrao Ghatge

Delve Publishing
4164 Lakeshore Road
Burlington, ON L7L 1A4
Canada
www.delvepublishing.com
Email: orders@arclereducation.com

ISBN: 978-1-77956-440-5 (Hardcover)

ABOUT THE EDITOR

Dr. Pravin Udhavrao Ghatge has total of 16 years working experience in teaching, research and extension with specialization in food quality analysis, food instrumentation, food processing and product development along with food safety management. During his service he work in different capacities as senior research assistant, Assistant Professor and Scientist in the College of Food Technology. He has significant contribution in 06 recommendations accepted by Govt. of Maharashtra for the benefits of food processors, cottage entrepreneurs and farmers. He has successfully guided 04 PG students and co-guided 22 PG students and 32 UG students in the discipline food science and technology. Dr. Pravin Ghatge has many publications to his credit to mention : 43 Research Papers published in the International and National abstracted generals, 12 papers presented in seminar, 18 popular articles. Best poster presentation awards in 2014 and 2022 in the National Conferences. Significant member contribution in Choudhary Singh award for AICRP on arid legumes. Member contribution in Centre for Excellence in Quality awarded to AICRP arid legumes project in 2007 . Presently working as Chairman of Vasantrao Naik Marawada Krishi Vidyapeeth Employees Credits Society, Parbhani. Life member of association of Food Scientist and Technologist, Mysuru .

TABLE OF CONTENTS

LIST OF FIGURES

LIST OF TABLES

PREFACE

In recent years, the concept of *functional foods* has gained widespread recognition for its potential to improve health and well-being. Unlike conventional foods, functional foods are enhanced with bioactive compounds that offer additional health benefits beyond basic nutrition. These foods are now seen as a key strategy for enhancing immune response, preventing disease, and promoting long-term health. As the global community faces unprecedented challenges, particularly with emerging diseases and an aging population, the role of functional foods in supporting the immune system has never been more critical.

This book, *Functional Foods in Enhancing Immune Response*, explores the multifaceted relationship between functional foods and the immune system. Across 10 chapters, this book highlights the nutritional, biochemical, and physiological aspects of functional foods, providing readers with a comprehensive understanding of how these foods can optimize immune function and promote overall health.

Chapter 1 provides a foundational overview of the field. It introduces readers to the concept of functional foods, highlighting their origins, definitions, and the criteria that distinguish them from traditional food products. This chapter sets the stage by discussing the evolution of functional foods in the global market and how scientific advancements have led to the development of these specialized products. Additionally, it outlines the regulatory frameworks that govern the production and marketing of functional foods, giving readers a holistic view of the industry's landscape.

In Chapter 2, the focus shifts to the core connection between diet and immune health. This chapter examines how various nutrients, including vitamins, minerals, and antioxidants, directly affect immune function. It discusses the role of a balanced diet in maintaining a healthy immune system and explores the mechanisms through which certain nutrients bolster immune responses.

Chapter 3 offers an in-depth look at two key components of functional foods: *hydrocolloids* and *lipids*. These compounds have unique properties that contribute to both the structural integrity of food products and their nutritional value.

The gut is often referred to as the "second brain" due to its significant influence on overall health, particularly immune function. Chapter 4 explores the gut–immune connection in detail. This chapter emphasizes the importance of gut health in maintaining a robust immune system and explains how functional foods containing probiotics, prebiotics, and other bioactive ingredients can fortify the gut microbiome.

In Chapter 5, the discussion shifts to one of the leading causes of death worldwide—*coronary heart disease* (CHD). This chapter examines the role of functional foods in reducing the risk of CHD through mechanisms such as cholesterol regulation, anti-inflammatory effects, and improved cardiovascular function. It highlights specific bioactive compounds, such as omega-3 fatty acids and plant sterols, which have been shown to positively impact heart health.

Chapter 6 explores the potential of functional foods in *cancer prevention*. The chapter provides an overview of how bioactive ingredients, such as *polyphenols*, *flavonoids*, and *carotenoids*, can inhibit cancer cell growth, enhance detoxification pathways, and promote apoptosis. In addition, the chapter addresses how functional foods can support overall health promotion and reduce the risk of chronic diseases.

The targeted application of functional foods in *viral defense* is the focus of Chapter 7. This chapter is particularly relevant in the context of recent global health crises, where the immune system's ability to fight off viral infections has been paramount. The chapter explores specific functional foods and bioactive compounds that can enhance antiviral immunity, highlighting their potential to serve as immune fitness boosters in vulnerable populations.

Chapter 8 shifts attention to the *skin*, the body's largest organ and a key component of the immune system. This chapter discusses how plant-based diets rich in antioxidants, vitamins, and minerals can promote skin health and reduce the risk of inflammatory skin diseases, such as eczema and psoriasis. It provides practical insights into how functional foods can be utilized to enhance skin resilience and overall well-being.

In Chapter 9, the focus turns to the *elderly population*. Aging is often associated with a decline in immune function and an increased risk of conditions such as osteoporosis. This chapter explores functional foods that are high in antioxidants and chicken eggshell calcium, which can help strengthen bones and reduce the risk of osteoporosis in older adults. It also discusses how these foods can support immune function in aging populations.

Chapter 10 wraps up the book by exploring the future of *plant-based functional foods* as immune enhancers. This chapter looks at emerging research on plant-based ingredients that have the potential to be developed into powerful immune boosters. It also addresses the growing trend of plant-based diets and their implications for health and wellness.

Functional Foods in Enhancing Immune Response provides a comprehensive exploration of the role of functional foods in supporting and enhancing immune function. Each chapter offers unique insights into how these foods can be applied to specific health challenges, making this book a valuable resource for nutritionists, healthcare professionals, and anyone interested in the potential of food to promote better health.

CHAPTER 1

INTRODUCTION TO FUNCTIONAL FOODS

Contents

INTRODUCTION

A food that has extra ingredients added to it, usually for the purpose of preventing disease or promoting health, is referred to as a functional food. The phrase can also refer to features intentionally incorporated into already-existing edible plants, such as purple or gold potatoes with higher levels of carotenoid or anthocyanin, respectively. Functional foods often resemble conventional foods in appearance and can be included in a regular diet. They can also be made to have physiological benefits and/or lower the risk of chronic disease beyond simple nutritional needs.

Because they can be used as a dietary supplement and are a significant source of important nutrients, functional foods are important for human health. Naturally, a variety of fruits, vegetables, grains, seafood, dairy products, and meat products are thought to be the main sources of functional food; additionally, each of these foods can nourish the host body. Research has indicated that tea and chocolate are considered functional foods because they contain active ingredients. A balanced diet is important for a person's overall health since it can sustain the body's normal functions. It also plays a significant role in the consumer's life. Functional food has a positive impact on an organism's body; it can meet the body's basic needs, shield the host body from malnutrition-related illnesses, and flush the body of numerous dangerous substances. According to studies, a food product can be considered functional if it helps to raise the body's metabolic rate without impairing its regular processes or stabilizing the body's outward appearance. Functional foods are the source of energy for an individual's body in this modern era, enabling smooth body growth and development. Functional foods can modulate the immune system, which may reduce the risk of cancer, osteoporosis, obesity, and cardiovascular problems. This could have positive impacts on a person's overall health. Vitamins, fortified foods, fiber, minerals, peanuts, fruits, and grain seeds are a few instances of functional foods.

The body needs functional foods just like it does other necessities. The key to keeping the human body healthy and active is eating a balanced diet. The results of the current studies indicate that people's busy schedules and negligent behavior prevent them from eating a balanced diet. A well-balanced diet has the potential to yield substantial health benefits for individuals. Reduced use of functional foods and a false lifestyle can harm one's health and increase the risk of several diseases, such as type 2 diabetes, high blood pressure, and cardiovascular disease (CVD). An imbalanced diet could also be a factor in memory loss and poor psychological health.

1.1. FUNCTIONAL FOODS

Functional foods have become more and more well-liked in the wellness and health communities in recent years. Functional foods, also referred to as nutraceuticals, are extremely nutrient-dense and linked to several potent health advantages. They might, for instance, guard against illness, stop nutrient shortages, and encourage healthy growth and development.

Foods classified as functional have health benefits that go beyond simple nourishment. Functional food advocates claim that their products enhance overall health and lower the risk of illness.

Oatmeal is a common example of a functional food because it contains soluble fiber, which has the potential to lower cholesterol. Certain foods have been altered for health reasons. Orange juice fortified with calcium to support bone health is one example.

The claims that manufacturers are permitted to make regarding the nutritional value of functional foods and their impact on illness, wellness, or bodily functions are governed by the Food and Drug Administration. Make informed choices if you wish to try functional foods. Furthermore, bear in mind that although functional foods might support wellness, unhealthy eating practices cannot be corrected by them.

1.1.1. Examples of Functional Foods

Conventional and modified foods are the two main classifications for functional foods.

Natural, whole-food components that are abundant in vital nutrients like vitamins, minerals, antioxidants, and heart-healthy fats make up conventional foods.

In the meantime, to boost a food's health benefits, modified foods have been fortified with extra ingredients like vitamins, minerals, probiotics, or fiber.

The following are some instances of traditional functional foods:

- **Fruits:** oranges, bananas, pears, peaches, apples, kiwis, and berries.
- **Veggies:** spinach, kale, broccoli, cauliflower, and zucchini.
- **Nuts:** almonds, cashews, pistachios, macadamia nuts, Brazil nuts.
- **Seeds:** chia, flax, hemp, and pumpkin seeds.
- **Legumes:** lentils, navy beans, chickpeas, and black beans.
- **Whole Grains:** couscous, brown rice, barley, buckwheat, and oats.
- Seafood includes cod, mackerel, anchovies, sardines, and salmon.
- **Fermented Foods:** tempeh, kombucha, kimchi, kefir, sauerkraut

- **Spices and Herbs:** cayenne pepper, ginger, cinnamon, and turmeric.
- **Drinks:** black tea, green tea, and coffee.

Here are some examples of modified functional foods:

- fortified juices
- fortified dairy products, such as milk and yogurt
- fortified milk alternatives, such as almond, rice, coconut, and cashew milk
- fortified grains, such as bread and pasta
- fortified cereal and granola
- fortified eggs

1.1.2. Potential Benefits

Functional foods have many possible health advantages.

1.1.2.1. Prevent Nutrient Deficiencies

Important nutrients including vitamins, minerals, fiber, and healthy fats are often abundant in functional foods.

Consuming a diverse range of functional foods, such as fortified and conventional foods, can help prevent nutrient deficiencies and guarantee that you get the nutrients as you need.

In fact, the prevalence of nutrient deficiencies has become much less common worldwide since the advent of fortified foods.

For example, rates of iron deficiency anemia in children were almost halved in Jordan following the introduction of wheat flour fortified with iron.

In addition to rickets, goiter and birth defects, fortification has also been used to prevent other conditions brought on by nutrient deficiencies.

1.1.2.2. Protect Against Disease

Functional foods offer vital nutrients that can aid in disease prevention.

Antioxidants are particularly abundant in many. By neutralizing dangerous substances known as free radicals, these molecules aid in the prevention of cell damage and several chronic illnesses, such as diabetes, cancer, and heart disease.

Omega-3 fatty acids, a good kind of fat that has been demonstrated to lower inflammation, improve brain function, and support heart health, are also abundant in some functional foods.

Some varieties have high fiber content, which can help improve blood sugar regulation and guard against diseases like diabetes, obesity, heart disease, and stroke. Additionally, fiber may help avoid digestive issues like diverticulitis, hemorrhoids, stomach ulcers, and acid reflux.

1.1.2.3. Promote Proper Growth and Development

For infants and children to grow and develop properly, a few nutrients are necessary.

Ensuring the nutritional needs are met can be facilitated by including a variety of nutrient-rich functional foods in a balanced diet. It may also be advantageous to incorporate foods that have been fortified with specific nutrients that are crucial for development and growth.

Cereals, grains, and flours, for instance, are frequently fortified with B vitamins, such as folic acid, which is critical for the development of the fetus.

Neural tube defects can impact the brain, spinal cord, or spine and are more common in cases of low folic acid. An estimated 50–70% reduction in the prevalence of neural tube defects could be achieved by increasing folic acid consumption.

Other elements that are frequently present in functional foods, such as calcium, iron, zinc, omega-3 fatty acids, and vitamin B12, are also important for growth and development.

1.1.3. Functional Food Consumption

Functional foods are not necessary for a healthy diet and do not by themselves treat or prevent diseases. For functional foods to fully realize their potential, a balanced diet is necessary. Their inclusion in the diet may meet the following criteria:

- To include functional foods in the diet, which provide advantages based on each person's unique physiological condition (e.g., pregnancy, lactation, physical activity, etc.).
- Eat functional foods if you have an intolerance to any specific food (like fish, nuts, or milk), have suboptimal intakes of some foods (like fruits and vegetables), or have low intakes of some foods (like fish, milk, and dairy products). In these situations, adding foods high in omega-3 fatty acids, vitamins, or calcium to milk and dairy products, or taking probiotics, may help to meet each person's specific nutritional needs.
- Certain nutrients may become more necessary in chronic illnesses

(osteoporosis, cardiovascular disease (CVD), obesity, diabetes, metabolic syndrome, etc.), and certain bioactive ingredients found in functional foods may have positive health effects.

- The final point to consider is the potential health benefits or preventive roles of functional foods, which when incorporated into a balanced diet, may help achieve optimal nutrition. However, this would necessitate a significant amount of additional research with carefully monitored methodology.

The obvious connection between our health and the food we eat has long been recognized. The idea of functional foods, as we know it today, emerged from the gradual realization that eating a diet rich in nutrients leads to a healthy diet and from the discovery of the mechanisms through which food affects metabolism and overall health. Our primary goals when we eat are to meet our body's nutritional needs and fulfill our metabolic demands. Nonetheless, certain food groups have additional health benefits in addition to their nutritional qualities. These foods are referred to as "functional foods," which are any foods that, in addition to being nutrient-dense, have a positive effect on a person's physical or mental well-being.

Other additional considerations have been proposed to define a functional food, including three additional conditions in particular:

- It is a food (not a capsule, tablet, or powder) derived from natural ingredients;
- It can and should be consumed as part of the daily diet;
- It has a particular function when eaten, serving to regulate a particular body process, such as:
- Enhancement of biological defense mechanisms
- Prevention of specific diseases
- Recovery from specific diseases
- Control of physical and mental disorders
- Slowing of the aging process.

It is also crucial to keep in mind that ongoing research is necessary to learn more about the functional qualities of food by studying how it affects biomarkers, as a food item can only be categorized as functional after its potentially positive impact has been demonstrated by carefully planned and carried out intervention studies in humans. For instance, it has been reported that plasma triacylglycerol concentrations serve as a functional indicator of n-3 polyunsaturated fatty acid (PUFA) because n-3 PUFA exerts a consistent, dose-dependent, and persistent hypotriacylglycerolemic effect; or that oligofructose and inulin alter several biomarkers, such as colonic microflora, gastrointestinal transit time, and experimentally induced neoplasia, suggesting that these nondigestible carbohydrates are naturally occurring dietary components that may enhance both human and animal health and resistance to disease.

1.2. PROBIOTICS IN FUNCTIONAL FOODS

Food is no longer thought of as a necessary component of nutrition. Most people believe that a person's health is directly impacted by their diet. Because of this, the presence of active ingredients that may alter to benefit the body is already one of the consumer's requirements for the quality of their food. Because of this, adding value-added food functionality in addition to maintaining food quality throughout the supply chain is crucial for the successful marketing and uptake of novel foods. Foods are therefore no longer only thought to be able to satisfy hunger; they also have the potential to enhance mental and physical well-being and prevent nutritional disorders in consumers. As the population ages and life expectancies rise, so do healthcare costs, which has led to a growing demand for so-called functional foods from health-conscious consumers (Li et al., 2017).

The term "functional" is frequently used to describe food items that are thought to offer specific physiological advantages or to be beneficial to the human body in some other way. Unlike pills or capsules, functional foods need to show that they are healthy when eaten in a well-balanced diet. However, there is not a single, accepted definition of what makes a functional food. Moreover, this term lacks a formal definition, and conventional and functional foods are distinguished in most nations. Certain functional foods are created by adding chemicals to food or changing the way it is processed. When it comes to bioactive ingredients, a product's characteristics can be specifically designed to improve overall health or treat a specific ailment.

The inclusion of various phytochemicals, such as antioxidants, vitamins, fibers, minerals, omega-3 fatty acids, etc., may provide the food's functional component. It might also be live "good" microorganism cells. These are referred to as probiotics. More accurately, they are defined as live bacteria that, in sufficient amounts, can offer health benefits by preserving the microbiological balance in the human gastrointestinal tract (GIT), averting illnesses, and enhancing the host immune system. However, the live probiotic count in most commercial probiotic meals varies, and their label claims regarding strain type are frequently not met. The health benefits of ingesting probiotic microorganisms differ depending on the strain or species, but they nevertheless offer benefits to the host. Most of the advantages associated with probiotics pertain to the maintenance of healthy intestinal flora, which exhibits the capacity to survive and proliferate in the intestinal region in addition to being able to pass through the upper GIT.

Adopting consistent standards for classifying microorganisms as probiotics is necessary due to their increasing use in food formulations. There needs to be general agreement on terminology among scientists, food manufacturers, regulators, and consumers. To classify a particular strain as a probiotic, a few procedures need to be followed. The candidate needs to be completely described and safe to eat. It must be alive during the production and storage of the product, and it must have at least one known positive effect on human health. Under the Qualified Presumption of Safety (QPS) concept, the European Food Safety Authority (EFSA) has been keeping a list of species presumed safe for human consumption in foods since 2007. Before being introduced to the market in the United States, a strain in Europe must adhere to the Novel Food regulation if it does not belong to a QPS species. The phrase "generally recognized as safe" was adopted by the Food and Drug Administration (GRAS). Only when a substance's general recognition of safety is supported by the opinions of professionals who are qualified to assess the substance's safety can it be classified as GRAS. It should be noted that only encapsulating materials that have been approved for human consumption may be used in the food industry when it comes to probiotics that are encapsulated (Arenas-Jal et al., 2020).

The probiotic microorganisms that have been studied and used the most are yeasts, lactic acid bacteria, and bifidobacteria. There are also strains of *Enterococcus* and *Streptococcus* in use. Probiotics can be found in a variety of food items, such as bread products, dairy products, meat, drinks, cereals, fruits, and vegetables. Probiotics must be kept at high levels of viability and activity, though, as these are crucial to their efficient operation. Probiotic viability is significantly reduced when exposed to harsh environmental conditions, such as food processing and storage. However, there is currently no consensus regarding the safety profile of using live microorganisms, and the issue is still delicate. In this regard, more trustworthy *in vivo* research is required. Thus far, research has indicated a strong correlation between *in vitro* and *in vivo* assays concerning strain specificity when predicting the immunomodulatory characteristics of lactic acid bacteria. However, research on maintaining the cell viability through the gastrointestinal system is needed concurrently.

Many technologies have been developed in the quest for suitable ways to preserve probiotic cells, the most important of which is microencapsulation. Numerous studies have documented the process of probiotic microencapsulation, which has garnered significant interest in encapsulation techniques employing a range of procedures, such as emulsification, extrusion, and spray drying. Nonetheless, a thorough evaluation of how every substance and environment affects the probiotic bacteria is necessary.

1.2.1. Probiotics in Food Products

Functional foods—especially those with probiotics or bioactive chemicals—have been created in response to the growing demand for a diet higher in nutrients. Probiotics have health benefits for the person consuming them when they are taken in sufficient quantities. Numerous advantageous impacts of these microorganisms have been identified; the majority are linked to immune system function and gastrointestinal health, but they also have positive effects on diabetes, obesity, hypercholesterolemia, and cancer, to mention a few. The development of innovative probiotic-rich food products and in-depth study on specific probiotic qualities and their effects on human health are the primary areas of interest for researchers and many food industry sectors interested in the global probiotic market. The American Gastroenterological Association's official report on the role of probiotics in the management of gastrointestinal disorders found that there are significant knowledge gaps, even though the aforementioned nearly meets the statute of postulate.

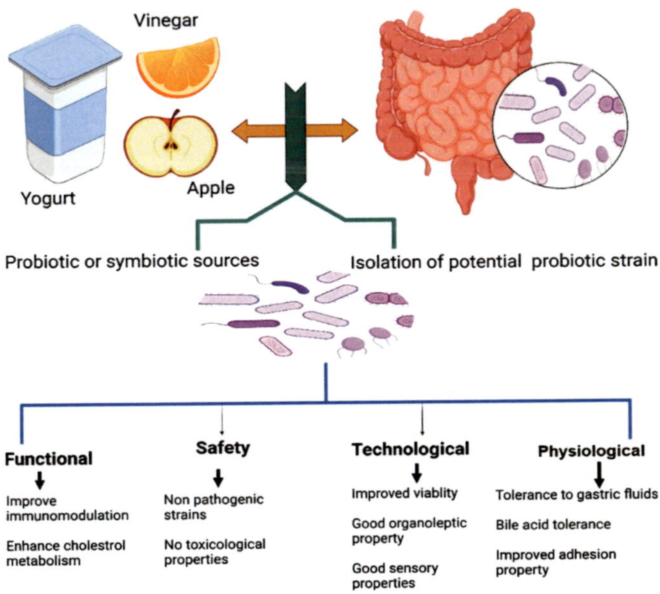

Figure 1.1. *Basic criteria for selecting suitable probiotic strains for specific applications.*

Source: Palanivelu J, Thanigaivel S, Vickram S, Dey N, Mihaylova D, Desseva I. (2022). Probiotics in functional foods: Survival assessment and approaches for improved viability. Applied Sciences; 12(1):455. https://doi.org/10.3390/app12010455.

The viability of probiotics in food is necessary to obtain the health benefits linked to them. Probiotic cultures are generally only beneficial to the host's health when determined by animal or clinical research. Therefore, the most practical quality assurance test is the viability test for probiotics. In most clinical studies, the presence of probiotics over viable cells daily has been associated with health benefits. Considering future developments, a recent study explored how food supplements, such as probiotics, could overcome current technological barriers and improve functionality and sensory acceptability. The fundamental standards for choosing probiotic strains that are appropriate for a given application are displayed in Figure 1.1. These criteria include technological, physiological, functional, and safe aspects. The primary sources may come from intestinal origin or from probiotic or symbiotic products.

1.2.1.1. Selection and Identification of Probiotic Strains

New probiotic strains must be identified, assessed, and characterized using a comprehensive, multi-step process. Probiotics are primarily chosen based on a variety of functional characteristics, including the capacity to deconjugate bile salts, bile and phenol toxicity, and tolerance to gastric acidity. Furthermore, the qualities of the strain should be taken into consideration when choosing new probiotics. Therefore, to choose and describe probiotic strains, new techniques are required. Molecular techniques are particularly significant in this regard. Probiotics' bioactive ingredients may help keep food from spoiling, boost its nutritional content, enhance its aroma and other sensory aspects, and detoxify it. Different species and strains have different bioactive synthesis processes. For instance, there are substantial differences in the synthesis of bile salt hydrolase (BSH) and exopolysaccharides (EPS) between species and strains within a single species. It is impossible to assume that all lactobacilli, or even all strains of a given species, produce the same amount of a given metabolite based on the results that are currently available. Determining functionality is the first step towards ensuring it.

1.2.1.2. Functionality of Probiotics

Numerous academic studies show that inactivated probiotic cells are beneficial to the digestive system. Piqué et al. reviewed the literature and compiled the most recent research on the use of heat-treated probiotic cells, cell-free supernatants, or purified key components as a source of primarily immunomodulatory effects, enteropathogen protection, and intestinal barrier integrity maintenance.

Most scientific publications continue to explore the benefits of viable probiotic cells in the GIT, alongside ongoing discussions about the safety of consuming live probiotics. Up until now, the direct effects of gastrointestinal disorders have received Most of attention. It is commonly reported that *Lactobacillus rhamnosus* lowers the risk of antibiotic-associated diarrhea in children. Significant anti-inflammatory potential was demonstrated by the strains of *Lactobacillus bulgaricus*, *Lactobacillus plantarum*, and *Streptococcus thermophilus*. Furthermore, probiotic bacteria have been shown to prevent radiation-induced diarrhea in cancer patients receiving chemotherapy and radiation therapy, according to a meta-analysis on the topic. Numerous studies concentrate on the potential advantages of probiotic treatment for conditions like skin disorders, cardiometabolic disorders, chronic kidney disease, and autism spectrum disorder. There is hope for new uses for these microorganisms as psychobiotics due to preliminary research on the relationship between probiotics and cognitive and behavioral processes (neurological, metabolic, hormonal, and immune signaling pathways).

1.2.1.3. Survival Assessment and Viability

When discussing probiotic survival, two key factors must be considered. The first is whether the food manufacturing process will not destroy the selective strains if they are initially able to survive through the GIT. It is not unexpected that milk and dairy products are now regarded as the primary functional food since Most of probiotics used in commercial applications were extracted from these foods. The two most crucial factors in choosing the probiotic strains are their high concentrations of bile salt and low tolerance to gastric pH. However, the conditions of subsequent processing and the selection of the food matrix are critical to their survival. Making sure probiotic food products have the minimum number of probiotics (10^6–10^7 CFU/mL product) in the finished product is also crucial. The number of viable and active probiotic cells per gram or milliliter of probiotic food products at the time of consumption determines their health effects, so this level of probiotics should be maintained throughout the product's shelf life. Probiotics must therefore adhere to strict technological requirements to continue functioning during the whole production and storage process. Because they are regarded as GRAS, members of the genera Lactobacillus and *Bifidobacterium* have a long history of use in food preparation.

The amount of active probiotic bacteria decreases by a fold factor of 10–100 or more during the manufacturing or storage process. The oxygen content, heating temperature, acidity, water activity, osmotic stress, moisture content, and packing material are all factors that affect the long-term survival of probiotics in meals. The most important factor affecting the functional food's effectiveness is the quantity of live, active cells per gram or milliliter of probiotic food products at the time of consumption. There are two general methods for determining how viable each individual bacterial cell is in food. The culture-dependent method is still the gold standard for figuring out how many colony-forming units (cfu/g) there are in one gram of original material. On the other hand, techniques based on DNA and flow cytometry are continuously enhanced to produce trustworthy and culture-independent screening methods.

The primary determinants of probiotic viability during production and ingestion are enumerated in Figure 1.2. The interactions between food ingredients such as sugars, salt, and antimicrobial agents can have a substantial impact on the viability of probiotics during processing. These are the primary requirements, along with the moisture content and acidity. The processing factors are also important because they typically involve different temperature regimes and the application of different mechanical forces. The latter is among the most important constraints on probiotic bacteria's ability to survive. Once more, the type of packaging material used, its acidity and moisture content, all matter during storage. The environmental conditions of the various GIT compartments are applied after the functional food has been consumed.

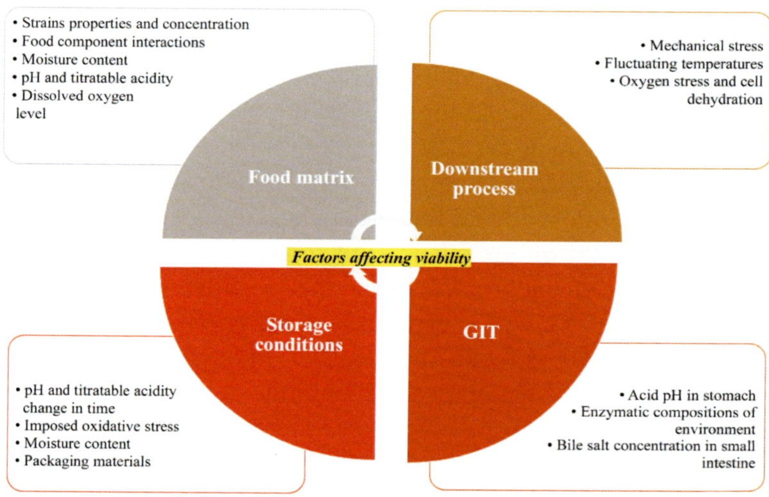

Figure 1.2. *Factors influencing the viability of probiotics.*

Source: Palanivelu J, Thanigaivel S, Vickram S, Dey N, Mihaylova D, Desseva I. (2022). Probiotics in functional foods: Survival assessment and approaches for improved viability. Applied Sciences, 12(1):455. https://doi.org/10.3390/app12010455.

1.2.1.4. Role of Food Ingredients and Additives

The growth and survival of probiotic bacteria are strongly influenced by how well they mesh with food products. Depending on the component, probiotics may have positive, neutral, or negative effects. Additionally, food additives like sugars, salts, sweeteners, aroma compounds, and preservatives have a big impact on how long probiotics survive. It has been demonstrated that certain strains of lactic acid bacteria interact with aroma compounds during the production of cheese, modifying the final product's flavor. The development and survival of probiotics may be adversely affected by food ingredients and compounds. For instance, polyphenolic compounds, which are widely found in foods derived from plants, are a significant class of bioactive compounds. They are often linked to health advantages and have antioxidant qualities. However, research shows that they prevent lactic acid bacteria from growing.

A class of substances has been shown to promote the development of probiotic cultures. The term "a substrate that is selectively utilized by host microorganisms conferring a health benefit" refers to what are known as prebiotics. These are food ingredients that probiotic bacteria metabolize instead of the body doing the digestion. Probiotics are thought to grow poorly or not at all in diets other than fermented milk products because they lack adequate glycolytic and proteolytic activity and have a high requirement for specific nutrients like B-group vitamins and non-protein nitrogen. Oligosaccharides are an excellent illustration of prebiotic substances that promote the growth of bifidobacteria and lactobacilli.

1.2.2. Progress to Enhance Viability

Several methods have been employed to increase the viability of probiotics. Using these functional ingredients to create stable products with active properties that hold true during production, storage, and consumption is a major challenge for the food industry.

The process of encapsulation involves the confinement of an active chemical by a wall material with a size ranging from nano to micro to millimeters. Micro- and nano-encapsulation are quickly becoming available to address possible bioavailability problems, providing special answers to the challenges faced by the food industry in the process. In general, colloidal particles with diameters between 10 and 1000 nm are known as nanoparticles and those with dimensions between 3 and 800 nm are known as microparticles.

Microencapsulation of probiotic cells, also called probiotic cell encapsulation, is the process of encasing probiotic cells in tiny sealed capsules to keep them isolated from their surroundings. Under the influence of different processing and environmental variables in the gut media, the capsules release their contents over an extended duration at a controlled rate. It appears to be the best available strategy for defending against microorganisms. However, the process of microencapsulating probiotics takes time. Many methods have been developed for microencapsulating probiotics or other beneficial compounds.

Most of these processes involve coating probiotics in emulsions, prebiotics with dry impacting and enteric coatings on the surface of food products, or encasing them in polymers like starch, alginate, and carrageenan. Probiotics are dried as part of encapsulation processes that have been thoroughly researched and used. It consists of vacuum, freeze, and spray drying. Moreover, gel microspheres are employed. One promising method that enables mild encapsulation at the nanoscale is electrospinning.

1.3. THE SCIENCE OF FUNCTIONAL FOOD

Functional foods must meet all safety standards outlined in the current food regulations because they are food. However, in many instances, to evaluate the risks associated with functional foods, new theories and methods will need to be created and proven. Certain functional foods in Europe, though by no means all of them, will be categorized as "novel foods," necessitating the safety decision tree evaluation outlined in the EU Novel Food Regulation.

It is important to note, though, that neither the nutritional value nor the physiological effects of these novel foods are covered by this regulation. It is unquestionably a safety requirement. Any development of a functional food must first satisfy the safety requirement. In fact, functional foods do not fall under the risk versus benefit paradigm—a concept that pharmacologists who are creating new drugs are familiar with—unless there are extremely specific circumstances for reducing the risk of a disease for which there is a lot of compelling scientific data. According to the European consensus document, the creation of functional foods is a major problem and a scientific challenge that should be based on fundamental research on the target functions and how food ingredients may affect them.

Local factors would need to be considered when implementing a food-based approach, as functional foods are not universally accepted. On the other hand, functional food is universally approached from a scientific standpoint. The scientific basis serves as the cornerstone of the function-driven approach, which aims to provide a more comprehensive understanding of the relationships between nutrition and health. Next, rather than focusing only on lowering the risk of disease, emphasis is placed on the significance of how food components affect specific, well-defined target bodily functions that are related to well-being and health-related concerns.

Drawing from fundamental concepts in nutrition and associated biological sciences, this kind of development necessitates the identification and, to some extent, comprehension of the mechanism(s) through which a prospective functional food or food component can modify the target function(s) that are/are acknowledged or demonstrated to be significant to the state of health and well-being and/or the mitigation of disease risk. If available, epidemiological data indicating a biologically meaningful and statistically supported association between the consumption of a given food component and a specific health benefit will be highly beneficial. A new generation of human nutrition studies aimed at demonstrating that relevant (in terms of dose, frequency, duration, etc.) intake of the specified food will be associated with improvements in one or more target functions, either directly or indirectly in terms of a valid marker of an improved state of well-being and health and/or reduced disease risk, will conclude that first step with the demonstration of a functional effect. A recommendation for new or improved dietary guidelines may come from the conclusion if it is well-supported by solid scientific evidence.

Although the next generation of human nutrition studies ought to be hypothesis-driven, they will frequently diverge significantly from what are traditionally known as clinical studies. The primary distinctions are that, while nutrition studies may focus on large, at-risk target groups or the general population, they are not symptom- or diagnostic-based, and they are not designed to assess a risk versus benefit approach. Instead, they test the effects of foods as part of a regular diet. Most of these studies will use changes in relevant and validated markers to show that consuming the potentially functional food over an extended Timeframe positively modulates target functions. Generally, a (double) blind design using parallel groups instead of a crossover design will be appropriate. These studies' data should be gathered and managed following best practices for data management, and data analysis should demonstrate the biological and statistical significance of the findings. Lastly, close observation of the long-term effects of any interactions between functional foods and bodily functions will be necessary.

1.3.1. Markers: Key to the Development of Functional Foods

Most of the time, measurements of "markers" will be necessary for the development of functional foods. The predictive value of these markers for potential benefits to a target function or the risk of a specific disease has to be determined and validated. "Indicators" are markers of related events, while "factors" are markers of an event that is directly related to the process. Indicators and even factors that are associated with the risk of disease may, in certain cases, be comparable to "surrogate markers," which are biological observations, outcomes, or indices that indicate the likelihood of developing chronic disease.

The identification and appreciation of these markers will become increasingly refined as more is understood about the mechanisms underlying health outcomes. The markers ought to be practicable, legitimate, repeatable, sensitive, and focused. They may have physiological, behavioral, psychological, or biochemical roots. On the other hand, single-point or static measurements may not be as useful as dynamic responses, if not more so. A battery of markers may often be required to build a decision tree from several tests.

These markers, most of which still need to be identified and validated, will relate to:

- Exposure to the food component under investigation is determined by measuring the concentration of an endogenous molecule that is directly impacted by the consumption of the food component, or by measuring the levels of the food component or its metabolite(s) in the serum, feces, urine, or tissue.
- The intended function or responses, such as alterations in the concentration of a particular protein, enzyme, hormone, metabolite, or change in serum or other bodily fluids, are examples of biological responses.

These first two markers are either indicators or factors:

- An appropriate endpoint of an improved state of well-being and health and/or reduction of disease risk. Such a marker is likely to be a factor rather than an indicator.
- Individual susceptibility or genetic polymorphism controlling the metabolism and/or the effect of the food component under study.

A state-of-the-art literature review will be required to find, define, and describe possible markers to further develop these markers. Additionally, a review of the fundamental scientific knowledge underlying these markers will be conducted. The evaluation of their significance for physiological function, general health, and well-being, and ultimately disease risk will be the next stage. After that, a validation for the methodology and biological significance will be required. Ultimately, possible dietary modifications will be illustrated and classified as an indicator or factor. Novel methods, akin to those employed by molecular and cellular biologists, will prove beneficial in pinpointing target populations that stand to gain from the ingestion of specific functional foods.

1.4. FUNCTIONAL FOODS AND HEALTH

The terms "designer food," "functional food," and "fortified food" are interchangeable and describe food that has been enhanced or fortified with additional nutrients or content that is complementary to it. Alternatively, they can be described as "foods that are specifically designed to satisfy any need in terms of convenience, nutrition, functionality, and therapeutic aspects." To provide the intended function, they are prepared by modifying the formulations, engineering them genetically, or using other traditional methods. Over the past few decades, there has been a significant focus on designer foods, which are primarily created to deliver nutrition. Nutraceuticals and functional foods also offer a way to continuously prevent illness and lower the rising costs of the healthcare system. Beginning in the early 1990s, the interest in functional foods has grown quickly, making it one of the fastest-growing segments of the global food industry.

Numerous health benefits of consuming functional foods have been shown or at least hypothesized by epidemiological studies and randomized clinical trials conducted around the globe. These benefits include a lower risk of cancer, improved heart health, enhanced immune system performance, a reduction in menopausal symptoms, improved gastrointestinal health, anti-inflammatory effects, lowered blood pressure, antibacterial and antiviral activity, and a reduction in osteoporosis.

1.4.1. Foods for Improved Gastrointestinal Health

The human body's gastrointestinal (GI) organ system is a crucial conduit between food and its subsequent health benefits. It is known that the GI tract contains more than 70% of our immune system. Any disruption to the delicate balance between the intestinal microflora and the host organism can result in a variety of chronic disorders, including inflammatory bowel syndrome (IBD), peptic ulcers, colon cancer, and acute gastroenteritis. The gut microflora is influenced by a variety of factors, primarily diet, but also by age, stress, lifestyle, and medication. Therefore, it is important to give priority to dietary management techniques that support the maintenance or even enhancement of the normal GI microflora. Probiotics are a well-known way to target GI microbes that have been shown to have the ability to prevent or cure disease. Probiotics that are viable, like *Lactobacilli* and *Bifidobacteria*, can persist in large enough quantities to aid in the GI tract's transition to a metabolically active state. Clinical trials have validated their therapeutic effects, and they have been successfully incorporated into the formulation of specific functional and nutritious foods. Probiotics mainly affect the immune system by acting as an antimicrobial, promoting the growth of immune-defense cells, controlling specific metabolic enzymes, and halting the aging process. The precise processes underlying the advantageous effects differ depending on the microorganisms and target group. Probiotics, prebiotics, and synbiotics are among the food items known as "colonic foods" that help to improve GI health.

The beneficial microflora in the colon helps the mucosa of the colon get all the nutrients it needs, such as vitamins, amino acids, polyamines, growth factors, and antioxidants. These include short-chain fatty acids (SCFAs). For mucosal cells, the normal nutritional needs are satisfied by colonic foods. Prebiotics are made in part by the probiotic bacteria from a wide range of raw materials, which they use as food. In addition to acting as prebiotics, complex carbohydrates, such as dietary fibers, resistant starch, and oligosaccharides, also have positive physiological effects, such as relieving constipation, preventing the absorption of cholesterol, and boosting the bioavailability of micronutrients. Known as "New age fiber," oligosaccharides are a common component of both plant and animal cellular constituents and have been linked to many health benefits. Furthermore, novel functionalities like water binding, gelation, and emulsification are exhibited by dietary fibers, resistant starch, and oligosaccharides. These functionalities can be leveraged in the creation of low-fat versions of probiotic products.

1.4.2. Important Functional Food Components of Plants

1.4.2.1. Carotenoids

Carotenoids are fat-soluble pigments that are mostly present in plants, fruits, flowers, algae, and bacteria that use light for photosynthesis. They can also be found in certain molds, yeasts, and non-photosynthetic bacteria. Although the structure of carotenoids ultimately determines their potential biological function(s), they are important for human health. The protective properties of carotenoids against degenerative eye disease, cancer, and heart disease have been established. This has led to extensive research on the functions of carotenoids as antioxidants and immune response regulators (Donaldson, 2004).

1.4.2.2. Dietary fibers

Dietary fibers are classified as soluble or insoluble and include cellulose, hemicellulose, polyfructoses, galactooligosaccharides, gums, mucilages, pectins, lignin, and resistant starches. Dietary and functional fiber consumption offers numerous potential health benefits, including the reduction of constipation and irritable bowel syndrome (IBS), the prevention of obesity and diabetes, the avoidance of colon cancer, the enhancement of breast cancer survival and the lowering of cholesterol and the incidence of coronary and cardiovascular heart diseases.

1.4.2.3. Flavonoids

Plants produce a broad family of polyphenolics called flavonoids, which are further subdivided into numerous groups, each of which contains hundreds of different compounds. Flavonoids have antiviral, antitoxic, antifungal, antibacterial, anti-allergic, anti-inflammatory and antioxidant properties that may be beneficial to human health.

1.4.3. Functional Foods for Infant and Weaning Purposes

Malnutrition and deficiency diseases in children are more common in India than in other countries. An estimated 50% or more of babies are born with low birth weights, which stunts their growth. The incidence of anemia and infectious diseases in children was caused by the deficiency of essential nutrients and bio-protective components during infancy. Although nursing is frequently not feasible, infants must be fed infant formula instead of their mother's milk, which is thought to be the ideal food provided by nature. The best example of designer food is infant formula. Cow's milk is used to make most infant formulas, but significant changes must be made to the recipe to make it more like breast milk. These changes include an increase in carbohydrates, a decrease in protein and minerals, and an addition of vitamins and trace elements.

Recent research suggests that infants may not be able to synthesize taurine and carnitine as well, necessitating the need for dietary sources. Long-chain fatty acids must be transported into cells by carnitine for β-oxidation and energy production to occur. A combination of different fat sources is preferred because the fatty acid profiles of individual fat sources do not match the complexity of breast milk at maturity. Most producers use a blend of vegetable oils. The essential fatty acids α-linolenic acid (C18:3, ω-3) and linoleic acid (C18:2, ω-6) must also be present in the fat source. It is suggested that the ratio of ω-6: ω-3, which is found in breast milk, be 5:1. Due to their ease of absorption and assimilation, medium- and short-chain fatty acids should also be present in sufficient amounts. However, before the ideal ratio and its effects on growth are assessed, more short- and long-term studies are required. The precursors of the very-long-chain (C20–C22) polyunsaturated fatty acids (LCPUFA), arachidonic and DHA, are linoleic and α-linolenic acids. Human milk contains LCPUFA, which are involved in the neural and vascular development of the fetus and neonates.

The non-protein nitrogen found in human milk contains nucleotides, which may be crucial for healthy immune system operation. Clinical trials suggest that adding nucleotides to infant formula may be helpful; however, more investigation is required before routine nucleotide supplementation of infant formula can be recommended. Numerous formulations have been developed because of the commercially prepared infant formulas' success, and there are currently hundreds of different proprietary infant formulas available worldwide. Additionally, there are unique dietary foods available that contain special formulas for use in clinical settings, for premature infants, or for infants with specific inborn errors of metabolism.

Infants' GI tracts are dominated by Bifidobacteria, which have health-promoting and protective qualities such as immune system activation, pathogen inhibition through the secretion of substances that are directly inhibitory towards multiple bacteria, pH lowering through the production of acids like lactic and acetate, which creates an antibacterial environment, vitamin production, and the production of digestive enzymes like casein phosphatase and lysozyme. For these reasons, it would seem ideal to boost the quantity of Bifidobacteria in formula-fed infants' intestinal flora. The best method for increasing the amount of Bifidobacteria in the intestine seems to be the administration of probiotic supplements and prebiotic oligosaccharides. The primary cause of breast milk's bifidogenic effects is human milk oligosaccharides. To provide formula-fed infants with a predominance of Bifidobacteria in their intestinal flora, several commercial formulations have been developed. Nevertheless, before being approved, the use of such unusual ingredients in baby formula formulations must undergo extensive research.

In many developing countries around the world, malnutrition-related issues are frequently caused by inadequate nutrition during the first two to three years of life. After six months, a child's nutritional needs—which go beyond what breast milk can supply—increase, making complementary nutrition essential for their continued normal and healthy growth. Furthermore, the food preparations eaten as weaning foods do not contain the right amount of nutrients that kids need. Cereal is typically the basis for traditional infant feeding practices in nations like India. To improve digestibility and improve the nutritional profile of these foods, grains are frequently germinated, fermented, processed, and cooked in different ways. Additionally, they are often combined with oilseeds or animal products. However, due to the higher proportion of anti-nutrients in these complementary foods, it has been reported that they are less safe for children to eat and less energy-dense. Weaning foods are typically prepared with a combination of cereals and milk solids. Complementary foods based on milk, cereal, and millets seem special in that they can give kids a variety of nutrients and work well together. The effectiveness of the resulting complementary food is determined by the proper method of incorporation, effective concentration, and necessary technological inputs. These goods might be a desirable choice for widespread child-feeding initiatives.

1.4.4. Specialized Foods with Plant Bioactive

Plant molecules have a well-established nutritional value, and a diet low in fruits and vegetables has been linked to an increase in cancer cases, heart disease, diabetes, and many other chronic diseases. Plants are obviously richer in vitamins, fibers, and other nutrients, and researchers are working feverishly to identify the precise plant components that may protect against which ailments. An increasing number of hitherto unidentified plant compounds with enigmatic names are being discovered. However, more research is needed to determine their precise metabolic function and potential application in designer foods.

The number of physiologically recognized disorders has skyrocketed in the last few years, and a diet high in plants can lower the risk of developing certain chronic diseases, according to copious evidence from epidemiological, *in vivo*, *in vitro*, and clinical trials. The role of phytochemicals in promoting better health is becoming increasingly acknowledged by medical professionals. The primary process linked to the medicinal properties of plant bioactive is their capacity to function as antioxidants.

Other substances found in plant-based diets that have a major positive impact on health include dietary fibers, tocotrienols, phenolic derivatives, and plant fatty acids. De novo synthesis of DHA, one of the most crucial structural elements of the brain and retina, is extremely uncommon. Since low plasma DHA concentrations have been linked to an increased incidence of several significant chronic diseases, including depression, attention deficit disorders, and Alzheimer's dementia, the decline in DHA intake could have major health effects on the general public. The marine algae strain *Cryptodinium cohmii* is utilized to produce DHA-rich oil on a commercial scale. Known as the "wonder algae" *spirulina* is a rich source of high-quality protein, omega-3 fatty acids, and numerous other medicinal compounds.

Plant polyphenols are widely distributed secondary metabolites found in higher plants. Nutritionists have historically viewed polyphenols as anti-nutrients because some of them, like tannins, have harmful effects that include lowering the activities of digestive enzymes, energy, protein and amino acid availability, mineral absorption, and other toxic effects. The understanding of the health advantages offered by numerous polyphenols has been reoriented with the discovery of these compounds' antioxidant properties. A large class of substances derived from plants known as phytoestrogens structurally resemble endogenous 17 beta-estradiol. From a nutritional and health standpoint, lignans (flaxseed) and isoflavones (soybeans) are two important phytoestrogens. These substances oppose or compete with the effects of estradiol. The precise biochemical mechanism that involves the activity of CYP3A monoxygenase when phase I enzyme inducers, like dixamethane, are present. An additional significant terpene subclass is phytosterols. Plants synthesize two types of sterol molecules: β-sitosterol and its glycoside. These two compounds have anti-inflammatory, anti-neoplastic, anti-pyretic, and immune-modulating properties in animals. Phytosterols help the body rid itself of cholesterol by competing with cholesterol in the intestine for absorption. When it comes to lowering the body's cholesterol levels, saturated phytosterols seem to be more beneficial than unsaturated ones. Some designer foods, such as yogurt with phytosterol, dairy drinks rich in β-glucan, infant foods with DHA, etc., have already attained the commercialization stage.

1.4.5. Milk Proteins and Peptides Based Nutraceuticals

Dietary proteins have biological, functional, and nutritional qualities. These qualities are frequently impacted by the technological procedures used in food processing and manufacturing. In recent years, there has been a growing recognition of the role that proteins play in the diet as physiologically active constituents. These proteins, or their precursors, can be found naturally in unprocessed food ingredients. They can function physiologically directly or through enzymatic hydrolysis *in vivo* or *in vitro*. A variety of dietary proteins can serve as a source of peptides that are physiologically active. These inactive peptides are still part of the parent protein and are released during food processing or gastrointestinal digestion. The bioactive peptides may serve various purposes *in vivo* or *in vitro* after they are released.

Enzymatic hydrolysis is required for the release of bioactive peptides from the parent protein. Both microbial fermentation and the use of isolated enzymes can accomplish this. The pharmaceutical industry is particularly interested in biologically active peptides due to their demonstrated physiological functions, which include opioid-like activity, antimicrobial, immunomodulatory, and antihypertensive properties. Enzymes from microbes or the digestive system can hydrolyze such peptides and release them. It has been shown that lactic acid bacteria's microbial enzymes can separate these peptides from milk proteins in a variety of fermented dairy products.

Bioactive peptides can have an impact on the digestive, immune, nervous, and cardiovascular systems when taken orally. Because of this, there has been a lot of interest in science over the past few years regarding the potential of specific peptide sequences to lower the risk of chronic diseases or increase natural immune protection. Certain peptide sequences that have been shown to have antimicrobial, antioxidant, antithrombotic, antihypertensive, and immunomodulatory properties may be responsible for these positive health effects. The most significant source of bioactive peptides is thought to be milk proteins, and fermented dairy products and milk protein hydrolysates are progressively revealing their bioactive peptide contents.

Many studies have been conducted worldwide in the past few years to clarify the bioactivity of milk proteins and their derivatives. These ingredients can be used to create functional foods or they can be used as nutraceuticals by the pharmaceutical industry. The majority of the physiological properties of milk proteins and their derivatives that have been proposed have been tested in animal or *in vitro* models; human validation of these properties is still pending. Due to their distinct amino acid composition and bioactivity, whey proteins are increasingly being used as a key ingredient in a variety of functional and health foods. Commercially available food products made with whey proteins include dairy desserts, sports supplements, medical foods, infant formulas, and geriatric foods. Foods and beverages for heart patients may be developed with antihypertensive bioactive peptides.

1.4.6. Other Prospective Designer Foods

Due to their high nutritional content, beverages represent another product category with enormous market potential for the Indian food industry. Similarly, it appears that milk beverages made from millets and small cereals are profitable items for school feeding programs. In certain nations, fortifying liquid milk with vitamins A and/or D is required. However, the sensory and processing quality characteristics of the milk were typically compromised by the fortification. Another significant worry is the bioavailability of nutrients that have been fortified. Recent studies revealed the possibility of adding calcium and iron to liquid milk fortification. These days, whey-based beverages and soups continue to garner a lot of interest. These show that both manufacturers and consumers are becoming more conscious of the huge potential these presented for broadening the product line. Additional designer foods include foods that are fun, low in sodium, low in calories or fat, etc.

1.5. EXERCISE AND FUNCTIONAL FOODS

Sufficient nutrition is necessary for optimal exercise performance. Correct nutrition is especially crucial for enhancing athletic performance, conditioning, recuperating from exercise-induced fatigue, and preventing injuries. A balanced basic diet is necessary for athletes, but since athletes need more nutrients than what is recommended daily, certain nutritional components are hard to get enough of from a typical diet. Thus, dietary supplements that are rich in proteins, carbs, vitamins, and minerals have been used extensively in a variety of sports, in part due to the ease with which they can be taken before, during, or following exercise. Numerous naturally occurring food ingredients have also been demonstrated to have physiological effects, and some of these are thought to be beneficial (when consumed frequently or in large amounts) for enhancing athletic performance or preventing the disruption of homeostasis caused by intense exercise. Food ingredients that have physiological effects have recently been dubbed "functional foods," and their effects have been studied by scientists. Here, we present a few functional foods, such as essential nutrients, which have been shown to positively impact the physiological alterations brought on by exercise.

1.5.1. Replenishment of Water

Water is the primary component of all living things and is vital to energy metabolism, circulation, waste product excretion, plasma volume and body temperature regulation, and waste product elimination. Sweating happens to radiate heat when the body temperature rises as a result of intense exercise or a high outside temperature. This causes significant water and electrolyte loss. This bodily fluid loss affects the circulatory system and thermoregulation, which lowers athletic performance. Therefore, replenishing water and electrolytes before, during, and after exercise is crucial to maintaining homeostasis and athletic performance. Drinking isotonic fluid, which contains electrolytes like sodium, potassium, and chloride at concentrations similar to those in bodily fluids, is generally thought to be beneficial. Another theory is that because hypotonic fluid is quickly absorbed from the small intestine along an electrochemical gradient, it may have a comparable or even faster effect on the body's ability to replenish water. Additionally, sweat has a lower sodium concentration (and osmolality) than extracellular fluid, which means that sweat loses a lot more water than electrolytes, which raises the plasma osmotic pressure.

On the other hand, during extended exercise that results in high sweat rates, replenishing water alone is unlikely to maintain body fluid homeostasis. Drinking only water while exercising for an extended period causes hyponatremia, lowers the osmotic pressure of bodily fluids, and inhibits the release of antidiuretic hormone, which suppresses water intake and increases urine output (spontaneous dehydration). Latzka and associates recommended that, to maintain carbohydrate oxidation and endurance performance during extended exercise lasting longer than 90 minutes, a fluid drink containing electrolytes and carbohydrates should be taken into consideration, rather than just water. Additionally, several studies have indicated that glycerol loading is one strategy recommended to avoid both excessive heat and dehydration during exercise. When taken orally, 1.0–1.2 g/kg BW glycerol combined with water temporarily increases bodily fluid by 300–700 ml and enhances endurance performance when compared to a placebo. As an osmolyte in bodily fluids, glycerol causes an increase in plasma osmolality. As a result, one of the mechanisms of the effect is thought to be the increase in water reabsorption in the kidney and the decrease in urine excretion (Buono et al., 2000).

1.5.2. Improvement of Endurance

The primary sources of energy used during exercise are fats and carbohydrates, so controlling how these two substrates are metabolized is crucial for increasing endurance. Glycogen, an energy substrate for muscle contraction, gradually runs out during endurance exercise, making it harder to continue. Increasing the amount of glycogen stored in the liver and skeletal muscle before exercise is a useful strategy for enhancing endurance. Glycogen synthetase activity temporarily increases in response to tissue glycogen stores being depleted, which increases glycogen storage through conversion from carbohydrates. A low-carb diet for three days starting six days before the competition, followed by a high-carb diet for the next three days, has been shown to increase glycogen stores by 1.5 times the amount of glycogen that would normally be stored. The inhibition of glycolysis caused by citrate will result in an additional increase in glycogen stores when taken in conjunction with a high-carb diet. Refueling the glycogen stores after a workout is crucial for athletes to ensure they have enough energy for the next training session or competition. A diet high in carbohydrates may be useful for quickly replenishing glycogen stores. When it comes to quickly restoring muscle glycogen after exercise, consuming protein in addition to carbohydrates may be more beneficial than taking supplements of carbohydrates on their own (Tarnopolsky et al., 1997).

Consuming carbohydrates either before or during an extended physical activity, like a marathon, is another useful strategy for enhancing endurance. Because they are quickly absorbed and delivered to the peripheral tissues, mono- or oligosaccharides are preferable for the athlete to consume in such circumstances. On the other hand, by increasing insulin secretion, eating carbohydrates prevents fat, another energy substrate, from breaking down. As a result, the lipid metabolism pathway that produces energy is impaired, and the glycolysis pathway, which produces energy alternatively, speeds up. Muscle contraction may be hampered because of the increased consumption of muscle glycogen and the resulting drop in intramuscular pH caused by increased lactic acid production. Consuming carbohydrates that will not impede lipid metabolism is therefore essential. It has been proposed that fructose supplements, as opposed to common carbohydrates like glucose and sucrose, may be more effective at increasing endurance because they stimulate insulin secretion less and are less likely to inhibit lipolysis. Moreover, concurrent consumption of citrate is anticipated to enhance lipid-based energy consumption by inhibiting glycolysis. This will prevent the synthesis of lactic acid and preserve glycogen, delaying the weakening of muscle contraction. It has been shown that the amino acid arginine can delay the depletion of glycogen during exercise and modulate hormones that control blood glucose levels without impairing lipid metabolism. To enhance energy metabolism and provide the best energy sources for extended exercise, it may be beneficial to consume both citrate and arginine along with carbohydrates that do not stimulate insulin secretion.

The changes in energy consumption patterns from primarily glucose- to lipid-based energy consumption may improve endurance by preserving glycogen stores and preventing the drop in intramuscular pH caused by lactate production during exercise. Numerous writers have discussed different elements that can promote lipid metabolism, but the effectiveness of these factors is not well supported by data. An intracellular enzyme called carnitine facilitates the β-oxidation of fatty acids and is necessary for the transport of fatty acids across the mitochondrial membrane and into the mitochondria. It is anticipated that supplementing with carnitine will cause the skeletal muscles' glycogen stores to be spared while simultaneously triggering lipid metabolism. A study found that giving participants in aerobic training 2–4 grams of carnitine either daily or prior to exercise increased their maximum oxygen consumption, or anaerobic threshold), and prevented lactate from building up after exercise. Studies have also been conducted on the impact of caffeine on endurance. By encouraging catecholamine release and raising hormone-sensitive lipase (HSL) activity, caffeine inhibits phosphotidiesterase. This results in an increase in circulating free fatty acids and further enhancement of endurance. Capsaicin, which is derived from spicy red peppers, may improve fat metabolism by influencing the lipolytic hormone balance and encouraging skeletal muscle fat oxidation (Yoshioka et al., 1995).

1.5.3. Enhancement of Muscle Strength

It is common knowledge that a muscle's cross-sectional area determines how strong it is, and that building more muscle mass is essential to increasing strength. Proteins (like actin and myosin) and water make up most of the muscle tissue, so when building muscle mass, it is critical to raise the protein content by adjusting protein metabolism. Put another way, either enhancing protein synthesis or preventing protein degradation can lead to an increase in muscle mass and strength. Resistance training increases the production and secretion of growth hormone and other growth factors and is intended to increase the bulk of muscle. Considering this, resistance training more strongly encourages protein synthesis and an increase in muscle mass than aerobic training. Put another way, either enhancing protein synthesis or preventing protein degradation can lead to an increase in muscle mass and strength. Maintaining the muscular pool and blood levels of different amino acids—substrates for the synthesis of muscle proteins—is crucial to maximizing the benefits of resistance training. To achieve this, increasing dietary protein intake is required to maintain a positive nitrogen balance. Strength training athletes have higher protein requirements than sedentary people, according to several studies. For resistance exercise, the estimated daily recommended protein intake is 1–8 g/kg when the intake of calories and carbohydrates is sufficient; however, for endurance athletes, excluding an elite minority, 1–0 g/kg of protein is generally sufficient. While maintaining such a high dietary protein intake may be challenging, taking protein supplements can be helpful. Powdered protein supplements can be made from a wide range of raw materials, and products made from sȯybeans, eggs, or whey (milk protein) are sold commercially. Each of these goods frequently receives an amino acid score of 100 and has an excellent ratio of essential amino acids. Specifically, whey protein is thought to be the best kind of protein for muscle growth because it is quickly absorbed and easily digested, raising blood levels of amino acids. Whey protein also contains a high concentration of glutamine and branched-chain amino acids (BCAA), which aid in the synthesis of muscle protein. To effectively build muscle, the timing and amount of protein intake are crucial factors. When it comes to increasing muscle mass, eating a meal right away following resistance training may be more beneficial than waiting a few hours to eat. Additionally, eating carbohydrates along with protein can quicken the synthesis of muscle protein because insulin promotes protein synthesis while preventing its breakdown.

Furthermore, it has been suggested that consuming peptides and amino acids is advantageous. Since free amino acids and peptides do not require digestion, quick absorption is to be anticipated. Certain amino acids have a range of physiological effects in addition to being used in the synthesis of muscle protein. The effects of BCAA, which are known to be present in relatively high concentrations in both muscle and food proteins, have drawn attention. These amino acids include valine, leucine, and isoleucine. The liver metabolizes most amino acids, but certain processes allow the muscles to metabolize BCAAs. Branched-chain-α-keto acid dehydrogenase (BCKDH) complex activation enhances the oxidation of BCAAs, which are used as energy substrates. Moreover, BCAAs influence how muscles use protein by promoting synthesis and preventing degradation, which gives muscles an anabolic boost. It has also been observed that glutamine increases muscle growth by preventing the breakdown of proteins. It is the most prevalent free amino acid in muscle tissue, and consuming it causes an increase in myocyte volume, which in turn stimulates the growth of new muscle. Glutamine plays a significant role in homeostasis and is present in many other human tissues at relatively high concentrations. As a result, glutamine is released from skeletal muscle into the plasma during catabolic states like exercise, where it is used to maintain the glutamine level in other tissues. An injection of arginine stimulates the release of growth hormone, which may result in an increase in muscle mass and strength. Arginine is a precursor to nitric oxide and creatine. While there is conflicting evidence regarding the impact of oral arginine on protein synthesis, current research suggests that arginine supplementation with other substances enhances exercise capacity.

Studies have also been conducted on a variety of other food ingredients to ascertain their impact on muscle mass and strength. A comprehensive review of research conducted between 1967 and 2001 endorsed the use of two dietary supplements to increase lean body mass and strength during resistance training: creatine and β-hydroxy-β-methylbutyrate (βHMB). Over 100 g of creatine are found in the human body, nearly all of which are kept as creatine phosphate in the skeletal muscles. Raising creatine stores is expected to improve anaerobic metabolism since this is used to produce ATP through degradation to creatine in anaerobic conditions. Consuming creatine also promotes the synthesis of proteins and water retention. A daily intake of ≥3 g of creatine has been shown to improve muscle strength and endurance, particularly during high-power output activities like resistance training or short-distance running. It also raises the intramuscular content of creatine phosphate. Furthermore, it has been noted that taking creatine during resistance training quickens the gains in lean body mass and muscle strength. βHMB is a metabolite of leucine, a branched-chain amino acid. It promotes muscle growth by preventing protein degradation through its impact on branches-chain amino acid metabolism. A higher increase in muscle mass and power was observed when 1.5 to 3.0 g/day of βHMB was consumed for 3 to 8 weeks, as opposed to when a placebo was consumed.

1.5.4. Prevention of Injury and Fatigue

Excessive exercise or strenuous physical activity can lead to muscle damage, the release of muscle protein, and pain in the muscles. Though the exact mechanism causing the delayed muscle damage following intense physical activity is unknown, it has been proposed that phagocyte infiltration, which is caused by an inflammatory response triggered by excessive mechanical stress, increased intracellular Ca^{2+} concentration, and oxidative stress, is the cause of this delayed injury. Since oxidative products are found in significant amounts in the blood and in the muscles after exercise, along with other markers of delayed-onset muscle damage, numerous studies have investigated whether antioxidants can lessen the damage to the muscles. Consuming antioxidants, such as vitamins C and E, carotenoids, or polyphenols, not only during exercise but also on a daily basis, can help prevent oxidative injury following acute exercise. On the other hand, many studies have shown that antioxidants do not affect the inflammatory response and muscle damage brought on by intense exercise. One explanation for the discrepancies in the results could be that the impact of antioxidants is probably due to variations in exercise settings, such as the degree of mechanical stress and oxygen uptake. The beginning of the muscle damage may be connected to reactive oxygen species (ROS). During exercise, ROS are produced by mitochondria and endothelium through an increase in myocyte oxygen uptake and the ischemia-reperfusion process. Following exercise, phagocytes invade the muscles through a redox-sensitive inflammatory cascade. As a result, if the production of ROS during exercise is reduced, even in the case of endurance prolonged exercise rather than resistance exercise, the inflammatory response may be inhibited. Furthermore, it is preferable to take multiple antioxidants concurrently because different antioxidant types—such as lipid- or water-soluble compounds—affect different organelles and can exchange electrons to keep a change in pro-oxidant status from occurring.

Two compounds that help to protect the joints are chondroitin and glucosamine. An amino acid called glutamate is produced by the body and is found in ligaments, tendons, and synovial fluid in the joints. Because of its hygroscopic action, chondroitin—which is mostly found in cartilage, tendons, and the skin's connective tissues—acts as a significant shock absorber. While the impact of supplementation during exercise is unclear, supplemental oral intake of these substances is suggested to be effective for either preventing or promoting recovery from osteoarthritis associated with exercise and aging.

Numerous factors, including the depletion of glycogen and buildup of lactic acid during exercise, as well as the hyperactivation of the sympathetic nervous system following exercise, can be identified as indicators of the fatigue condition brought on by exercise. As was already mentioned, a diet high in carbohydrates encourages the recovery of glycogen storage in muscle. Simultaneously, it is more efficient to consider the timing of carbohydrate intake and a factor that inhibits glycolysis, like citrate. Additionally, the buildup of lactate in muscles reduces their ability to contract, which is linked to a drop in muscle pH and may be a sign of fatigue. Therefore, dietary supplements that control lactate production or clearance may be useful. Dipeptides with a pH-buffering effect include carnosine and anserine, which are found in large amounts in skeletal muscle. Through these buffering effects, supplementing with dipeptides can also prevent the intramuscular pH from declining as a result of exercise.

1.5.5. Maintenance of Immunity

While excessive physical activity suppresses the immune system and increases the risk of inflammatory and allergic disorders, moderate exercise is generally thought to improve immunocompetence and be useful in preventing infections, cancer, and inflammatory diseases. Increased production of immunosuppressive factors, such as adrenocortical hormones and anti-inflammatory cytokines, is thought to be the cause of susceptibility to infections after excessive physical activity. This results in a decrease in the quantity and activity of circulating T cells and natural killer (NK) cells as well as a lower concentration of IgA in saliva. As a result, athletes who engage in high-intensity training run the risk of developing compromised immune systems. Consuming carbohydrates while engaging in prolonged, submaximal exercise reduces the rise in plasma cortisol and cytokine levels following exercise, which may prevent immunosuppression. In addition to being necessary for T cell differentiation and maintenance of T cell function, vitamins C and E have immune-boosting properties. On the other hand, the effects of vitamin supplementation on immune function in connection to exercise are not well-established. Glutamine serves as a vital source of energy for neutrophils, macrophages, and lymphocytes. It is also a necessary amino acid for the development and proliferation of these cells. Exercise that is too intense lowers the concentration of plasma glutamine, which could be linked to immunosuppression. Castell and associates found that after a marathon, athletes who consumed glutamine had a lower infection rate than the placebo group. Additionally, they showed that consuming glutamine raised the ratio of T-helper to T-suppressor cells. Moreover, glutamine suppresses the synthesis of cytokines linked to immunosuppression or inflammation and increases the activity of intestinal enterobacteria.

1.6. COMMUNICATING WITH FUNCTIONAL FOOD

In the Canadian market, the functional food sector has grown and developed quickly. Several factors have contributed to the growth of the functional food industry, such as advancements in food science and technology, an aging population with escalating health concerns, changes in the regulatory landscape permitting food claims to be made for health benefits, and increased promotion of functional food items. Beyond the traditional focus on the prevention of nutrient deficiency diseases, the evolution of consumer and scientific interest in nutrition's ability to optimize health and prevent chronic disease has been fundamental to the advancement of functional foods.

Foods that have been shown to have physiological advantages and/or lower the risk of chronic disease beyond their basic nutritional roles are examples of this evolution in food and health. Functional foods' long-term commercial success, however, is reliant on consumer acceptance as well as attitudes and perceptions about them. Correlated to this, it has been determined that consumers' acceptance of functional foods can be influenced by the efficient dissemination of nutrition and health information; however, more investigation is required to clarify how consumers view these topics.

Given that the demographic stands to gain significantly from including functional foods in their diets, it is particularly interesting to learn about consumer awareness of and attitudes toward these foods among older adults. The population segment of older adults is growing quickly; estimates show that the number of Canadians over 65 will rise from 4.2 million to 9.8 million between 2005 and 2036 and that by 2041, it could account for up to 25% of the country's total population. The healthcare system is heavily burdened by the concurrent rise in chronic age-related diseases like cancer, osteoporosis, age-related macular degeneration, and CVD. With their bioactive ingredients, functional foods offer a potential tactic to reduce the higher risk of chronic illness in the elderly population. To provide appropriate guidance regarding functional foods and health, it is necessary to understand older adult awareness and perceptions of these foods, as they interact with healthcare providers more frequently.

By highlighting the advantages of functional food products for health, nutrition and health information, as well as the source of this information, can affect consumer acceptance of these products.

According to consumer research, if consumers believe that functional food products are healthier than conventional foods, they are more likely to replace conventional foods with functional foods. Contrary to sensory aspects of a food product, consumers cannot directly perceive the health benefits of functional foods; therefore, perceptions of functional food products can be influenced by information about these benefits and how it is communicated. Another important factor influencing the acceptance of functional foods is consumer confidence and trust in the source of nutrition information. Consumer research has shown that health professionals—such as doctors and dietitians—are regarded as reliable sources of information about nutrition and health. Customers have also indicated that reliable sources of information about nutrition and health are family and friends.

It is important to identify the preferred information sources and perceived need for additional information regarding functional foods among older adult consumers, as nutrition and health information can impact consumer acceptance of functional food products.

Sources of nutrition information on food labels, especially those of health claims, are an important consumer education tool that may affect the acceptability of functional foods. Studies on the effects of health claims on functional food labels on consumers have shown that these claims positively impact consumers' perceptions of the healthfulness of functional foods and lead to more positive attitudes toward them. On the other hand, other studies have revealed a lack of trust among consumers concerning the validity of health claims made about functional foods, the scientific proof supporting these claims, and the laws that govern them. Studies have looked at consumer attitudes toward functional food health claims as well as the kinds of health claims that consumers prefer. Studies have indicated that consumers tend to favor health claims about food products that link to a physiological benefit or lower risk of disease rather than those that link to psychological benefits like less fatigue or stress. Additionally, it has been discovered that among Belgian consumers, nutrient function claims are more appealing and persuasive than nutrient content claims, and both are preferred over claims about lowering the risk of disease. Evidence also suggests that since consumers are already aware of the component's health benefits, they view nutrient content claims—which state that the product contains a well-known component—as being just as beneficial as nutrient function or disease risk reduction claims. In the end, additional study is required to identify the kind of health claim that different consumer segments prefer to inform food manufacturers and enable the provision of targeted information.

The awareness and comprehension of this information among older adults, a major beneficiary of functional foods, has not yet been investigated, despite evidence that nutrition and health information may have an impact on acceptance of functional foods. Although consumer needs, interests, and perceptions of nutrition and health information vary greatly, there has not been enough focus on these differences in the current research on consumer perceptions. The current study's goal was to gather data on the consumption of functional foods among a sample of older adults by investigating the sources of such data as well as participant awareness, perceptions, and comprehension of the health claims made on functional food products.

1.6.1. A Communication Challenge

The question of how to effectively communicate the unique benefits of functional foods is becoming more and more important as the public's acceptance of the link between nutrition and health grows. Improving public health and developing functional foods require the public to be informed about the benefits of these products through intermediaries like educators, media, health professionals, and the food industry. It is also crucial for preventing issues related to consumer misunderstanding of health messages. Out of all the various communication formats, claims—whether stated directly on food product labels or packages, or indirectly through supplementary data—continue to be a subject of intense debate.

According to CB Hudson, consumers will not be able to understand the connections between nutrition science and food product development until the necessary channels of communication are established.

However, the communication of health benefits and other physiological effects of functional foods remains a major challenge because science should remain the driving force:

- Messages—claims—must be based on sound, objective, and appropriate evidence; and
- Evidence must be consistent, able to meet established scientific standards and plausible.

Furthermore, information on nutrition often originates from a variety of sources, some of which are contradictory, giving the impression of disarray. Furthermore, disorganized information is easily misinterpreted and frequently leads to ignorance.

1.6.2. The Scientific Challenge

The claims connected to specific food products are the best way to inform consumers about functional foods. Any claim made following the fundamental principle must be accurate, not deceptive, supported by science, unambiguous, and understandable to the consumer. These fundamental ideas must be upheld, though, so as not to impede the development of functional foods or the public's acceptance of them. Distinct types of claims exist, and the term "health claims," which is traditionally used to communicate the benefits of foods, is defined differently in different parts of the world. This makes difficult to communicate the benefits of functional foods, even though a general definition of "claim" is widely accepted in the field of nutrition as "any representation, which states, suggests or implies that a food has certain characteristics relating to its origin, nutritional properties or any other quality."

Recently, Codex Alimentarius defined and categorized four distinct categories of claims, omitting the term "health claim" in the process of seeking clarification:

- Relate to dietary guidelines.
- Relate to nutrient content.
- Are comparative (reduced, less, more).
- Describe nutrient function.

These assertions touch on the idea of adequate nutrition as well as recognized nutrients and their roles in normal functions, growth, and development. They do not allude to any specific outcome beyond what would be anticipated from consuming a balanced diet, even though they are founded on well-established and generally acknowledged knowledge. Therefore, these claims do not really help to convey the unique advantages of functional foods. In fact, the scientific categorization of markers (indicators and/or factors) for target functions and the effects on these markers should serve as the foundation for the claims made for functional foods. A claim for an enhanced function will be made if the effect is one that goes beyond what is expected given the established role of diet in growth, development, and other normal bodily functions and affects a target function or biological activity without specifically mentioning a disease or pathological process. However, claims for disease risk reduction will be made if the benefit is evidently a decrease in the likelihood of a disease or pathological process. According to the European consensus on scientific concepts of functional foods, these two types of claims—which are unique to functional foods—are referred to as type A and type B claims, respectively.

Type A claims are comparable to "structure/function claims." while type B claims are comparable to "health claims" in the United States. The Swedish "health claim" and the type B claim are equivalent. Codex Alimentarius has included type A and type B claims and defined them as follows in its most recent proposed draft recommendations for the use of health claims.

- Type A or claims that, without mentioning nutrient function, speak to the specific advantages of consuming foods and their ingredients on biological, psychological, or physiological processes. Such claims are related to improving a function, maintaining, or enhancing health, or making a positive contribution to health or a health-related condition.
- Type B, or "risk of disease reduction claims," are those that discuss how consuming a food or a food ingredient in the context of a daily diet may help lower the risk of a particular disease or condition.

Particularly about these two kinds of claims, one of the main unresolved issues is the biological threshold at which evidence is deemed to "satisfactorily demonstrate" improved function or decreased risk of disease. Every piece of information that falls into one of three categories should be the basis for this evidence:

- biological observations;
- epidemiological data; and
- intervention studies, mostly based on markers.

There may be insufficient or no evidence from all three areas to support improved function or decreased risk of disease for a particular food product. However, all the supporting evidence ought to be:

- consistent in itself;
- able to meet accepted scientific standards of statistical as well as biological significance, especially dose–effect relationship, if relevant;
- plausible in terms of the relationship between intervention and results, especially in terms of mechanism(s) of action; and
- provided by several sources (including obligatory human studies) that give consistent findings able to generate scientific consensus (Donaldson, 2004).

REFERENCES

1. Akalın, A. S., Karagözlü, C., & Ünal, G. (2008). Rheological properties of reduced-fat and low-fat ice cream containing whey protein isolate and inulin. *European Food Research and Technology, 227*(3), 889–895.
2. Alzamora, S. M., Salvatori, D., Tapia, S. M., Lopez-Malo, A., Welti-Chanes, J., & Fito, P. (2005). Novel functional foods from vegetable matrices impregnated with biologically active compounds. *Journal of Food Engineering, 67*(3), 205–214. https://doi.org/10.1016/j.jfoodeng.2004.05.067.
3. Arenas-Jal, M., Suñé-Negre, J. M., & García-Montoya, E. (2020). An overview of microencapsulation in the food industry: Opportunities, challenges, and innovations. *European Food Research and Technology, 246*(6), 1371–1382.
4. Buono, M. J., & Wall, A. J. (2000). Effect of hypohydration on core temperature during exercise in temperate and hot environments. *Pflügers Archiv–European Journal of Physiology, 440*(5), 476–480.
5. Buriti, F. C. A., & Saad, S. M. I. (2014). Chilled milk-based desserts as emerging probiotic and prebiotic products. *Critical Reviews in Food Science and Nutrition, 54*(2), 139–150.
6. Donaldson, M. S. (2004). Nutrition and cancer: A review of the evidence for an anti-cancer diet. *Nutrition Journal, 3*(19). https://doi.org/10.1186/1475-2891-3-19.
7. Foligne, B., Nutten, S., Grangette, C., Dennin, V., Goudercourt, D., Poiret, S., Dewulf, J., Brassart, D., Mercenier, A., & Pot, B. (2007). Correlation between *in vitro* and *in vivo* immunomodulatory properties of lactic acid bacteria. *World Journal of Gastroenterology, 13*(2), 236–243.
8. Friedman, M. (2007). Overview of antibacterial, antitoxin, antiviral, and antifungal activities of tea flavonoids and teas. *Molecular Nutrition and Food Research, 51*(1), 116–134. https://doi.org/10.1002/mnfr.200600209.
9. Gotova, I., Dimitrov, Z., & Najdenski, H. (2017). Selected *Lactobacillus bulgaricus* and *Streptococcus thermophilus* strains from Bulgarian yogurt demonstrate significant anti-inflammatory potential. *Acta Microbiologica Bulgarica, 33*(2), 137–142.
10. Grangette, C., Nutten, S., Palumbo, E., Morath, S., Hermann, C., Dewulf, J., Pot, B., Hartung, T., Hols, P., & Mercenier, A. (2005). Enhanced anti-inflammatory capacity of a *Lactobacillus plantarum* mutant synthesizing modified teichoic acids. *Proceedings of the National Academy of Sciences, 102*(29), 10321–10326.
11. Hojsak, I., Szajewska, H., Canani, R. B., Guarino, A., Indrio, F., Kolacek,

S., Orel, R., Shamir, R., Vandenplas, Y., Van Goudoever, J. B., et al. (2018). Probiotics for the prevention of nosocomial diarrhea in children. *Journal of Pediatric Gastroenterology and Nutrition, 66*(1), 3–9.

12. Li, H., Inoue, A., Taniguchi, S., Yukutake, T., Suyama, K., Nose, T., & Maeda, I. (2017). Multifunctional biological activities of water extract of housefly larvae (*Musca domestica*). *PharmaNutrition, 5*(3), 119–126.
13. Malmo, C., Giordano, I., & Mauriello, G. (2021). Effect of microencapsulation on survival at simulated gastrointestinal conditions and heat treatment of a non-probiotic strain, *Lactiplantibacillus plantarum* 48M, and the probiotic strain *Limosilactobacillus reuteri* DSM 17938. *Foods, 10*(2), 217.
14. Mantegazza, C., Molinari, P., D'Auria, E., Sonnino, M., Morelli, L., & Zuccotti, G. V. (2018). Probiotics and antibiotic-associated diarrhea in children: A review and new evidence on *Lactobacillus rhamnosus* GG during and after antibiotic treatment. *Pharmacological Research, 128*, 63–72.
15. Tarnopolsky, M. A., Bosman, M., Macdonald, J. R., Vandeputte, D., Martin, J., & Roy, B. D. (1997). Postexercise protein-carbohydrate and carbohydrate supplements increase muscle glycogen in men and women. *Journal of Applied Physiology, 83*(6), 1877–1883.
16. Williamson, M. P., McCormick, T. G., Nance, C. L., et al. (2006). Epigallocatechin gallate, the main polyphenol in green tea, binds to the T-cell receptor, CD4: Potential for HIV-1 therapy. *Journal of Allergy and Clinical Immunology, 118*(6), 1369–1374. https://doi.org/10.1016/j.jaci.2006.08.025.
17. Yoshioka, M., Lim, K., Kikuzato, S., Kiyonaga, A., Tanaka, H., Shindo, M., & Suzuki, M. (1995). Effects of red pepper diet on the energy metabolism in men. *Journal of Nutritional Science and Vitaminology, 41*(5), 647–656.
18. Zhou, L., Li, D., Wang, J., et al. (2007). Antibacterial phenolic compounds from the spines of *Gleditsia sinensis*. *Natural Product Research, 21*(4), 283–291. https://doi.org/10.1080/14786410601110728.

CHAPTER 2

THE RELATIONSHIP BETWEEN NUTRITION AND THE IMMUNE SYSTEM

Contents

INTRODUCTION

The relationship between food, nutrition, and health is strong, and the health effects of consuming particular nutrients are significant. Inappropriate nutrient consumption is linked to the development of major human diseases because it impairs the immune system. The quantity and type of nutrients consumed are closely linked to the metabolic stage and the immune system's health.

Nutrition has a strong influence on the inflammatory mechanisms that make up the innate immunity, and when this interaction is disrupted, it can have a significant impact on the development of disease. Using both innate and adaptive immune cells, as well as antibodies that are unique to each pathogen, the immune system can eliminate antigens. Studies about how nutrition affects the immune system are becoming more and more common. The effects of micro and macronutrients were primarily the focus of the early research on nutritional modulation of immune function. The list of nutritional-immuno-modulators has expanded recently to include a wide range of phytochemicals and other chemical biocomponents present in nutrients. While they have an impact on immune function, these biocomponents are not essential for preserving regular cell metabolism and function. Several phytochemicals have been shown to have strikingly beneficial effects on the immune system.

Given the potent effects that nutrients have on the immune system—which we will outline in the paragraphs that follow—it is evident that a diet high in nutrients is strictly necessary to preserve a sufficient level of health. This is on top of the fact that nutrients are essential for survival because they support immune system function, tissue and organ growth, cell division, specialization, energy supply, and growth of tissues and organs.

To make sense of everything, it is necessary to comprehend what constitutes a sufficient diet and, in addition, to be aware of the negative effects processed foods have on the immune system (Figure 2.1).

Furthermore, a weakened host's ability to fight off infections and a compromised immune response are directly linked to nutritional deficiencies. On the one hand, malnutrition is still a significant health issue in less developed areas because it is linked to higher rates of morbidity and mortality, which are typically brought on by the higher incidence of bacterial and parasitic infectious diseases in these areas. In contrast, industrialized nations deal with excessive calorie intake and insufficient diets that offer little nutritional value. Therefore, malnutrition resulting from undernutrition or from consuming poor diets lacking in macro- and micronutrients decreases the immune system's effectiveness. This reduces the immune system's ability to both deteriorate immunity and effectively eliminate pathogens, leaving people vulnerable to a wide range of diseases (Popkin et al., 2012).

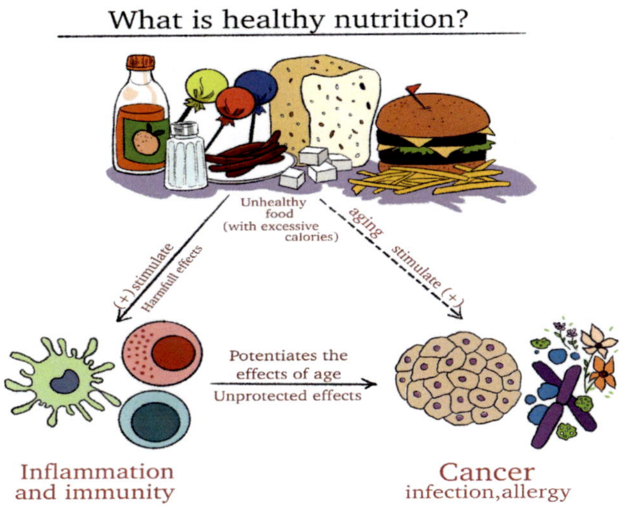

Figure 2.1. *The importance of understanding the meaning of a healthy diet and concomitantly recognizing the harmful effect of pro-inflammatory foods that impact the immune system.*

Source: Munteanu C and Schwartz B (2022). The relationship between nutrition and the immune system. Front. Nutr. 9:1082500. doi: 10.3389/fnut.2022.1082500.

Apart from dietary intake, a crucial inquiry that emerges concerns the bioavailability of nutrients. Could we treat certain deficiencies using supplements if we are aware of them? Or should we try to build an adequate and complex menu that ensures the proper and desired bioavailability? Another important question refers to where does the absorption of nutrients take place? It is critical to understand that pathogenic agents, including bacteria, bacterial products, some toxic food particles, and the intestinal microbiota, frequently cause an immune response. This is critical because the intestinal epithelial barrier plays a major role in maintaining pathogen prevention. Because of its lymphoid tissue, the gastrointestinal tract (GIT) plays a vital role and is a vital component of the immune system. It has been shown that cells within the epithelial barrier present dendritic cells (DCs) in the lamina propria with antigens. The CD103 + CX3CR1-DCs are the most significant. They leave an imprint on the intestinal lymphocytes that promote the growth of regulatory T cells, IgA production, and DCs that produce Th17 cells and TNFα. Additionally, by interacting with Toll-like receptors (TLRs), proteins known to be important in the regulation of the innate immune system, nutrients can regulate the expression of pro- and anti-inflammatory cytokines. They are found in cells like dendritic and macrophage cells, and as such, they use the right signals and crosstalk to regulate the activity of immune cells. Consequently, the enzymatic activity of immune cells is impacted, which in turn causes molecular and chemical alterations associated with inflammation and oxidative stress, ultimately impacting immune function. The majority of the activities have an impact on neutralized cytotoxicity through oxidative reactions.

Except for vegetarians and vegans, humans typically consume vegetables, eggs, milk, dairy products, and meat in their diets. From a biochemical standpoint, the foods can be transformed into the macro- and micronutrients needed for the organism to function properly.

To highlight the significance of certain micro- and macronutrients in preserving human health, we provide a partial summary of the vast body of research that has been gathered thus far on the modulation of immune function.

2.1. THE EFFECT OF NUTRITION ON OPTIMAL IMMUNE RESPONSE

Nutrition is crucial for maintaining the best possible immune response because it gives immune cells the right nutrients at the right concentrations. The immune system can launch potent defenses against infections in this way. Nutrients from the diet play a major role in triggering this rapid response, which helps prevent chronic inflammation. Immunity is weakened and the supply of these components to immune system cells is greatly spared when dietary nutrients are inadequate or ineffective.

Certain macronutrients, like certain amino acids, and certain micronutrients, like vitamins and minerals, have been shown to have a very significant and unique effect on immune modulation. L-arginine and L-tryptophan are two examples of amino acids that are essential to the proper immune function of macrophages. Macrophages exhibit variability in their polarization and plasticity in reaction to alterations in the intracellular milieu. They can change into distinct subtypes in response to various signaling molecules and intracellular microenvironments. L-arginine is linked to a well-known immunoregulatory pathway that M2 macrophages take advantage of. The mechanism is mediated by arginase 1, which simultaneously promotes M2 and consumes L-arginine and the genes that cause M1 inhibition. Furthermore, the synthesis of polyamines is controlled by the combination of methionine and arginine. They promote cell proliferation because they can preserve DNA homeostasis and the integrity of the cell membrane. Furthermore, a plethora of research indicates that these specific amino acids play a role in both the metabolic pathways driving tumor cell growth and the immune system's response against cancer. These types of amino acids provide chemical precursors for several biological reactions through their breakdown. Insulin growth hormone and insulin-like growth factor-I can both be used as potent secretagogues of arginine. Type 2 diabetes and metabolic syndrome are two serious global health issues that are closely linked to diet. Many amino acids are involved in the synergistic stimulation of pancreatic β-cells that release insulin. The ability of arginine to depolarize the plasma membrane at a neutral pH in the presence of glucose is one of the most well-known mechanisms (Figure 2.2). We refer to this gating mechanism as cationic (Newsholme et al., 2005).

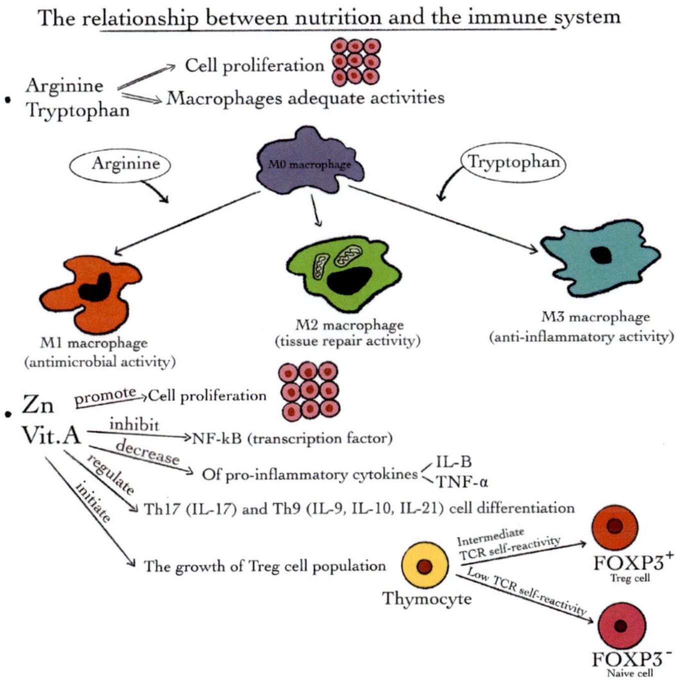

Figure 2.2. *The relationship between nutrition and the immune system. Macronutrients such as arginine and tryptophan are involved in cell proliferation and macrophages' adequate activities. Micronutrients like Vitamin A and Zinc can promote cell proliferation; inhibit the nuclear factor kappa-light-chain-enhancer of activated B cells (NF-κB) pathway; decrease the pro-inflammatory cytokines IL-1β and tumor necrosis factor-α (TNF-α); regulate the Th17 and Th9 cell differentiation; initiate the growth Treg cell population.*

Source: Munteanu C and Schwartz B (2022). The relationship between nutrition and the immune system. Front. Nutr., 9:1082500. doi: 10.3389/fnut.2022.1082500.

A potent immunosuppressive enzyme called indoleamine 2,3-dioxygenase 1 (IDO1) catalyzes the initial and rate-limiting step of the catabolism of L-tryptophan. Tumor necrosis factor (TNF), IL-1, and interferon-γ (IFN-γ) are produced when IDO1 activates the storage of L-tryptophan. Dendritic cells (DCs) express high levels of IDO1 and exhibit catalytic activity in response to IFN-γ. NAD + is produced as a result of tryptophan metabolism and is a cofactor that can participate in redox reactions. The initiation of the nuclear factor kappa-light-chain-enhancer of activated B cells (NF-κB) pathway may be determined by arginine catabolism in immune tolerance. In addition to acting as a substrate for cell growth and survival, arginine is also essential for proper gene expression and differentiation.

The endoplasmic reticulum (ER) becomes stressed due to a decrease in protein metabolism that is connected to a drop in the concentration of some amino acids. Consequently, pro-inflammatory cytokine-producing T cells are activated. There is a correlation between decreased T cell capacity to induce tumor immunity and Arg deficiency.

In addition to their roles in cell division and proliferation, vitamins, and minerals like zinc (Zn) also play a role in immune modulation. For instance, these micronutrients can alter the rate at which antibodies are synthesized.

Vitamin A plays a crucial role in the biosynthesis of carotenoids and retinyl esters, two molecules that have been shown to influence healthy immune function. If it binds to retinoic acid receptors (RARs), it can also function as a transcription factor. By controlling the expression of particular genes, it can therefore be in charge of lipid homeostasis, cell division, growth, and specialization. Immune system dysfunctions caused by vitamin A deficiency include decreased neutrophil function, inhibition of natural killer (NK) cell activity and number, and impaired macrophage phagocytosis ability. Furthermore, it might have an impact on B cell development and differentiation. The likelihood of contracting an infection can rise in this way (Prentice, 2017).

Another member of the micronutrient group is zinc. Zinc inhibits the transcription factor NF-κB. Additionally, alteration of the Toll-Like Receptor 4 (TLR4) signaling pathway may lead to a suppression of the production of the pro-inflammatory cytokines IL-1β and tumor necrosis factor (TNF-α). Moreover, zinc can moderate the Th17 and Th9 cell differentiation pathway, which is specific to pro-inflammatory responses. Since zinc administration can increase the number of Treg cells, zinc is thought to be important for the development of immune cells. Natural killer cell activity, delayed-type hypersensitivity (DTH) response, and reduced lymphocyte proliferation are some of the specific effects. As we have previously discussed, immunity and diet have robust and dynamic relationships that are critical to preserving health (Munteanu and Schwartz, 2022).

2.2. NUTRIENTS INVOLVED IN MEDIATION OF PRO-AND ANTI-INFLAMMATORY RESPONSES

The cells that make up the immune system are from the innate and adaptive mechanisms, which are the two categories of immune responses. The innate immunity system's cells mediate the body's initial response to pathogens once they enter. Phagocytes, DCs, neutrophils, eosinophils, mast cells, and a few other cells make up this system. Here, the immune system responds swiftly. The innate immunity system differs from the adaptive response in that it is less effective and less specialized. T cells, on the other hand, are crucial for antigen identification because the adaptive response can identify any pathogen and can also recall whether it has been encountered previously. They also play a role in controlling the immune response. Additionally, there are two types: T4 helper cells (CD4 inducer), which help regulate the responses of other cells, and cytotoxic T cells/T8 (CD8 inducer), which are directly implicated in the death of tumor cells and infected cells. Additionally, Th1, Th2, and Th17 are some of the subtypes of T helper cells that produce cytokines. Th1 cells are in charge of battling viruses and bacteria. These cells' primary function is to produce IL-2 and interferon γ (IFNγ). Like IL2, IFNγ is produced by both innate and adaptive immune cells, including Th1, T8 lymphocytes, NK, and innate lymphoid cells. Th2 cells also stimulate the immune system at the same time. They can make additional interleukins (ILs). Both cytotoxic $CD8^+$ T cells and activated macrophages can induce cell death. It is interesting to note that the other immunity regulatory T cells contribute to the immune response's inhibition or blocking of cytokine secretion. Given this knowledge, they play a critical role in the development and maintenance of peripheral tolerance.

Part of the adaptive immune system, B lymphocytes are also responsible to produce antibodies. Like T cells, they are able to react differently to every antigen. Antigens can cause harm to the tissue they target. It is understandable that the pathogens in the surrounding tissues and tissues encourage an inflammatory response. Its primary function is to restore the damaged tissue in a way that reduces the antigens' extension and gets rid of their effects. Following this process, a few physiological alterations take place, which are in charge of raising the phagocyte count in the area where this process is taking place. This leads to the delivery of pro- and anti-inflammatory cytokines, prostaglandins, and complements, particularly through phagocyte activation. All these modifications lead to an increase in the inflammatory response. Now that the cells involved in each inflammatory response pathway have been identified, it is possible to clarify the impact of particular nutrients on each of these processes.

To achieve this, we decided to talk about the impact of a few different vitamins, including A, B1, B2, B3, B12, C, and D; minerals, like zinc and selenium; and specific amino acids, like tryptophan and arginine; as well as some fatty acids.

2.2.1. Vitamin A

By activating nuclear RARs, all-trans retinoic acid, 9-cis retinoic acid, or other metabolites, vitamin A regulates both innate and cell-mediated immunity as well as antibody responsiveness. The immune system is significantly regulated by vitamin A and related retinoid metabolites. When there is a vitamin A deficiency and an increased susceptibility to infections, this crucial role is demonstrated. A vitamin A deficiency impacts the release of cytokines and the production of antibodies in the right amounts. Furthermore, low vitamin A levels are linked to decreased generation of NK cells, monocytes, or macrophages, as well as compromised T- and β-lymphocyte maturation and proliferation. A vitamin A deficiency reduces the ability of neutrophils, macrophages, and NK cells to function normally and prevents the normal regeneration of mucosal barriers damaged by infection. Supplementing with vitamin A reduces morbidity and mortality in a range of infectious diseases. Children are most affected by vitamin A deficiency because it affects the integrity of the mucosal epithelium, which increases the accessibility of various pathogens to the GIT and other organs. Severe vitamin A deficiency in children leads to the nearly complete disappearance of goblet cells, which are found in the upper layer of the epithelial line. As a result, these cells are unable to produce mucus, which strengthens the bacterial adherence to the epithelial lining and is a major factor in the development of bacterial diseases. Furthermore, decreased phagocytic activity and macrophage oxidative breakdown that occurs during the inflammatory process, as well as a decrease in the quantity of NK cells, are linked to vitamin A deficiency.

It has been shown that vitamin A promotes Th1 and Th2 cell growth and differentiation. Thus, by suppressing the production of IL-12 and IFNγ by Th1 lymphocytes, vitamin A can enhance the Th2 anti-inflammatory response. Furthermore, some research indicates that vitamin A is positively correlated with pro-inflammatory cytokines (IFN-γ) and anti-inflammatory cytokines (IL-10), which are induced by mitogens. It is crucial to understand that there are three different forms of vitamin A: retinol, retinoic acid (RA), and retinal. Numerous biological activities involving RA have been demonstrated. As stated by Rampal et al. in situations where inflammation is present, RA may either maintain or exacerbate intestinal inflammation. Moreover, RA can influence the activity of macrophages by releasing specific cytokines, including nitric oxide (NO), IL-1, IL-6, and IL-12. Vitamin A administration lowers diarrheal disease-related mortality in children with hypovitaminosis. The antitumor effects of vitamin A on human pancreatic cell lines may be the cause. All-trans-RA (ATRA) appears to have a comparable effect on renal carcinoma metastases. ATRA is a nutrient that the human body synthesizes from the A vitamin and is needed in trace amounts. When treating acute promyelocytic leukemia (APL), ATRA is a highly effective treatment. Furthermore, they can lengthen life expectancy in conjunction with arsenic trioxide (ATO). About 95% of cases of this disease recover as a result of this combination. After vitamin A was administered, cytotoxic effects were observed in breast cancer patients, but healthy cells were unaffected. Consequently, vitamin A has the ability to mitigate certain adverse effects of chemotherapy. The next important question to consider is whether or not pro- and anti-inflammatory responses are mediated by additional nutrients.

2.2.2. Vitamins B1, B2, B3, and B12

The eight water-soluble vitamins that make up the B-vitamin group are responsible for carrying out crucial, interconnected tasks that are necessary for proper cellular functioning. These vitamins are necessary cofactors for numerous significant cellular metabolic pathways and function as effective co-enzymes in a wide range of catabolic and anabolic enzymatic reactions. Thus, it would be impossible to discuss life or cellular life without mentioning the B vitamins. In this context, certain cofactors, such as nicotinamide adenine dinucleotide (NAD) (B3), flavin mononucleotide (Flavin Adenine Dinucleotide/B2), and thiamine pyrophosphate (B1), are used by key enzymes that regulate essential cellular processes. Regarding B vitamins, however, it is crucial to remember that, in this case of the human body, the activity of the gut microbiota determines a significant source of these vitamins, with the possible exception of those that may be consumed through diet. The large and small intestines are the two different intestinal regions where B vitamins are absorbed. For the most part, the large intestine serves as the primary site of absorption for B vitamins produced by bacteria. In addition, the absorption of dietary B vitamins occurs in the small intestine. It is speculative to assume that the two distinct absorption sites cause two distinct immune responses. Since the population of gut immune cells varies, we hypothesize that the immune activities at the two distinct locations differ (Rodionov et al., 2019).

Thiamine, also known as vitamin B1, protects neutrophil surface sulfhydryl groups, which in turn has an antioxidative effect. Consequently, there is an additional blockage in the synthesis of cytokines by macrophages. Thiamine, along with other substances, can prevent and inhibit PMNL's oxidative system in relation to the stimulation of antimicrobial oxidative reactions myeloperoxidase (MPO), H_2O_2, and a halide (HRP/H_2O_2/Nal) determined by the activation of polymorphonuclear leukocytes. Thiamine inhibits the NF-κB pathway, which is involved in controlling oxidative stress. This function is emphasized by inhibiting inhibitory kappa B's (IκB) phosphorylation and catabolism, which in turn prevents the transcription factor-sensitive redox NF-κB from translocating to the nucleus. Based on biochemical and immunological perspectives, it can be inferred that thiamine derivatives modulate immune cell activity, thereby influencing immune metabolism. These characteristics stem from its role in preserving the balance between the TCA cycle and glycolysis. They serve as cofactors for the enzymes involved in these pathways, as we have previously discussed. The primary source of resting macrophages, T-regulatory cells, and naïve T cells is the TCA energy cycle. It is interesting to note that not enough energy is obtained from TCA for activated T helper cells; instead, they require energy from aerobic glycolysis. B1 deficiencies have major side effects because thiamine has a significant impact on these pathways. One of the adverse effects is associated with the induction of pro-inflammatory cytokines (TNF-α, IL-6, and IL-1), as well as neuro-inflammation. Lastly, inhibition of CD 40 and CD 40L regulation may result in neuronal death. B1 has been shown to have a potential role in the treatment of neurodegenerative diseases by inhibiting the pro-oxidative activity of microglial cells. In addition, we should be aware of the role that B vitamins play in oncogenesis. It is also critical to distinguish between people who are well and those who are not. Because B1 functions as a cofactor in energy pathways and proliferation, both of which are critical for the growth of tumor cells, there are some theories about how B1 may be involved in this disease. It will take more investigation to definitively identify thiamine's potential carcinogenic effects.

As a cofactor, riboflavin, also known as vitamin B2, is essential for the metabolism of energy. Additionally, it is a significant anti-inflammatory and antioxidant modulator, particularly in the lungs. It is important to discuss a few particular details about the relationship between the major histocompatibility complex (MHC) and B2 bacterial compounds. Invariant T cells are stimulated as a result of this function on the innate mucosa. The majority of human innate-like T cells, known as mucosa-associated invariant T cells (MAIT), are selectively activated by riboflavin and its precursors. It is unclear how they are synthesized and how they are connected to the MHC via the major histocompatibility complex protein (MR1). It was found that genes encoding the enzymes needed to form intermediate compounds during the synthesis of bacterial riboflavin are necessary for the activation of MAIT cells.

Because these cells produce IL-17 and IFN-γ, they are well-known for their role in the inflammation and defense activity in the gut mucosa. The activity of riboflavin may increase the growth of neutrophils and monocytes as well as the stimulation of neutrophil and macrophage activities. The pro-inflammatory factor Kappa B (NF-κB) is activated through the catabolism of inhibitory kappa B (IκB). The inflammatory signaling pathway is then triggered by this catabolic pathway. TNF-α and ILs, two pro-inflammatory cytokines, are activated at the end of this signaling pathway. Vitamin B2 functions as an anti-inflammatory suppressor in this signaling process and has the potential to prevent NF-κB activation. Moreover, vitamin B2 may lessen oxidative stress by overexpressing catalase and nitric oxide synthase (iNOS).

Niacin, i.e. vitamin B3, is a precursor to NAD and NADP. It is a cofactor for many different enzymes involved in multiple metabolic pathways, just like all B vitamins are. Unlike other B vitamin groups, human cells have separate pathways for synthesizing NAD and NADP cofactors. Niacin and the resultant cofactors are involved in redox reactions from a biochemical perspective. Genomic balance is maintained by NAD, and its mode of action may involve epigenetic regulation.

Furthermore, a positive correlation has been observed between elevated NAD concentrations and the inhibition of ROS synthesis. Furthermore, because of its deacetylation and inhibitory effects on the NF-κB pathway, NAD can be regarded as an anti-inflammatory micronutrient. Additionally, it inhibits animal tumor cells as well as inflammatory cytokines. NAD is regarded as an effective anti-inflammatory agent due to its ability to decrease the release of specific cytokines from alveolar macrophages.

Cobalamin, or B12, influences both pro- and anti-inflammatory reactions. There is evidence of a negative correlation between TNF-α and vitamin B12. It has been shown that an elevation in TNF-α causes the antioxidants involved in the defense against free radicals to become exhausted. Pro-inflammatory cytokines and a few other pro-inflammatory substances are consequently activated. It is interesting to note that, in contrast to healthy individuals, there is a decrease in the number of $CD8^+$ T cells in human anemia caused by cobalamin deficiency. On the other hand, compared to healthy individuals, patients with cobalamin deficiencies showed an increase in the number of $CD4^+$ T cells. The $CD4^+$ / $CD8^+$ ratio is pathologically higher in these situations. Furthermore, there is a decrease in NK cell activity in these patients. It is interesting to note that vitamin B12 deficiency causes hyperhomocysteinemia, which increases inflammatory processes and can result in chronic illnesses like insulin resistance and coronary heart disease. Since aberrant TNF-α activity is linked to vitamin B12 deficiency, insulin resistance may also result from it. In terms of cancer activity, Cheng et al. discovered no link between B12 and specific cancers, including squamous cell carcinoma, prostate, breast, and colorectal cancer, when viewed through a genetic lens. Administration of vitamin B12 was not regarded as a risk factor in the case of lung cancer. Conversely, a large meta-analysis of cancer patients revealed that a higher B12 intake was associated with risk for many cancer types (Arendt et al., 2019).

2.2.3. Vitamin C

Considering that humans are unable to synthesize it, vitamin C is regarded as an essential micronutrient. When compared to other species that can synthesize vitamin C, humans absorb more of it. Vitamin C has a function as a regulator of cell signaling and is involved in the modulation of many immune functions. In addition to its involvement in hydroxylation reactions, vitamin C is also involved in gene transcription. It gained the ability to protect the body from reactive oxygen species (ROS), which are brought on by the activity of toxins and pollution, through the course of its primary role as an antioxidant.

Vitamin C is in charge of blocking the start of the NF-κB reaction and stopping the pro-inflammatory cytokine actions. When vitamin C is administered to peripheral blood cultures stimulated with lipopolysaccharide (LPS), it has been observed that IL-10 is enhanced and TNF-α and IFN-γ are reduced. Furthermore, vitamin C induces neutrophils to relocate into infected areas as a result of ROS buildup in microbial infections. Furthermore, in cases of severe infections, vitamin C may be helpful as a cofactor in the pathways leading to the synthesis of norepinephrine and vasopressin. When the pathological state poses a risk, this has a discernible impact on the cardiovascular system's infection response. Because external factors, especially pollutants, work in concert to combat ROS, it seems that vitamin C can act as an antioxidant protector for the skin. In this instance, administering vitamin C in addition to vitamin E has a more noticeable effect. Ellulu et al. have shown that adults with hypertension and/or diabetes experience a reduction in inflammation after receiving C vitamins, which is accompanied by a moderate drop in inflammatory markers like IL-6 and high-sensitivity C-reactive protein (hs-CRP). Additionally, vitamin C plays a role in controlling the activity of hypoxia-inducible factor 1-alpha (HIF-1α), which allows neutrophils to survive in hypoxic environments and postpones neutrophil apoptosis. Moreover, it is believed that vitamin C contributes to the immune system's ability to combat tumor cells by boosting the production of NK cells.

2.2.4. Vitamin D

As vitamin D receptors are expressed in various organs throughout the human body, vitamin D has a multitude of anti-inflammatory functions. The effects that are most well-known and proven are related to the metabolism of minerals and bone. Wöbke and associates have shown that the effects of vitamin D are mediated by pro-inflammatory components involved in both nuclear and cytosolic signaling control pathways. When vitamin D binds to its receptors (VDR), a complex of vitamin D-VDR is formed. This complex can either form a heterodimer compound with the nuclear retinoid X receptor (RXR) or a homodimer with another VDR. Furthermore, the nuclear role is shown after heterodimers with steroid hormone receptors are formed. When vitamin D is bound to VDR or RXR, it can pass through the nuclear membrane, bind to a response element, and activate the expression of that response gene to initiate its particular gene regulation action.

As a result of the high expression of VDRs by immune cells like B and T cells, vitamin D appears to be involved in adaptive immunity. From an immunological regulatory perspective, vitamin D can prevent monocytes from secreting pro-inflammatory cytokines like TNFα or IL-6. In addition, prostate cells have been shown to exhibit the same behavior. The inhibition of P-38 MAP kinase, a subclass of mitogen-activated protein kinase, in response to pro-inflammatory cytokines is what causes these effects. Furthermore, the connection between the activation or inhibition of the MAP kinase signaling pathway and VDR/RXR is intriguing. The outcomes are intimately linked to the effects of triggering factors, the response of cells, and their specificity. Additionally, the complex VDR/RXR can bind to other substances that are involved in transcription, like NF-κB and the glucocorticoid receptor (GCR). Because of these interactions, vitamin D may be regarded as an anti-inflammatory micronutrient.

As a result of becoming active, the vitamin D-bound VDR inhibits NF-κB, another heterodimer compound. Furthermore, certain research indicates that the anti-inflammatory properties of vitamin D are also facilitated by the suppression of particular pro-inflammatory Th1 cell cytokines, including TNF-α, IFN-γ, IL-6, IL-2, and IL-17. Furthermore, due to an increase in Th2 cell activity, vitamin D can raise the concentration of cytokines like IL-10, IL-4, and IL-5. It may also result in a decrease in Th17 cell counts and an increase in Treg cell amplification.

From a medical standpoint, vitamin D significantly impacts the lung defense mechanism against microbiological infections. This function is the outcome of the activation expression of antimicrobial peptides in monocytes, neutrophils, NK cells, and respiratory tract epithelial cells. Higher infection risks are associated with lower serum levels of vitamin D. In particular, vitamin D administration causes a decrease in acute respiratory infections.

The main effects mediated by calcitriol are the stimulation of apoptosis, the suppression of cancer cell proliferation, and the associated delayed tumor development in cancer. Calcitriol is the biologically active molecule of vitamin D that mediates the anti-cancer effect of vitamin D in tumor cells. At the molecular level, calcitriol is regarded as an anti-inflammatory nutrient because it inhibits the activity of prostaglandins (PG) and pro-inflammatory cytokines while also blocking the NF-κB signaling pathway. According to this theory, calcitriol has potential applications in cancer treatment and prevention.

2.2.5. The Minerals-Zinc and Selenium

Chronic inflammation occurs when levels of inflammatory cytokines are sustained at elevated levels. The activity of certain minerals is intimately related to this process. Thus, it is critical to investigate zinc's function in this crucial process. Many signaling pathways that are activated by certain modifications made by antigens and their metabolites mediate the process. Because it plays a part in cell proliferation, cell apoptosis, and the release of specific cytokines like IL-6, IL-8, and IL-1β, the primary compound involved in inflammatory responses is mediated by the activity of the NF-κB factor. Zinc's function in this context is debatable. Studies conducted *in vitro* show that zinc has both pro- and anti-inflammatory properties. Certain intracellular membrane chelators, such as TPEN (N, N, N,' N'-tetrakis (2-pyridinylmethyl)-1,2-ethanediamine), can inhibit the NF-κB signaling pathway. After the chelator and heavy metals bind, the apoptosis effect is visible from one angle. On the other hand, several investigations show a substantial correlation between zinc and the start of LPS-induced NF-κB. Furthermore, zinc can prevent pro-inflammatory effects following a decrease in IL-1β release.

Moreover, it has been documented that the production of cytokines is contingent upon the level of zinc, and this is strongly linked to persistent inflammation. In this context, it has been noted that IL-1β, IL-1α, and IL-6 genes are overexpressed in obese individuals with low zinc plasma concentrations. T lymphocyte differentiation and proliferation are positively impacted by zinc. Cross-sectional studies have shown a strong correlation between a high number of cytokines and a decrease in zinc plasma levels in infections and trauma-associated conditions. Genes involved in cytokine production have been found to be upregulated in patients with severe head injuries. Furthermore, a decrease in plasma zinc concentration is associated with increased cytokine production in critically ill patients. Additionally, the antioxidant properties of zinc support the body's defenses against ROS and reactive nitrogen species (RNS). Zinc has the ability to trigger the start of $CD8^+$ T cell proliferation. Additionally, zinc can maintain the integrity of mucous membranes and skin. In summary, zinc plays a critical role in the growth and multiplication of immune system cells, including T lymphocytes and $CD8^+$ T cells, which are renowned for their rapid turnover. It has been noted that a lack of this mineral has a detrimental effect on health by lowering resistance to dermatitis, growth disorders, infectious diseases, and genetic disorders.

Selenium is another essential micronutrient that is required for the proper operation of the thyroid metabolism, cardiovascular system, immune system, and prevention of cancer. Even if these stages are uncommon in the human body, there are still disagreements from a cellular perspective about the precise dosage that may result in deficiency or toxicity.

It is commonly known that selenium, when present in the amino acid selenocysteine, can regulate specific metabolic reactions that can result in the synthesis of lipoxygenase, which can ultimately contribute to the production of inflammatory mediators. Selenium may help regulate cellular immunity in mice by increasing the activity of T-cell receptor complexes (TCRs) and converting T0 cells into Th1 cells. Because of how zinc affects redox signaling activities, it can also help the body fight off infections. Recent research has shown that zinc and selenium both play protective roles in COVID-19 patients and are linked to improved survival rates. It has been shown that selenium administration may enhance the response during the COVID-19 vaccination, as well as the titer antibody response. Selenium is thought to play a role as a cofactor in the vaccination-mediated immune response.

Selenium treatment has also been shown to decrease IL-1 and TNF-α gene expression in women with polycystic ovary syndrome who are infertile and for whom *in vitro* fertilization has been advised. This result points to selenium's potential anti-inflammatory function in the human body. Additionally, selenium supplementation raised IgA and IgG antibody titers as well as neutrophil counts in cancer patients. It is possible to conclude that selenium regulates the synthesis of inflammatory mediators and may also stimulate phagocytic cell activity. Th1 cell-mediated immune response and T cell-based stimulation can both be improved by selenium. There is a positive correlation between selenium and B cell count. The administration of selenium may enhance the innate immune system. Cellular immunity has been shown to be affected similarly. Supplemental selenium has been shown to increase antibody titers, which may amplify the benefits of vaccinations. Following selenium infusion, the brain exhibits an increase in both neurogenesis and hippocampal neural precursor cells.

2.2.6. Macronutrients: Amino Acids Arginine and Tryptophan

In addition to micronutrients, macronutrients like proteins and amino acids are crucial for immune system function. Amino acids are building blocks of other proteins, such as cytokines and antibodies, which are characteristic immune system proteins. These building blocks are known as proteins.

In macrophage cells, arginine aids in the synthesis of nitric oxide. The production of nitric oxide by arginine, which is then processed by iNOS, controls the cytotoxicity of macrophages in the defense of pathogenic bacteria and parasites. Additionally, arginine is used by M1 macrophages to generate NO. Although arginine was once thought to be a non-essential amino acid, research published a decade later has shown that it is necessary for ontogenetic fetal development, constant hemodynamics and vascular parameters, and embryonic outliving. Furthermore, there is a connection between the arginine degradation pathway and the induction of the NF-κB pathway. As previously discussed, arginine can enhance the release of insulin from pancreatic β cells via a mechanism that is dependent on cations. Furthermore, arginine raises Ca^{2+} concentration in β-pancreatic cells via electron transport, which is reliant on the mCAT2A transporter amino acid. Following the opening of Ca^{2+} voltage-dependent channels and an increase in calcium intracytoplasmic concentration, depolarization of the membrane stimulates the release of insulin. Clinical assessments, however, have revealed that the beneficial effects of arginine administration are minimal. This is most likely because arginine is rapidly converted by epithelial cells into ornithine or citrulline. Additionally, balanced levels of membrane, mRNA, and DNA are maintained by the polyamines, which are substances also derived from the breakdown of arginine. Polyamines can therefore regulate the rate at which cells proliferate. Polyamines have the ability to alter the inflammatory process *in vitro*. Moreover, it has been shown that increased intracellular polyamine concentrations may alter the *in vitro* cytotoxicity that macrophage cells control. Polyamines have a significant role in the regulation of inflammation and pathogen identification due to their mode of binding to receptor-ligand complexes.

Over the past few decades, research has shown that arginine is more advantageous than previously thought. For instance, arginine has been shown to reduce intestinal inflammation and oxidative stress. In humans and other mammals with intestinal diseases, arginine can reduce damage to the gut and restore the balance of the mucosal immune system. The NF-κB pathway mediated by IL-1β can be inhibited by arginine in an *in vitro* intestinal system in Caco-2 cells. It is still unclear how to stop the inflammatory pathways, though. It might have something to do with Arginase-1 (Arg-1) activity, which is in this case stimulated by the substrate L-arginine. The enzyme arginase-1 helps to convert l-arginine into l-ornithine and urea at the end of the urea cycle. According to research, these anti-inflammatory properties help treat several inflammatory illnesses. On the other hand, research has demonstrated that tumor cells' increased arginine metabolism in combination with their unique environment produce conditions that are both intermediary and essential for the survival and growth of cancer cells. These activities translate into appropriate immunosuppression. There is no denying the contentious nature of the arginine-cancer cell relationship. On the one hand, the data point to a possible correlation between arginine deficiency and a postponement of certain tumor cell development.

However, arginine can also have antitumoral effects, which are demonstrated by an increased immune response. Additionally, Al-Koussa et al. have revealed that arginine is necessary for the growth of some types of cancer cells. Accordingly, arginine deficiency may inhibit the migration of cancer cells, according to some writers. The process of movement is beneficial for immune system and embryonic growth under physiological conditions. That being said, things are different with regard to cancer cells. This occurs because specific types of cancer cells are able to take advantage of this characteristic to promote metastasis. Thus, it is possible to lower the activity of metastasis in cancer cells by depriving them of arginine. Regretfully, the precise mechanism is still a mystery.

It is evident that tryptophan (Trp) is necessary for the immune system to function. Trp becomes essential for cell division and development since it is required for protein synthesis. Trp must be obtained through diet because the human body is unable to synthesize it. Trp is a substrate that is used in the synthesis of kynurenine (Kyn), indoles, and serotonin (5-HT). The Kyn route, which is linked to the synthesis of kynurenic acid and NAD, is the most beneficial and active Trp metabolism. Naturally, as with all pathways, this kind is caused by the participation of two different kinds of enzymes: tryptophan 2,3-dioxygenase (TDO) and indoleamine 2,3-dioxygenase (IDO and IDO2). In addition to the metabolism of Trp, we also presented data regarding its function in controlling inflammation via its initiators, beginning with IDO, which has negligible effects on normal and healthy conditions. Certain cytokines, such as interferons, which are the end product of an inflammatory process that has been triggered, alter things. The extensively utilized and extremely powerful cytokine interferon-gamma (IFN-γ). It can express itself in a variety of cell types and is connected to the promotor region of IDO. Though it can also be seen in other tissues like connective tissues and epithelium, DCs, and macrophages exhibit the highest expressive grade. The serum ratio Kyn/Trp may serve as a marker for inflammation, which is linked to IDO activity, concerning cancer, infections, auto-immune diseases, or cardiovascular issues. Trp biochemistry controls both chronic immune tolerance and inflammation, as we have previously discussed. This is impacted by the capacity of IDO to alter the Kyn/Trp ratio. By influencing specific metabolic pathways and immune signaling, the Kyn/Trp equilibrium can be used by IDO-competent cells to initiate an anti-inflammatory response. NAD +, the final step metabolite of the Kyn pathway, is recognized for starting the programmed death of $CD8^+$ and $CD4^+$ lymphocytes. Amino acid metabolism is a key component of metabolic reprogramming in tumor cells. According to some authors, there is a strong correlation between the two in the case of gliomas because certain clinical characteristics and survival may be predicted by the metabolic amino acid pathway. As previously discussed, malignant cell control and microenvironment modification are attributed to amino acids and their metabolites. The outcomes are thus translated into an improvement in the malignancy state and immunosuppression.

The metabolism of kynurenine has the ability to activate a pathway that resists oxidative stress, which can lead to modifications in the tumor microenvironment that promote tumor growth. On the other hand, 5-methoxytryptophan (5-MTP), a different tryptophan metabolite, has the power to prevent tumor growth and the spread of cancer cells to other tissues. Wu and associates ponder the possibility that cyclooxygenase-2 (COX-2) suppression is the cause of these effects. This particular kind of inflammation-associated enzyme is highly prevalent in tumor cells and aids in the progression of cancer (Munteanu and Schwartz, 2022).

2.3. THE ROLE OF CHOLESTEROL IN THE IMMUNE RESPONSE

The functioning of cellular membranes is significantly impacted by cholesterol, particularly in the cell's plasma membrane where it is present in higher concentrations. Its unique position within the lipid bilayer promotes the best possible interaction with other lipids and plays a big part in membrane fluidity. All the cholesterol's structure is directed toward the lipid bilayer, except the hydroxyl group, which faces the outside environment. As a result, the hydrocarbon chains and steroid rings of adjacent lipids are close by. For the plasma membrane to perform its various physiological functions, cholesterol is essential. Cholesterol molecules allow the cells to maintain their proper lipid bilayer functionality; without them, the microenvironment, endocytosis, signaling pathways, and other processes would be affected.

Bilayer fluidity and receptor arrangement are caused by cholesterol, which also plays a role in membrane integrity. We will give a schematic description of the cholesterol synthesis and pathway so that you can better understand the information that will be provided later. Although the process of cholesterol biosynthesis is complicated, the pathways involved have been well-defined. More than 20 metabolic-specific processes are involved in its synthesis, including enzymatic reactions from the cholesterol synthesis pathway and the mevalonate pathway. The majority of the enzymes involved in the biosynthesis of cholesterol are found in the ER membranes. Several tightly regulated molecular reactions aim to target these enzymes and prevent any damage to the cells. Nonetheless, the plasmalemma does not distribute cholesterol in an even manner. The primary pool and storage in a cell is represented by the plasma membrane. It has been noted that every pool in the physiology of the plasma membrane corresponds to a distinct function.

The concentration gradient from areas of high cholesterol concentration to areas of low cholesterol concentration or loss makes it evident that a transport mechanism is involved in cholesterol equilibrium. Because of its hydrophobic conformation, cholesterol cannot be transported through the blood without the help of proteins. As a result, cholesterol binds to various proteins to form unique lipoprotein compounds, including high-density lipoprotein (HDL) and low-density lipoprotein (LDL). As one might anticipate, distinct regulatory pathways underlie the synthesis of every lipoprotein. As most human cells are unable to effectively break down cholesterol, any excess can either be transported through the efflux process or deposited as intracellular lipid droplets. Lipid droplet plaque deposition results in the release of inflammatory cytokines into the bloodstream, which in turn triggers an inflammatory process. This event's effect is linked to inflammation brought on by the cytokine interleukin-1β (IL-1β). Moreover, IL-1β is thought to be a significant marker of the inflammatory process.

We will draw attention to the cholesterol signaling pathway because it is involved in the immune response. The protein known as sterol response element-binding protein, or SREBP, plays a crucial part in the cholesterol signaling pathway. These proteins are normally found in the ER membrane, which can bind to two other complex proteins, the insulin-inducible genes (INSIGs) and the cleavage-activating protein and generator (SCAP). One important aspect of SCAP's function is its transition from the ER to the Golgi apparatus. SREBP2 is the most significant of the three SREBP protein variants, which are SREBP1a, SREBP1c, and SREBP1. One protein complex structure called SREBP2 appears to have the ability to control the expression of every enzyme involved in the biosynthesis of cholesterol. Its primary function is its unique reaction to elevated sterol concentrations, which effectively inhibit the synthesis of cholesterol.

The function of SREBP2 is fulfilled when the concentration of sterols decreases. The complex SCAP will subsequently be transferred from the ER to the Golgi apparatus as a result of this change in sterol concentration brought on by SREBP2 activity. This organelle has a modified SCAP molecule. Proteases sit 1 and 2 cut this complex once the SCAP reaches the Golgi apparatus. Consequently, the production and stimulation of the transcription factor (TF) occurs. The TF then makes its way into the nucleus, where it controls the activity of the enzymes involved in the cholesterol synthetic pathway. In terms of immune-related processes, the T cell receptor (TCR) and B cell receptor (BCR) of immune cells may activate SREBP2 via their signaling pathways. The flux of cholesterol biosynthesis may be stimulated by all these pathways (Figure 2.3).

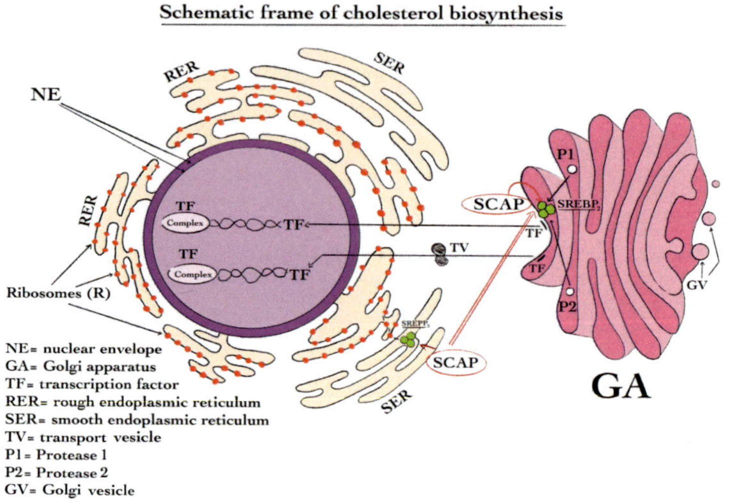

Figure 2.3. *Schematic frame of cholesterol biosynthesis.*

Source: Munteanu C and Schwartz B (2022) The relationship between nutrition and the immune system. Front. Nutr. 9:1082500. doi: 10.3389/fnut.2022.1082500.

The fact that all cholesterol-related pathways, including synthesis, influx, efflux, and esterification, occur through mechanisms that are closely related to one another and enable well-adjusted whole mechanistic biochemical pathways, should also be mentioned. These meticulously regulated systems elucidate the pivotal function of cholesterol in longevity and elucidate the possible hazards that may arise from deviations from the ideal range of concentrations. Regarding this, Luo et al. have shown a connection between many metabolic diseases, including alterations in cholesterol metabolism and Schnyder corneal disease and familial hypercholesterolemia.

Furthermore, because of the consequences of inflammation, the biochemical equilibrium of cholesterol is significantly disrupted in several diseases, including allergies, infections, and different forms of cancer. About the correlation between macrophages and cholesterol, counter-regulatory mechanisms work against both the accumulation of cholesterol in cells and the inflammation caused by macrophages. Certain sterols are produced when cellular cholesterol concentration rises. The transcription factors RXR and liver X receptor (LXR) are activated with their assistance. Controlling the expression of ATP-binding cassette transporters (ABC transporters), namely ABC subfamily A member 1 (ABCA1) and ABCG1, is one of these anti-inflammatory functions of heterodimers. They also play a role in promoting cholesterol outflow from macrophages. They can inhibit the TLR signaling activation caused by the elevated intracellular cholesterol concentration in this manner. LXR genes are inhibited when TLRs are activated, which lowers macrophage cholesterol efflux. Apolipoprotein A1 (APOA1), which forms HDL and starts the process of transferring cholesterol back to the liver through blood vessels and lymphatics, is how ABCA1 and ABCG1 activate cholesterol efflux. Thus, cholesterol homeostasis can be changed by the immune system to enhance the inflammatory response. Disturbances in the regulation of cholesterol biosynthesis, leading to elevated cholesterol levels, may result in metabolic disorders like dyslipidemia and atherosclerosis. Both the innate and adaptive immune systems can control this phenomenon in some of these situations.

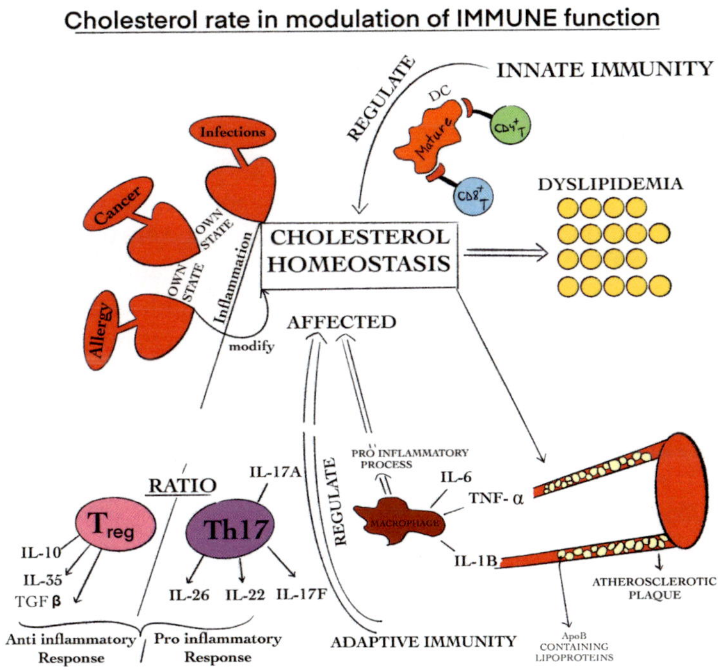

Figure 2.4. *Cholesterol rate in the modulation of immune function.*

Source: Munteanu C and Schwartz B (2022) The relationship between nutrition and the immune system. Front. Nutr.9:1082500. doi: 10.3389/fnut.2022.1082500.

Thus, lipoproteins containing ApoB are produced right after atherosclerosis causes damage. These occur in the endothelial compartment where they are produced, developed, and stored. Fascinatingly, these molecules cause oxidation and acetylation, and most notably, they cause aggregation with other molecules to cause pro-inflammatory effects. The changes brought about by the buildup of lipoproteins containing ApoB in the endothelium cause immune cells to become more adherent, tenacious, and mobile in this area. In conclusion, the production of ROS and specific cytokines like TNF α, IL6, and IL1β contribute to the inflammation. Regarding the inflammatory phase, $CD8^+$ Th and $CD4^+$ T cell inhibition by Treg cells has an anti-inflammatory effect. Th1 and Th17 both contribute to the pro-inflammatory process concurrently. The reduction in the Treg/Th17 ratio is closely associated with atherosclerosis development and chronic inflammation (Figure 2.4).

2.4. THE POLYUNSATURATED FATTY ACIDS EFFECTS ON IMMUNOMODULATION

Essential fatty acids with multiple double bonds in their backbone are known as polyunsaturated fatty acids (PUFAs). The two primary categories of PUFAs are omega-3 and omega-6. The location of the cis double bonds indicates the structural chemical difference between the two groups. PUFAs and cholesterol play a crucial role in the development, integrity, and upkeep of cell membrane homeostasis. Additionally, they promote cell division and are utilized by specific cell structures. Calder and Grimble showed that modifications to the fatty acid composition of membranes can impact membrane fluidity. This is primarily because these modifications affect the enzymes' affinity for their substrates, which in turn can alter signaling pathways. The immune system's sensitivity can be changed in this way. The three omega-3 fatty acids eicosapentaenoic acid (EPA), alpha-linolenic acid (ALA), and DHA are the most representative forms of PUFA. Since they are involved in numerous vital and essential processes—and, more intriguingly, immunomodulation pathways—they are the subject of extensive research. Furthermore, the significance of ALA stems from its role as a precursor to other fatty acids. By lowering pro-inflammatory eicosanoids, omega-3 PUFAs contribute to immunomodulation. They serve as a substrate for the enzymes involved in the AA cascade, which produces specific prostanoids and leukotrienes. Precursors of some lipid mediators, like maresins, are omega-3 PUFAs. They encourage the inflammatory process to come to an end. ALA cause cell proliferation inhibition and apoptosis activation in human breast cancer cells. ALA improves insulin sensitivity and corrects abnormalities in the metabolism of fat and glucose in diabetic rats. The inability of the human body to synthesize ALA makes it essential. In this context, linoleic acid (LA), a member of the omega-6 group, is regarded as essential. This fatty acid is rapidly changed into arachidonic acid (ARA) after consumption of LA, which is what gives the cell membrane its flexibility and fluidity. Through the stimulation or inhibition of their functions, free ARA plays a role in the modulation of ion channels, enzymes, and receptors. Both *in vivo* and *in vitro*, free, unesterified

ARA exhibits antitumoral activity. It is a medication with anti-cancer properties. Furthermore, ARA can induce the death of tumor cells by suppressing their proliferation, which in turn triggers the neutral sphingomyelinase (nSMase) mechanism.

Both omega-3 and omega-6 have a role in immunomodulation. As stated by Simonetto and associates, they produce antagonistic results. By inhibiting ARA from the membrane, which is the primary precursor for pro-inflammatory eicosanoids, omega-3 from the PUFAs group contributes to anti-inflammatory responses. They can control the duration and intensity of immune and inflammatory responses. Pro-inflammatory effects include, on the one hand, fever, vasodilatation, and increased pain. However, by preventing lymphocyte proliferation and NK activity, they may have anti-inflammatory effects. Furthermore, they have the ability to suppress TNF-α, IL-2, and IL-6. What matters most, though, is the ratio of the two PUFA groups. In an attempt to shed some light on the matter, Simopoulos suggested that women's lower risk of breast cancer is due to a low omega-6/omega-3 ratio. She also came to the conclusion that a lower ratio is linked to a general decline in highly prevalent chronic diseases in Western society (Walls et al., 2016; Munteanu and Schwartz, 2022).

REFERENCES

1. Adepoju, A., & Allen, S. (2019). Malnutrition in developing countries: Nutrition disorders, a leading cause of ill health in the world today. *Paediatrics and Child Health, 29*(4), 394–400. https://doi.org/10.1016/j.paed.2019.06.005.
2. Al Tanoury, Z., Piskunov, A., & Rochette-Egly, C. (2013). Vitamin A and retinoid signaling: Genomic and nongenomic effects: Thematic review series: Fat-soluble vitamins: Vitamin A. *Journal of Lipid Research, 54*(7), 1761–1775. https://doi.org/10.1194/jlr.R030833.
3. Arendt, J., Sørensen, H., Horsfall, L., & Petersen, I. (2019). Elevated vitamin B12 levels and cancer risk in UK primary care: A THIN database cohort study. *Cancer Epidemiology, Biomarkers & Prevention, 28*(5), 814–821. https://doi.org/10.1158/1055-9965.EPI-17-1136.
4. Bold, R., Ishizuka, J., Townsend, C., Jr., & Thompson, J. (1996). All-trans-retinoic acid inhibits growth of human pancreatic cancer cell lines. *Pancreas, 12*(2), 189–195. https://doi.org/10.1097/00006676-199603000-00014.
5. Bozonet, S., Carr, A., Pullar, J., & Vissers, M. (2015). Enhanced human neutrophil vitamin C status, chemotaxis, and oxidant generation following dietary supplementation with vitamin C-rich SunGold kiwifruit. *Nutrients, 7*(4), 2574–2588. https://doi.org/10.3390/nu7042574.
6. Brasky, T., Ray, R., Navarro, S., Schenk, J., Newton, A., & Neuhouser, M. (2020). Supplemental one-carbon metabolism related B vitamins and lung cancer risk in the Women's Health Initiative. *International Journal of Cancer, 147*(5), 1374–1384. https://doi.org/10.1002/ijc.32913.
7. Brieger, A., Rink, L., & Haase, H. (2013). Differential regulation of TLR-dependent MyD88 and TRIF signaling pathways by free zinc ions. *Journal of Immunology, 191*(4), 1808–1817. https://doi.org/10.4049/jimmunol.1301261.
8. Chen, X., Andresen, B., Hill, M., Zhang, J., Booth, F., & Zhang, C. (2008). Role of reactive oxygen species in tumor necrosis factor-alpha induced endothelial dysfunction. *Current Hypertension Reviews, 4*(4), 245–255. https://doi.org/10.2174/157340208786241336.
9. Corbett, A., Eckle, S., Birkinshaw, R., Liu, L., Patel, O., Mahony, J., et al. (2014). T-cell activation by transitory neo-antigens derived from distinct microbial pathways. *Nature, 509*(7500), 361–365. https://doi.org/10.1038/nature13160.
10. Cox, S., Arthur, P., Kirkwood, B., Yeboah-Antwi, K., & Riley, E. (2006). Vitamin A supplementation increases ratios of proinflammatory to anti-

inflammatory cytokine responses in pregnancy and lactation. *Clinical & Experimental Immunology, 144*(3), 392–400. https://doi.org/10.1111/j.1365-2249.2006.03082.x.

11. Cunningham-Rundles, C. (2002). Hematologic complications of primary immune deficiencies. *Blood Reviews, 16*(1), 61–64. https://doi.org/10.1054/blre.2001.0185.
12. Grohmann, U., Mondanelli, G., Belladonna, M., Orabona, C., Pallotta, M., Iacono, A., et al. (2017). Amino-acid sensing and degrading pathways in immune regulation. *Cytokine & Growth Factor Reviews, 35,* 37–45. https://doi.org/10.1016/j.cytogfr.2017.05.004.
13. Haryanto, B., Suksmasari, T., Wintergerst, E., & Maggini, S. (2015). Multivitamin supplementation supports immune function and ameliorates conditions triggered by reduced air quality. *Vitamins & Minerals, 4*(1), 1–15. https://doi.org/10.4172/2376-1318.1000128.
14. Hu, X., & Ivashkiv, L. (2009). Cross-regulation of signaling pathways by interferon-γ: Implications for immune responses and autoimmune diseases. *Immunity, 31*(4), 539–550. https://doi.org/10.1016/j.immuni.2009.09.002.
15. Huskisson, E., Maggini, S., & Ruf, M. (2007). The role of vitamins and minerals in energy metabolism and well-being. *Journal of International Medical Research, 35*(3), 277–289. https://doi.org/10.1177/147323000703500301.
16. Jasenosky, L., Scriba, T., Hanekom, W., & Goldfeld, A. E. (2015). T cells and adaptive immunity to *Mycobacterium tuberculosis* in humans. *Immunological Reviews, 264*(1), 74–87. https://doi.org/10.1111/imr.12274.
17. Jones, P. T., & Anderson, R. (1983). Oxidative inhibition of polymorphonuclear leukocyte motility mediated by the peroxidase/H O /halide system: Studies on the reversible nature of the inhibition and mechanism of protection of migratory responsiveness by ascorbate, levamisole, thiamine, and cysteine. *International Journal of Immunopharmacology, 5*(4), 377–389. https://doi.org/10.1016/0192-0561(83)90012-7.
18. Kitabayashi, C., Fukada, T., Kanamoto, M., Ohashi, W., Hojyo, S., Atsumi, T., et al. (2010). Zinc suppresses Th17 development via inhibition of STAT3 activation. *International Immunology, 22*(5), 375–386. https://doi.org/10.1093/intimm/dxq017.
19. Koithan, M., & Devika, J. (2010). New approaches to nutritional therapy. *Journal of Nurse Practitioners, 6*(10), 805–806. https://doi.org/10.1016/j.nurpra.2010.07.001.

20. Li, Y., Jiang, C., Xu, G., Wang, N., Zhu, Y., Tang, C., et al. (2008). Homocysteine upregulates resistin production from adipocytes *in vivo* and *in vitro*. *Diabetes, 57*(4), 817–827. https://doi.org/10.2337/db07-0617.
21. Lin, J., Selim, M., Shea, C., Grichnik, J., Omar, M., Monteiro-Riviere, N., et al. (2003). UV photoprotection by combination topical antioxidants vitamin C and vitamin E. *Journal of the American Academy of Dermatology, 48*(6), 866–874. https://doi.org/10.1067/mjd.2003.425.
22. Liu, M., Bao, S., Gálvez-Peralta, M., Pyle, C., Rudawsky, A., Pavlovicz, R., et al. (2013). ZIP8 regulates host defense through zinc-mediated inhibition of NF-κB. *Cell Reports, 3*(2), 386–400. https://doi.org/10.1016/j.celrep.2013.01.009.
23. Maggini, S., Beveridge, S., Sorbara, P., & Senatore, G. (2008). Feeding the immune system: The role of micronutrients in restoring resistance to infections. *CAB Reviews: Perspectives in Agriculture, Veterinary Science, Nutrition and Natural Resources, 3*(1), 1–21. https://doi.org/10.1079/PAVSNNR20083098.
24. Marshall, J., Warrington, R., Watson, W., & Kim, H. (2018). An introduction to immunology and immunopathology. *Allergy, Asthma & Clinical Immunology, 14*(1), 49. https://doi.org/10.1186/s13223-018-0278-1.
25. Maywald, M., Wang, F., & Rink, L. (2018). Zinc supplementation plays a crucial role in T helper 9 differentiation in allogeneic immune reactions and non-activated T cells. *Journal of Trace Elements in Medicine and Biology, 50,* 482–488. https://doi.org/10.1016/j.jtemb.2018.02.004.
26. McDole, J., Wheeler, L., McDonald, K., Wang, B., Konjufca, V., Knoop, K., et al. (2012). Goblet cells deliver luminal antigen to CD103+ dendritic cells in the small intestine. *Nature, 483*(7389), 345–349. https://doi.org/10.1038/nature10863.
27. Mellor, A. L., & Munn, D. H. (2004). IDO expression by dendritic cells: Tolerance and tryptophan catabolism. *Nature Reviews Immunology, 4*(10), 762–774. https://doi.org/10.1038/nri1457.
28. Molina, N., Morandi, A., Bolin, A., & Otton, R. (2014). Comparative effect of fucoxanthin and vitamin C on oxidative and functional parameters of human lymphocytes. *International Immunopharmacology, 22*(1), 41–50. https://doi.org/10.1016/j.intimp.2014.06.026.
29. Mullero, K. (2005). Malnutrition and health in developing countries. *Canadian Medical Association Journal, 173*(3), 279–286. https://doi.org/10.1503/cmaj.050342.
30. Munteanu, C., & Schwartz, B. (2022). The relationship between nutrition

and the immune system. *Frontiers in Nutrition, 9,* 1082500. https://doi.org/10.3389/fnut.2022.1082500.

31. Newsholme, P., Brennan, L., Rubi, B., & Maechler, P. (2005). New insights into amino acid metabolism, β-cell function, and diabetes. *Clinical Science, 108*(3), 185–194. https://doi.org/10.1042/CS20040290.
32. Nishikimi, M., Fukuyama, R., Minoshima, S., Shimizu, N., & Yagi, K. (1994). Cloning and chromosomal mapping of the human nonfunctional gene for L-gulono-gamma-lactone oxidase, the enzyme for L-ascorbic acid biosynthesis missing in man. *Journal of Biological Chemistry, 269*(17), 13685–13688.
33. Popa, C., Netea, M., Van Riel, P., Van Der Meer, J., & Stalenhoef, A. (2007). The role of TNF-α in chronic inflammatory conditions, intermediary metabolism, and cardiovascular risk. *Journal of Lipid Research, 48*(4), 751–762. https://doi.org/10.1194/jlr.R600021-JLR200.
34. Popkin, B., Adair, L., & Ng, S. (2012). Global nutrition transition and the pandemic of obesity in developing countries. *Nutrition Reviews, 70*(1), 3–21. https://doi.org/10.1111/j.1753-4887.
35. Prentice, S. (2017). They are what you eat: Can nutritional factors during gestation and early infancy modulate the neonatal immune response? *Frontiers in Immunology, 8,* 1641. https://doi.org/10.3389/fimmu.2017.01641.
36. Qureshi, A., Reis, J., Qureshi, N., Papasian, C., Morrison, D., & Schaefer, D. (2011). δ-Tocotrienol and quercetin reduce serum levels of nitric oxide and lipid parameters in female chickens. *Lipids in Health and Disease, 10*(1), 39. https://doi.org/10.1186/1476-511X-10-39.
37. Rampal, R., Awasthi, A., & Ahuja, V. (2016). Retinoic acid–primed human dendritic cells inhibit Th9 cells and induce Th1/Th17 cell differentiation. *Journal of Leukocyte Biology, 100*(1), 111–120. https://doi.org/10.1189/jlb.3VMA1015-476R.
38. Rodionov, D., Arzamasov, A., Khoroshkin, M., Iablokov, S., Leyn, S., Peterson, S., et al. (2019). Micronutrient requirements and sharing capabilities of the human gut microbiome. *Frontiers in Microbiology, 10,* 1316. https://doi.org/10.3389/fmicb.2019.01316.
39. Rosenkranz, E., Maywald, M., Hilgers, R., Brieger, A., Clarner, T., Kipp, M., et al. (2016). Induction of regulatory T cells in Th1-/Th17-driven experimental autoimmune encephalomyelitis by zinc administration. *Journal of Nutritional Biochemistry, 29,* 116–123. https://doi.org/10.1016/j.jnutbio.2015.11.010.
40. Rosenkranz, E., Metz, C., Maywald, M., Hilgers, R., Weßels, I., Senff,

T., et al. (2016). Zinc supplementation induces regulatory T cells by inhibition of Sirt−1 deacetylase in mixed lymphocyte cultures. *Molecular Nutrition & Food Research, 60*(3), 661–671. https://doi.org/10.1002/mnfr.201500524.

41. Shi, H., & Walker, W. (2015). Development and physiology of the intestinal mucosal defense. *Mucosal Immunology, 1*(1), 9–29. https://doi.org/10.1016/B978-0-12-415847-4.00002-1.
42. Shoeb, M., & Ramana, K. (2012). Anti-inflammatory effects of benfotiamine are mediated through the regulation of the arachidonic acid pathway in macrophages. *Free Radical Biology and Medicine, 52*(1), 182–190. https://doi.org/10.1016/j.freeradbiomed.2011.10.444.
43. Szefel, J., Danielak, A., & Kruszewski, W. (2019). Metabolic pathways of L-arginine and therapeutic consequences in tumors. *Advances in Medical Sciences, 64*(2), 104–110. https://doi.org/10.1016/j.advms.2018.08.018.
44. Valacchi, G., Muresan, X., Sticozzi, C., Belmonte, G., Pecorelli, A., Cervellati, F., et al. (2016). Ozone-induced damage in 3D-skin model is prevented by topical vitamin C and vitamin E compound mixtures application. *Journal of Dermatological Science, 82*(3), 209–212. https://doi.org/10.1016/j.jdermsci.2016.02.007.
45. Villamor, E., & Fawzi, W. W. (2005). Effects of vitamin A supplementation on immune responses and correlation with clinical outcomes. *Clinical Microbiology Reviews, 18*(3), 446–464. https://doi.org/10.1128/CMR.18.3.446-464.2005.
46. Walls, J., Sinclair, L., & Finlay, D. (2016). Nutrient sensing, signal transduction, and immune responses. *Seminars in Immunology, 28*(5), 396–407. https://doi.org/10.1016/j.smim.2016.09.001.
47. Yadav, U. C., Kalariya, N. M., Srivastava, S. K., & Ramana, K. V. (2010). Protective role of benfotiamine, a fat-soluble vitamin B1 analogue, in lipopolysaccharide-induced cytotoxic signals in murine macrophages. *Free Radical Biology and Medicine, 48*(10), 1423–1434. https://doi.org/10.1016/j.freeradbiomed.2010.02.031.
48. Zhou, E., Li, Y., Yao, M., Wei, Z., Fu, Y., & Yang, Z. (2014). Niacin attenuates the production of pro-inflammatory cytokines in LPS-induced mouse alveolar macrophages by HCA2 dependent mechanisms. *International Immunopharmacology, 23*(1), 121–126. https://doi.org/10.1016/j.intimp.2014.07.006.
49. Zuniga, E., Macal, M., Lewis, G., & Harker, J. (2015). Innate and adaptive immune regulation during chronic viral infections. *Annual Review of Virology, 2*(1), 573–594. https://doi.org/10.1146/annurev-virology-100114-055226.

CHAPTER 3

FUNCTIONAL AND NUTRITIONAL ASPECTS OF HYDROCOLLOIDS AND LIPIDS

Contents

INTRODUCTION

Recent years have seen a shift in people's preferences toward natural food products due to their growing desire to maintain their health through diet. Convenience foods are also necessary in this fast-paced world. For the two reasons listed above, i.e., supplying the public with wholesome, fresh, and ready-to-eat food has primarily caused the hydrocolloids market to expand. It is primarily influenced by consumer preferences, eating patterns, R&D endeavors, product innovation, and the usefulness of hydrocolloids in a range of food items. Food products such as sauces (thickening agents), puddings (gel-forming agents), ice cream (emulsifying agents), cheese (syneresis inhibitors), beer and wine (clarifying and foam stabilizing agents), fruits and vegetables (edible film or coating formers), and sausages (binding agents and fat replacers) are among those that use hydrocolloids. They are also known to prevent ice cream from crystallizing. Hydrocolloids are therefore seen as a low-risk and nutrient-supplying option, which has increased their worldwide business. The global market for hydrocolloids was valued at $8.8 billion in 2018 and is expected to grow at a compound annual growth rate (CAGR) of 5.03% from 2018 to 2023, reaching $11 billion by that time, according to the "Hydrocolloids Market" report.

Hydrocolloids are a type of colloidal system made up of a variety of long-chain hydrophilic polymers that are prone to swelling in water and can easily disperse, either partially or completely, in it. The hydrocolloids dissolve and the solution's physical characteristics change, allowing for thickening, coating, emulsification, and stabilization as well as a tendency to form gels. The ability of hydrocolloids to bind water and alter the product's rheological characteristics—a feature that is useful in modifying and improving sensory properties—is the main driver of their widespread use in the food industry. Hydrocolloids therefore have the potential to be a very important food additive. The use of hydrocolloids as a food additive could help create functional meals. The more difficult task for scientists is to create innovative nanoscale structures that could deliver hydrocolloids as a physiologically active ingredient to give consumers additional health benefits.

In addition, hydrocolloids have also attracted interest because of their nutritional properties. As per the most recent iteration of the Codex Alimentarius Commission's definition of dietary fiber, the majority of food hydrocolloids may be classified as such. Consequently, hydrocolloids have become a crucial component of a balanced diet in recent years. Eating food hydrocolloids can help with weight management, immune regulation, constipation relief, heart disease prevention, and colon cancer prevention, among other health benefits. In addition to giving food viscosity, hydrocolloids aid in the creation of highly satisfying foods. Hydrocolloids have the satiety effect because they slow down the effectiveness of enzyme action and/or delay the emptying of the stomach. Many studies suggested that consuming particular hydrocolloids could control the bacterial flora in the gut, favoring beneficial bacteria like Lactobacillus and *Bifidobacterium*. Due to the significant biological effects of short-chain fatty acids, which include the regulation of cell proliferation, apoptosis, and angiogenesis, this could result in both increased fermentative activity and short-chain fatty acid production (Viebke et al., 2014).

Thus, the purpose of this chapter is to discuss how lipids and hydrocolloids, both alone and together, are used in the food industry as food additives, ingredients, and—above all—health promoters. It will address the nutritional and functional aspects of lipids and hydrocolloids in food, a topic that has drawn a lot of attention from food scientists.

3.1. HYDROCOLLOIDS AND LIPIDS: CHARACTERISTICS AND FUNCTIONALITY IN THE FOOD SECTOR

The food industry makes extensive use of hydrocolloids for a variety of purposes, such as gelling, thickening, stabilizing foams, emulsions, and dispersions, and enabling flavor release that is controlled. Although hydrocolloids usually exist in small amounts at concentrations below 1%, their ability to control the rheology and texture of food systems is actually the primary reason for their routine application. These characteristics are intimately linked to the structure of hydrocolloids and molecular interactions. The application and creation of colloidal particles have the potential to greatly impact the nutritional value of the entire system in addition to enhancing the physicochemical characteristics of food. Similarly, hydrophilic colloids find extensive application in the delivery of bioactive compounds, fat substitution, glycemic reduction, salt reduction, and inhibition of food safety hazards. They also improve satiety, control obesity, adjust the glycemic index, modulate the immune system, and preserve colon health. Therefore, hydrophilic colloids, in terms of applications, structure, and other characteristics play a major role in determining the nutritional value and overall quality of food products.

In essence, hydrocolloids are materials that, when combined with water, gel. This characteristic of hydrocolloids has primarily been used to inhibit the growth and formation of ice and sugar crystals as well as to stabilize emulsions, suspensions, foams, and proteins. Because of their exceptional ability to bind water, they are also frequently used as a processing aid in a variety of food operations, helping to encapsulate and form films. Commercial hydrocolloids that are frequently used include methylcellulose (MC), carboxymethylcellulose, starch, pectin, alginate, and carrageenan.

3.1.1. Cellulose and Its Derivatives

The most common organic polymer on Earth, cellulose serves as the main structural element of plant cells. It is a linear polysaccharide with the chemical formula $(C_6H_{10}O_5)_n$ that is made up of numerous glucose monosaccharides. These are extremely insoluble substances because of the crystalline structure of cellulose or the extremely close packing of polymer chains. Because of this, alkali treatment can increase their water solubility by swelling the structure and causing them to react with methyl chloride, propylene oxide, or chloroacetic acid to form a variety of cellulose derivatives, including MC, carboxymethyl cellulose (CMC), hydroxypropyl cellulose (HPC), and hydroxypropyl methylcellulose (HPMC) (Bourtoom, 2008).

The FDA-approved HPMC is used in the food industry as a gelling and stabilizing agent because of its high-water permeability, which also restricts its use for coating. Their lipid and gas barrier qualities are exceptional. Because it is nontoxic and viscous, CMC is frequently used to stabilize emulsions in a variety of products, including ice cream, and as a thickening agent in food applications.

The most affordable and easily accessible hydrophilic water-soluble cellulose derivative is MC. In food applications, it is frequently used as an emulsifier and thickener. While MC and HPC films are effective at blocking the passage of oxygen, carbon dioxide, and lipids, they are not as effective at stopping the movement of water vapor. This issue could be resolved by incorporating hydrophobic materials, like lipids, to enhance the moisture barrier qualities of the films. They can also cover fried products with a coating that is formed through thermal-induced means. Because it prevents lipid migration, MC can also be used as a coating on confectionery products. On the other hand, because HPC is thermoplastic, it can be extruded into films and used to stabilize foams in whipped products.

Compared to synthetic polymers, cellulose derivatives such as cellophane have a higher water vapor permeability, which limits their application in the food packaging sector. However, they have good mechanical qualities and can be used to package candies, processed meat, baked goods, and cheese when coated with PVDC or nitrocellulose wax, which improves the barrier properties of the product.

3.1.2. Chitin and Chitosan

Chitin is the second most common naturally occurring biopolymer, and it resembles cellulose in almost every way. They come from a variety of fungi, other biological materials, and the exoskeleton of crustaceans, which includes the shells of shrimp and crabs. When chitin is deacetylated with an alkali, chitosan is produced. It is a naturally occurring polymer that is safe, biodegradable, and compatible with humans. It is soluble in a variety of acidic solvents, including diluted hydrochloric, acetic, and formic acids, but insoluble in water. In agricultural produce, chitosan and its derivatives can be used to create semi-permeable films and coatings with good mechanical properties and the ability to change the internal atmosphere, which delays ripening and lowers transpiration rates (Figure 3.1). Furthermore, chitosan's polycationic nature is thought to be the source of its reported antifungal and antimicrobial qualities. Additionally, they can absorb heavy metal ions, which may be helpful in lowering food oxidation caused by free metals (Chandra and Rustgi, 1998).

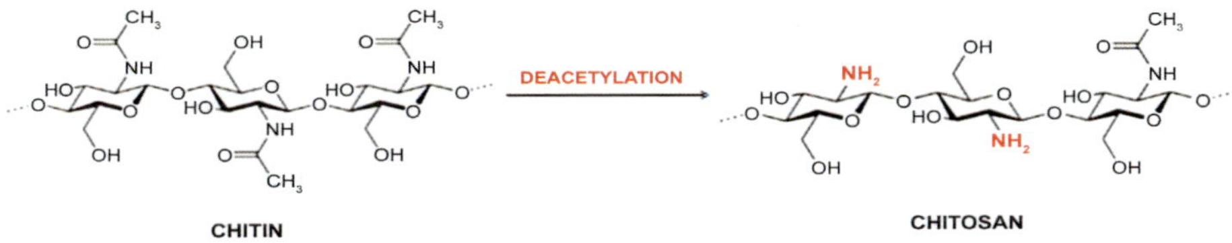

Figure 3.1. *Structures of chitin and chitosan.*

Source: Piekarska K, Sikora M, Owczarek M, Jóźwik-Pruska J, Wiśniewska-Wrona M. (2023). Chitin and chitosan as polymers of the future—obtaining, modification, life cycle assessment and main directions of application. Polymers, 15(4):793. https://doi.org/10.3390/polym15040793.

3.1.3. Starch

The most prevalent and biodegradable polysaccharide that makes up most of the carbohydrates in a human diet is starch. It is made up of two polymers: amylose, which is a linear chained molecule, and amylopectin, which is a branched molecule. A large number of glucose monomers are joined together by glycosidic bonds to form this complex. Amylopectin occasionally contains α-1,6-glycosidic bonds that cause branching, but amylose is entirely made up of D-glucose units connected by α-1,4-glycosidic linkages. When heated, starch dissolves in water after being insoluble in alcohol or cold water. Granules of starch tend to swell when they absorb water and may rupture, losing their semicrystalline structure. Subsequently, the amylose molecules may seep out of the granule, creating a network that retains water and raises the viscosity of the mixture. Due to their low cost, biodegradability, and renewability, starches are frequently used in the food industry. Starch-based films have also been found to have good mechanical qualities and the potential to replace plastic polymers completely or in part (Figure 3.2).

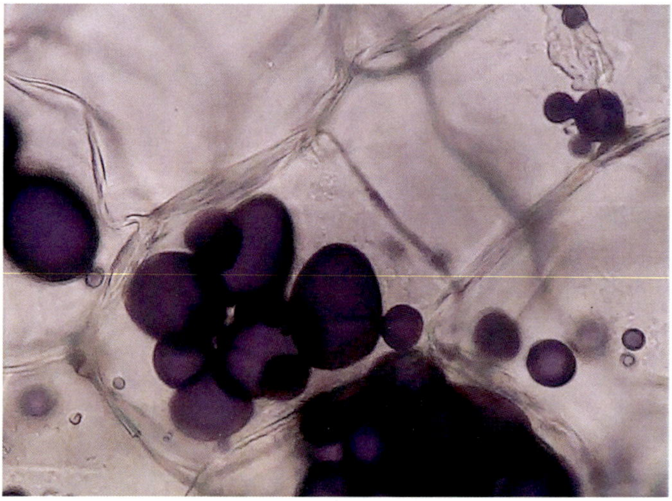

Figure 3.2. *Potato starch granules in cells of potato.*

Source: https://upload.wikimedia.org/wikipedia/commons/thumb/c/cc/Starch_granules_of_potato02.jpg/800px-Starch_granules_of_potato02.jpg.

3.1.4. Alginate

The term "alginate" is frequently used to refer to the sodium, calcium, and magnesium salts of alginic acid that are present in the cell walls of brown algae belonging to the Phaeophyceae class. Sodium dissolves in water, but the salts of calcium and magnesium do not. As a result, all the alginate salts are changed into sodium salt, which is then dissolved in water and filtered out. They differ in their molecular weight, structural arrangement, and degree of composition between D-mannuronic acid and L-guluronic acid monomers. The food industry has shown a great deal of interest in it lately as a texturizing and gelling agent because of its many qualities, including thickening, gel formation, film formation, and emulsion stabilization. The ability of alginates to react with di-valent and tri-valent cations—such as calcium, iron, or magnesium—that are added as gelling agents determines the formation of alginate coatings. They are also thought to be a possible component for a coating or film, which could also act as a vehicle for antioxidant and antimicrobial substances (Hassan et al., 2018).

Their use at concentrations higher than 10% w/w is discouraged because of pourability issues brought on by a sudden increase in the suspension's viscosity.

3.1.4.1. Structure

As a linear copolymer, alginic acid is composed of *homopolymeric* blocks of covalently linked β-*D-mannuronate* (M) and α-*L-guluronate* (G) residues, arranged in distinct sequences or blocks. Monomers can show up as consecutive G-residues (G-blocks), consecutive M-residues (M-blocks), or alternating M and G-residues (MG-blocks) in *homopolymeric* blocks. The C-5 epimer of β-*D-mannuronate* is α-*L-guluronate*.

3.1.4.2. Forms

Brown seaweeds are used to refine alginates. Many brown seaweeds belonging to the Phaeophyceae class are harvested worldwide and processed into sodium alginate. Numerous industries, including food, pet food, fertilizers, textile printing, and pharmaceuticals, use sodium alginate. Alginate is the gelling agent used in dental impression material. Food-grade alginate is a permitted component in manufactured and processed foods.

The size range of brown seaweeds is 20–40 meters for the giant kelp *Macrocystis pyrifera*, 2-4 meters for thick, leather-like seaweeds, and 30–60 cm for smaller species. Except for *Laminaria japonica*, which is grown for food in China and the excess material from its cultivation is used in the country's alginate industry, the majority of brown seaweed used for alginates is collected from the wild.

Alginates from various brown seaweed species have distinct chemical structures, which affect their physical characteristics. Certain species produce an alginate that gels strongly, while others produce a weaker gel. Some species can also produce an alginate that is cream or white in color. These species are best suited for technical applications where color is not a concern.

Alginate of commercial quality is extracted from *Macrocystis pyrifera* giant kelp, *Ascophyllum nodosum*, and various Laminaria species. Pseudomonas and Azotobacter, two bacterial genera that also produce alginates, were crucial in revealing the pathway of its biosynthesis. Alginates derived from bacteria can be used to create micro- or nanostructures that are appropriate for use in medicine.

- The sodium salt of alginic acid is called sodium alginate ($NaC_6H_7O_6$). A gum is sodium alginate.
- Sodium alginate is a gum.
- The potassium salt of alginic acid is known as potassium alginate ($KC_6H_7O_6$).
- The calcium salt of alginic acid is called calcium alginate ($CaC_{12}H_{14}O_{12}$). It is created by ion exchange, which involves switching out the sodium ion in sodium alginate for a calcium ion.

3.1.5. Pectin

Pectin is a complex anionic polysaccharide that occurs naturally and is soluble in water. It is primarily obtained from citrus fruits. Pectin is divided into two categories based on its "methoxyl" content: low methoxyl pectin (LMP), which has a methoxyl content of 0–5% and 7%, and high methoxyl pectin (HMP), which has a methoxyl content of 7–12%. The process of gelation used in the production of fruit juices is significantly influenced by the level of methylation. When calcium ions are added to LMP solutions, they have a tendency to gel similarly to how alginate solutions do. Pectin is primarily used in food as a thickening, gelling, or stabilizing agent in items like fruit juices, jams, milk and yogurt drinks, and ice cream. In recent years, studies on the gelling ability of pectin have been carried out to explore its potential application in the creation of edible films (Figure 3.3).

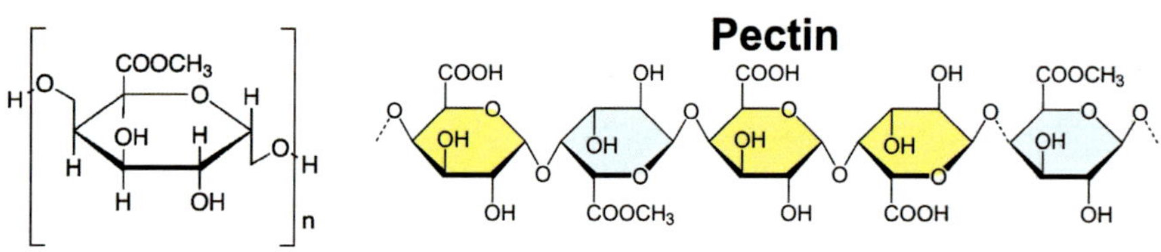

Figure 3.3. *Structure of pectin.*

Source: https://commons.wikimedia.org/wiki/File:Pectin.png.

Pectin is primarily used in food as a stabilizer, thickening agent, and gelling agent. Pectin can also be purchased in some nations as a blended powder, extract, or solution for use in jam-making at home.

The traditional use is to give jams or marmalades—which are otherwise sweet juices—a jelly-like consistency. Pectin also strengthens the gel of low-calorie jams and decreases syneresis in marmalades and jams. Pectin is a household ingredient that is diluted to the proper concentration with sugar and a small amount of citric acid to adjust the pH in gelling sugar, also called jam sugar.

Different pectin types can be identified for different food applications based on their characteristics, including their acidity, degree of esterification, relative number of methoxyl groups in the molecules, etc. To put it another way, HMPs are distinguished from LMPs by having a higher percentage of their carboxyl groups esterified with methanol:

- HMPs, which are characterized by a degree of esterification of 50 or higher, are commonly utilized in the production of traditional jams and jellies. These pectins form gels at high sugar concentrations and in acidic environments, giving them a smooth texture and making them ideal for use in confectionery and bakery fillings.

- LMPs can be either amidated or non-amidated, with a degree of esterification of less than 50. The efficacy of pectin is determined by the percentage level of substitution of the amide group, which is defined as the degree of amidation. Depending on the calcium concentration and calcium reactivity of the pectin selected, LMPs can provide a variety of textures and rheological properties; amidated LMPs can form gels that can melt and reform, while non-amidated LMPs can form thermostable gels that can withstand high temperatures. These characteristics make LMPs suitable for dairy products, low-sugar products, and stabilizing acidic protein drinks.

High-ester (high methoxyl) pectins are used in traditional jams and marmalades that have a sugar and soluble fruit solids content of more than 60%. Dietary products can be made with less sugar thanks to amidated and low-ester (low methoxyl) pectins. In Taiwan, Aiyu jelly is traditionally made from water extract of Aiyu seeds. The extract gels without heating because of the bivalent cations in the water and the low-ester pectins in the seeds.

To give confectionery jellies a good gel structure, a clean bite, and a good flavor release, pectin is used. Pectin can also be used as a fat substitute in baked goods, to enhance the mouthfeel and pulp stability in juice-based drinks, and to stabilize acidic protein drinks like yogurt.

Pectin levels in food additives typically range from 0.5 to 1.0%, which is roughly equivalent to the pectin content of fresh fruit.

In medicine, pectin is used to treat diarrhea and constipation by increasing the volume and viscosity of stool. It was one of the primary ingredients, along with kaolinite, in Kaopectate, a medication to treat diarrhea, until 2002. It has been applied to the delicate removal of heavy metals from biological systems. Moreover, pectin serves as a demulcent in throat lozenges. Due to its low cost and safety, pectin can be used in a variety of oral drug delivery platforms, such as gastro-retentive, colon-specific, controlled release, and mucoadhesive delivery systems. It was discovered that due to differences in molecular size and chemical composition, pectin from different sources provides varying gelling abilities. Similar to other natural polymers, one of pectin's main issues is sample variability, which can lead to poor reproducibility in drug delivery properties.

3.1.6. Carrageenan

A family of naturally occurring linear sulfated polysaccharides known as carrageenans or carrageenins is taken from red, edible seaweeds. The food industry makes extensive use of carrageenans because of their gelling, thickening, and stabilizing qualities. Because of their robust binding to food proteins, their primary use is in dairy and meat products. Because carrageenans resemble native glycosaminoglycans (GAGs), they have been a promising candidate in tissue engineering and regenerative medicine applications in recent years (Figure 3.4). Their primary applications have been in tissue engineering, covering wounds, and medication delivery.

Figure 3.4. *Manufacturers extract carrageenan from seaweed.*

Source: By StinaTano - Own work, CC BY-SA 3.0, https://commons.wikimedia.org/w/index.php?curid=18604544.

Carrageenans are anionic polysaccharides because they have a 15–40% ester-sulfate content. Based primarily on their sulfate content, they can be divided into three classes. Each disaccharide in kappa-, iota-, and lambda-carrageenan contains one, two, or three sulfate groups.

The dark red algae known as *Chondrus crispus*, or Irish moss, is a common red seaweed that is used to make the hydrophilic colloids needed to make carrageenan. It grows attached to rocks and resembles parsley. From around the 15th century, food additives have been made using gelatinous extracts from the seaweed *Chondrus crispus*. Carrageenan can be used in place of gelatin in confections and other foods because it is a vegetarian and vegan substitute for gelatin in certain situations. Carrageenan's safety as a food ingredient is not supported by clinical research, primarily due to the lack of sufficient information regarding what happens to it after digestion.

The initial commercial production of *Eucheuma* and *Kappaphycus* species on an industrial scale. Because carrageenan originated in the Philippines, the Philippines and Indonesia are the world's leading producers of carrageenan. Gulaman, a traditional jelly dessert from the Philippines, is made with carrageenan and agar.

3.1.6.1. Properties

Large, incredibly flexible molecules called carrageenans curl into helical shapes. They can now create a wide range of gels at room temperature as a result. They are extensively employed as thickening and stabilizing agents in the food and other industries.

Carrageenans, which are high-molecular-weight polysaccharides, are primarily composed of disaccharide repeating units, which are alternating 3-linked β-D-galactopyranose (G-units) and 4-linked α-D-galactopyranose (D-units) or 4-linked 3,6-anhydro-α-D-galactopyranose (DA-units).

There are three main commercial classes of carrageenan:

- In the presence of potassium ions, kappa reacts with dairy proteins to form strong, rigid gels. The primary source of it is *Kappaphycus alvarezii.*
- *Iota forms soft gels* in the presence of calcium ions. It is produced mainly from *Eucheuma denticulatum*.
- Lambda does not gel, and is used to thicken dairy products.

The number and location of ester sulfate groups on the repeating galactose units of kappa, iota, and lambda carrageenan are the main variations that affect their respective properties. Elevated ester sulfate concentrations cause gel inhibition (lambda carrageenan), which lowers the solubility temperature of the carrageenan and results in weaker gels.

Different kinds of carrageenans are produced by numerous red algal species at different stages of their development. For example, in its gametophytic stage, the genus Gigartina primarily produces kappa carrageenans, and in its sporophytic stage, lambda carrageenans. All of them dissolve in hot water, but only the lambda form and the sodium salts of the other two do so in cold water (Figure 3.5).

Carrageenan, when present as processed *eucheuma* seaweed, has the EU additive E numbers E407 or E407a when used in food products. Carrageenan is regarded as a dietary fiber in theory.

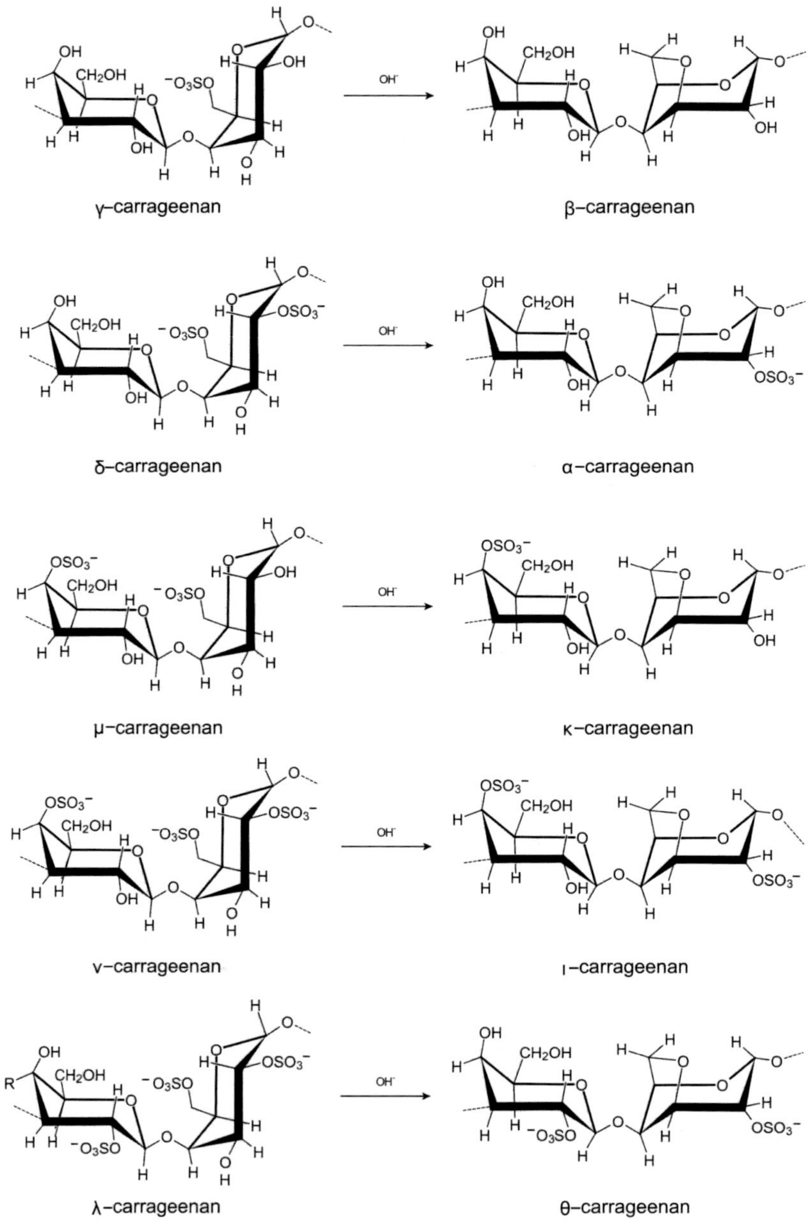

Figure 3.5. *Molecular structure of carrageenan.*

Source: By Roland Mattern - Roland1952, Public Domain, https://commons.wikimedia.org/w/index.php?curid=7623678.

3.1.7. Xanthan

Xanthan is the second commercially available microbial polysaccharide, following dextran, and it is fully soluble in both hot and cold water. They are made from *Xanthomonas campestris* and are marketed as xanthan gum. Because of their exceptional ability to bind water, they hydrate rapidly in the presence of water and produce very viscous solutions even at low concentrations. The resulting solutions are not impacted by variations in pH, temperature, or salt content. Because of this feature, xanthan works incredibly well as a thickener and stabilizer. Solutions containing xanthan gum are very pseudoplastic, i.e., even at elevated shear rates, the original viscosity is restored right away. This characteristic improves the mouthfeel and flavor release of food, among other sensory aspects. The commercially available xanthan may vary in terms of particle size, viscosity grades, dispersibility rate, pseudo-plasticity, and hydrating time because different strains of organisms and growth conditions are used in their production.

3.1.7.1. Uses

Around 1% xanthan gum has the ability to significantly raise a liquid's viscosity.

Xanthan gum is frequently found in sauces and salad dressings. Despite not being an emulsifier, it stabilizes the emulsion, which helps to prevent oil separation. Spices and other solid particles are also suspended by the use of xanthan gum. In many ice cream recipes, xanthan gum contributes to the desired texture. Xanthan gum is frequently added to toothpaste as a binder to maintain consistency. In addition, xanthan gum aids in thickening commercial egg substitutes formed from egg whites, which are meant to replace the yolks' fat and emulsifiers. Since it does not alter the color or flavor of food or beverages at normal use levels, it is also a recommended method of thickening liquids for people with swallowing disorders. Xanthan gum is used in gluten-free baking to provide the sticky texture that gluten would normally provide. It is used at concentrations of 0.5% or less in the majority of foods. Many different types of food products, including sauces, dressings, dairy products, baked goods, confections, meat and poultry products, and others, contain xanthan gum.

Large amounts of xanthan gum are used in the oil industry to thicken drilling mud. The solids that the drilling bit has cut are brought to the surface by these fluids. Improved low end rheology is provided by xanthan gum. The solids stay suspended in the drilling fluid when the circulation is stopped. The increased use of horizontal drilling is a result of its widespread application and the requirement for effective control of drilled solids. To boost viscosity and stop washout, it has been added to concrete that has been poured underwater.

3.1.7.2. Shear Thinning

Higher shear rates cause xanthan gum solutions to become less viscous. Pseudoplasticity or shear thinning is the term for this. This implies that any product that is sheared—by shaking, mixing, or chewing—will become thinner. The food will thicken once more after the shear forces are eliminated. When xanthan gum is added to salad dressing, the mixture becomes sufficiently thick in the bottle to maintain some homogeneity, but the shear forces created during shaking and pouring cause the mixture to thin out and make pouring easier. It sticks to the salad because the shear forces are released as it leaves the bottle and it thickens once more. Higher xanthan gum concentrations in water cause the rheology of xanthan aqua solutions to become Visco-elastic.

3.1.7.3. Concentrations Used

A liquid will get thicker the higher the xanthan gum concentration in the liquid. It is possible to form an emulsion with as little as 0.1% (by weight). Up to 1% xanthan gum, increasing the gum concentration results in a thicker, more stable emulsion. A teaspoon (approximately 2.5 grams) of xanthan gum can be used to concentrate one cup (250 ml) of water to 1%.

3.1.7.4. Preparation

The fermentation of glucose and sucrose yields xanthan gum. The xanthan polymer is generated extracellularly and added to the well-aerated, agitated medium. Isopropyl alcohol is added to the medium after one to four days, causing the polymer to precipitate out of it. The precipitate is then dried and ground into a powder that is easily soluble in brine or water.

It is made up of pentasaccharide repeat units, which have the molar ratio of 2:2:1 for glucose, mannose, and glucuronic acid.

It is possible to use *X. campestris*, a waste product of cheese production, to process whey because it can grow on lactose. For every 40 g/L of whey powder, this can yield 30 g/L of xanthan gum. Xanthan gum derived from whey is frequently found in a wide range of commercial products, including salad dressings and shampoos.

Glucose serves as the starting point for the synthesis of the sugar nucleotide precursors GDP-mannose, UDP-glucose, and UDP-glucuronate, which are needed to construct the pentasaccharide repeat unit. This connects the metabolism of carbohydrates and xanthan synthesis. Repeat units are assembled at lipid carriers anchored in the cytoplasmic membrane by undecaprenyl phosphate.

The nucleotide sugar xanthan precursors' sugar moieties are successively transferred to the lipid carriers by particular glycosyltransferases. The remaining acetyl and pyruvyl are added as non-carb decorations. Mature repeat units are exported and polymerized like Enterobacteriaceae-mediated Wzy-dependent polysaccharide synthesis mechanism. The repeat unit's synthesis, polymerization, and export are driven by the products of the gum gene cluster.

3.1.8. Gum Arabic

Gum Arabic is a naturally occurring gum derived from different acacia tree species. *Acacia senegal* is the primary supplier of Arabic gum used in commerce. Even at higher concentrations, gum Arabic dissolves in cold water and has extremely low viscosity—at low shear rates, 30% of gum Arabic has a viscosity of less than 1% xanthan gum or CMC. Unlike most other hydrocolloids, it behaves like a Newtonian, its viscosity being independent of shear rate. These aforementioned characteristics render it unsuitable for use as a thickening agent; however, when added in moderation, it can enhance the quality attributes of gluten-free products.

In addition to its extensive use in the food and pharmaceutical industries as a safe thickener, emulsifier, and stabilizer, gum Arabic is a rich source of dietary fiber and has a host of other health benefits that have been amply demonstrated by numerous *in vitro* and *in vivo* studies. Instead of breaking down in the stomach, gum Arabic ferments into a variety of short-chain fatty acids in the large intestine. It is thought to be a prebiotic that promotes the development and multiplicity of the advantageous gut microbiota; as a result, consuming it is linked to numerous advantageous health benefits.

These health benefits include:

- Improved absorption of calcium from the gastrointestinal tract
- Anti-diabetic
- Anti-obesity (gum Arabic lowers the body mass index and body fat percentage)
- Lipid-lowering potential (gum Arabic decreases total cholesterol, LDL, and triglyceride)
- Antioxidant activities
- Kidney and liver support
- Immune function via modulating the release of some inflammatory mediators
- Prebiotic functions like improving the intestinal barrier function, preventing colon cancer, and alleviating symptoms of irritable bowel diseases
- In rats, a protective effect on the intestine against the adverse actions of the NSAID drug meloxicam

In the food business, gum Arabic is used to bind the sweeteners and flavorings in soft drinks as well as stabilize, emulsify, and thicken frostings, fillings, soft candies, chewing gum, and other confections. Gomme syrup, a mixture of sugar and gum Arabic dissolved in water, is occasionally added to cocktails to give the sugar a smooth texture and stop it from crystallizing.

Gum Arabic, a soluble dietary fiber and complex polysaccharide, is widely accepted to be safe for ingestion by humans. Some persons who take large doses of 30 g (1 oz) or more per day may experience mild flatulence. It is a prebiotic (as opposed to a probiotic) that is fermented in the colon by microorganisms rather than broken down in the intestine. Regarding its calorie content, neither science nor regulations have come to a consensus. For rats, an upper limit of 2 kcal/g was established, but this does not apply to humans. In the beginning, the US FDA set a value for food labeling of 4 kcal/g, but no value for soluble dietary fiber was established in Europe. A 1998 review concluded that the only arbitrary value that can be used for regulatory purposes is the current state of scientific knowledge. The FDA responded to an application to lower the rated caloric value of gum Arabic to 1.7 kcal/g in 2008 with a letter of no objection.

3.1.9. Galactomannans (Guar Gum, Locust Bean Gum, Tara Gum, Cassia Gum)

Galactomannans are polysaccharides with a variable mannose-to-galactose ratio (M/G) that are made up of D-mannose and D-galactose. They are extracted from the seeds of fenugreek, tara, guar, and carob plants for commercial use. Galactomannans are found to interact synergistically to produce thicker, more viscous products than either of them alone because of their different M/G ratios. Since guar gum is a neutral polysaccharide, pH and ions have little effect on its solutions. When combined with xanthan, κ-carrageenan, or agar, it works synergetically to produce very high viscosities. It is frequently used as an emulsifier for flavor oils to be incorporated into beverages in the confectionery industry. When combined with carrageenans, locust bean gum greatly enhances texture, transparency, and gel strength while also preventing syneresis. It is frequently found in a wide range of food items, including cheese, ice cream, prepared foods, sauces, salad dressings, and baked goods. The less-used hydrocolloids are Tara, and cassia gum; however, there is ongoing discussion regarding the acceptability and utility of these ingredients in food. The most widely used galactomannan is guar gum, primarily because of its affordable and consistent price.

3.1.10. Lipids

A vast array of naturally occurring organic compounds known as lipids are hydrophobic, nonpolar compounds that are soluble in nonpolar organic solvents like acetone, benzene, ether, and chloroform but insoluble in water. For several centuries, lipid-based coatings have been used to coat fruits or confectionery items with wax. Their hydrophobic characteristic, which makes them superior moisture barriers and lowers respiration, extending the shelf life of coated products, is the main reason for their preference. The most widely used and successful lipids in coating formulations are surfactants, acetylated monoglycerides, vegetable and mineral oil, paraffin wax, beeswax, and carnauba wax. Because lipids have low polarity, they mainly prevent all forms of moisture transfer while forming thick, brittle, opaque films. They also add gloss, which improves the coated products' aesthetic appeal. Lipid-based films and coatings have not been widely used in food because of their poor film-forming ability and weak mechanical strength.

Because they may disrupt the polymer chain-to-chain interactions and improve the flexibility of the films and coatings, they must be used in conjunction with film-forming agents like proteins and polysaccharides. Apple slices have been coated with natural wax, such as carnauba wax, along with glycerol and cassava starch. The coating has been reported to have better mechanical qualities and act as a barrier against the migration of gas and moisture.

3.1.11. Essential Oil

An essential oil (EO) is a concentrated hydrophobic liquid that contains volatile plant-based chemical compounds that evaporate easily at room temperature. EOs are sometimes referred to as volatile oils, ethereal oils, aetheroleum, or just the plant's oil, like clove oil, from which they were extracted. An EO is considered essential when it contains the essence of the scent of the plant it comes from—that is, the distinctive scent of the plant from which it is extracted. The word "essential" in this context does not imply "useless" or "usable by the human body," unlike the terms "essential fatty acid" and "essential amino acid," which have this meaning because they are necessary for a living thing to function.

Usually, distillation is used to extract EOs, frequently with the use of steam. Expression, solvent extraction, absolute oil extraction, resin tapping, wax embedding, and cold pressing are additional procedures. They are used to flavor food and beverages, add fragrances to incense and household cleaning products, and enhance the scent of perfumes, cosmetics, soaps, and air fresheners, among other products.

Aromatherapy is a type of complementary medicine that attributes therapeutic properties to aromatic compounds. EOs are frequently utilized in this process. While aromatherapy can help promote relaxation, there is not enough proof to say that EOs can effectively treat any kind of illness. When EOs are used improperly, they can lead to adverse effects such as allergic reactions, inflammation, and skin irritation. Youngsters may be especially vulnerable to the harmful consequences of incorrect use. If consumed or absorbed through the skin, EOs have the potential to be toxic.

3.2. POTENTIAL APPLICATION IN THE FOOD INDUSTRY

Hydrocolloids are used as ingredients in a variety of food products. Among its uses are as an emulsifying and stabilizing agent in ice cream, yogurt, and butter; as a thickening and gelling agent in soups, salads, sauces, and jellies; as a fat substitute in meat and dairy products; and as a raw material for edible coatings that extend the shelf life and enhance the quality of fresh fruits and vegetables. They frequently serve as carriers for antimicrobial or antioxidant substances, such as EOs, which help extend the shelf life of food by protecting it from harmful bacteria, fungi, and deterioration. Table 3.1 briefly discusses the numerous uses of hydrocolloids in various food products.

3.2.1. Thickening Agent

The most common application for hydrocolloids is as a thickening agent in a variety of culinary processes. Because of the non-specific entanglement of conformationally disordered polymer chains, hydrocolloids' thickening property essentially confers increased viscosity. Over a critical concentration referred to as the overlap concentration (C*), thickening is thought to occur. Below this overlap concentration (C*), the hydrocolloid dispersions behave Newtonian, meaning that viscosity is independent of the shear rate, and above this concentration, they behave non-Newtonian. The kind of hydrocolloid used, its concentration, the pH level, and the temperature of the food system in which it is incorporated all affect the thickening effect that the hydrocolloids produce. Starch, xanthan, CMC, HPMS, or galactomannans like guar gum and locust bean gum are a few frequently used thickening agents.

Table 3.1. Aquatic Animals Derived Gelatin Compared

Source	Yield (Wet Basis)	Bloom/gel Strength
Atlantic salmon skin	4–11.3%	80–108 g
Atlantic cod skin	44.8% [c]	71 g
Bigeye snapper skin	6.5%	105.7 g
Bigeye snapper skin	40.3% [a]	138.6 g
Brownbanded bamboo shark	19.06–22.81%	56.53–217.26 g

Blacktip shark	21.17–24.76%	10.43–207.83 g
Black tilapia skin	5.39%	181 g
Bigeye snapper skin	6.5%	105.7 g
Channel catfish	19.2% [b]	252 g
Cod skin	17%	180 g
Cuttlefish skin	36.82% [c] (dorsal skin) and 59.69% [c] (ventral skin)	126 g (dorsal skin) and 137 g (ventral skin)
Giant catfish skin	20.1%	153 g
Giant squid inner and outer tunics	12%	147 g
Grass carp	11.3% [a]	N/E
Lumpfish skin	14.3%	N/E
Megrim skin	10%	360 g
Nile perch bone	2.4%	134–160 g
Nile perch skin	16%	134–229 g
Pollock skin	18% [b]	460 g
Red tilapia skin	7.81%	128 g
Shark cartilage	17.34%	111.9 kPa
Shortfin scad skin	7.25%	177 g
Sin croaker skin	14.3%	125 g
Silver carp skin	11% [a]	600 g
Tilapia skin	N/E	263 g
Tuna fin	1.25%	126 g
Yellowfin tuna skin	89.7%	426 kPa

[a]Based on the hydroxyproline content of the gelatin in comparison with that in the skin. [b]—based on the protein content of the gelatin in comparison with the wet weight of raw material. [c]—dry weight. N/E—not evaluated.

Source: Waraczewski R, Muszyński S, Sołowiej BG. (2022). An analysis of the plant- and animal-based hydrocolloids as byproducts of the food industry. Molecules; 27(24):8686. https://doi.org/10.3390/molecules27248686.

Across a broad pH and temperature range, xanthan gum exhibits very high low-shear viscosity, high-shear thinning, and viscosity maintenance in the presence of electrolytes. Although CMC has a high viscosity, it decreases when electrolytes are added or the pH drops. Conversely, MC and HPMC are not affected by pH or electrolytes, but they do increase viscosity with temperature. Galactomannans, which include gum Arabic and locust bean gum, have very high low-shear viscosity and strong shear thinning, while gum Arabic is a low viscosity gum and shows near Newtonian behavior, shear thinning at low shear rates. While the presence of electrolytes does not affect them, high and low pH can degrade them, and high temperatures can cause them to lose their viscosity. Ketchup is the most popular food item where the thickening property of hydrocolloids is used to control the viscosity, and starch is the most popular hydrocolloid used as a thickener because it is inexpensive and does not change the taste of the food product when used at low concentrations of 2–5%.

3.2.2. Gelling Agent

Hydrocolloids have the capacity to gel in addition to thickening food. In a gel, the liquid phase is discontinuous and the solid matrix is continuous, forming a colloidal dispersion. It creates a three-dimensional web of intertwined suspended particles, which limits the mobility of the particles in the dispersing medium. These gelling agents experience significant cross-linking upon hydration, which raises viscosity. A gel is defined as a viscoelastic system with a "storage modulus" (G') greater than the "loss modulus" (G') in rheological terms. Agar, pectin, alginate, carrageenan, gelatin, modified starch, and MC are a few hydrocolloids that do not gel. They are frequently found in food items such as puddings, jellies, jams, restructured meals, and baked goods.

Table 3.2. Health Benefits of Aquatic Animal Gelatin Hydrolysates and Gelatin-Derived Peptides with Used Enzymes

Fish or Aquatic Animals	Enzyme Used	Pro-Health Benefits
Alaska pollock skin	Pronase E	Antioxidant
Atlantic salmon skin	Flavorzyme	Dipeptidyl-peptidase IV enzyme inhibitory activity–type 2 diabetes, symptoms mitigation
Amur sturgeon skin	Alcalase	Antioxidant, cryoprotective benefit
Brownstripe red snapper skin	Trypsin-like prote-ases from pyloric caeca	Antioxidant
Blacktip shark skin	Papain, papaya latex crude enzymes	Antioxidants, hypertension pre-vention, human LDL cholesterol inhibition, DNA oxidation inhibi-tion, metal ion chelation

Chum salmon skin	Papain, Alcalase	Cell proliferation, cycle progression, apoptosis
Hoki skin gelatin	Trypsin	Antioxidant
Japanese flounder skin	Pepsin	Antioxidant
Jumbo squid skin	Trypsin	Antioxidant
Nile tilapia scale	Alcalase	Antioxidant
Pacific cod scale	Pepsin, trypsin, α–chymotrypsin	Antioxidant, antihypertensive benefit
Pacific cod skin	Papain	Antioxidant, ACE-inhibition (hypertension prevention)
Squid inner and outer tunics	Protamex, trypsin, neutrase, savinase, NS37005, esperase, alcalase	Antioxidant, hypertension prevention, anticancer benefit against lines MCF-7 and U87
Squid skin	Pepsin	Hypertension prevention
Tilapia skin	Properase E, multifactor neutral	Antioxidant, photoaging prevention

Source: Waraczewski R, Muszyński S, Sołowiej BG. (2022). An analysis of the plant- and animal-based hydrocolloids as byproducts of the food industry. Molecules; 27(24):8686. https://doi.org/10.3390/molecules27248686.

Certain hydrocolloids show a temperature-induced, reversible transition from sol to gel. The nature of secondary or non-covalent molecular forces like hydrogen bonds and hydrophobic interaction is a major determinant of this sol–gel transition. Syneresis is a common, unwanted phenomenon that happens in some products, like jelly, as a result of the gel structure shrinking as a result of calcium ions cross-linking pectin, carrageenan, or alginate. Conversely, syneresis might be advantageous in the process of making cheese. When other hydrocolloids are combined with gums that create brittle gels, such as locust bean gum added to κ-carrageenan-based gel, a more elastic and transparent gel can be created. This is a practical way to create synergy. Furthermore, combining gelling and non-gelling agents yields better food gels. For instance, alginate desserts can be made more resistant to textural deterioration during freezing and thawing by adding CMC, or agar-based marshmallows can be made softer by adding gum Arabic. Similarly, adding non-gelling hydrocolloids can enhance the texture of many products, including meringues, pie fillings, and marshmallows (Kiani et al., 2010).

3.2.3. Emulsifying Agent

By lowering the interfacial tension between the two phases, the emulsifier in an emulsion system encourages emulsion formation and preserves the mixing system's uniformity, while the stabilizer gives the emulsion long-term stability.

The texture, body, and gelation of milk-based formulations, such as flavored milk or yogurt drinks, are enhanced by the use of hydrocolloids. Additionally, they help to prolong the shelf life of chocolate milk and acidified fruit milk drinks by preventing the emulsion droplets from creaming or flocculating, whey formation, or settling scattered particles. Gum Arabic is the most widely used hydrocolloid emulsifier. It is used in soft drinks to emulsify flavor oils at low pH and high ionic strength, as well as when beverage coloring agents are present. The food industry also uses modified starches, galactomannans, and some pectins, which are hydrocolloid emulsifiers.

3.2.4. Surface Activity and Emulsification Property

The lack a flexible molecular structure, rigid hydrocolloids are generally not regarded as classical emulsifiers; instead, their hydrophilicity, high molecular weight, and gelation behavior make them suitable as stabilizers.

Excellent emulsifiers, like protein emulsifiers, are thought to have flexible molecular structures that enable quick adsorption and rearrangement at the interface to produce a coherent macromolecular protective layer. This is the conventional wisdom. Rigid hydrocolloids, therefore, are categorized as stabilizers rather than classical emulsifiers because of their hydrophilicity, high molecular weight, and gelation behavior.

However, it is well known that certain hydrocolloids can stabilize emulsions and adsorb on oil-water interfaces. They prevent foams and oil-containing gel systems from synergizing, delay the precipitation of dispersed solid particles and the coalescence of oil droplets, and prevent the aggregation of dispersed particles. Furthermore, the large molecular size and predominate hydrophilicity of a hydrocolloid emulsifier, as opposed to protein emulsifiers, cause a thicker stabilizing layer to form, which can shield droplets from aggregation under challenging circumstances like the addition of calcium salts and thermal shock treatment.

Regardless of esterification level, commercial pectins made from citrus peel are typically not used as emulsifiers. Surface activity has been reported for highly acetylated pectin from sugar beet, and good surface activity is also reported for pectins from citrus fruits and apples with low acetyl content ($<0.8\%$). Many surface-active cellulose derivatives, including hydroxypropyl (methyl) cellulose, can also produce good stabilization (Dickinson, 2009).

3.2.5. Fat-Replacing Agents

The fats are present in food, they often give it a desirable taste, texture, and appearance. However, because of their detrimental effects on human health, the food industry has been forced to create food products with lower fat content. Effective measures to reduce the fat content of certain emulsion-based food products, like ice cream, frozen desserts, puddings, sauces, and salad dressings, are necessary because their fat content is relatively high. But this fat loss could also result in a loss of other desirable sensory attributes.

Food products must therefore be designed with a lower fat content while maintaining sensory attributes that are comparable to those of their full-fat equivalents. The use of fat replacers or substitutes can help achieve this. An ingredient that replaces part or all of the functions of fat may or may not have nutritional value. This is known as a fat replacer. Pectin, inulin, guar gum, xanthan gum, gum tragacanth, locust Bena gum, κ-carrageenan, and sodium alginate are a few hydrocolloid-based fat replacers.

3.2.6. Bioplastics for Food Packaging

Food packaging has long used nonbiodegradable plastics, but with the rise in environmental awareness recently, bioplastics have shown to be an effective replacement for conventional plastic. A biodegradable hydrocolloid that can be used to make cellophane films is cellulose. The most widely used cellulose derivative for food packaging, particularly for baked goods and fresh produce, is cellulose acetate, one of several derivatives with good film-forming properties. Fruit, potato chips, and pasta are packaged in cellulose films by many major retailers, including Birkel and Wal-Mart.

Another hydrocolloid that is readily biodegradable and a crucial part of bioplastics is starch. They have been used in blow molding, injection molding, extrusion, and fil blowing on an industrial scale. Cornstarch trays have been used in milk chocolate by well-known retailers like Marks & Spencer Group and in organic tomato packaging by Iper. Bioplastics can also be made from hydrocolloid mixtures, such as starch and cellulose, pectin and chitosan, pectin and gelatin, or gelatin and sodium alginate.

3.3. HEALTH BENEFITS

Consumers' growing health consciousness has increased demand for low-fat foods enhanced with fiber. Dietary fiber is primarily composed of polysaccharides, which are thought to be structural elements of vegetables and cereals and are resistant to human endogenous enzymes. People all over the world generally agree that eating natural foods high in fiber promotes a healthy lifestyle. Good textural and fiber properties in hydrocolloids may support intestinal health. The overindulgence in high-calorie, high-fat foods has led to an alarming rise in chronic diseases. As a result, it becomes necessary to increase the intake of dietary fiber while concurrently reducing the amount of sugar consumed in the diet. It was noted many years ago that guar gum-containing diets improved colon health, delayed gastric emptying, lengthened the time it took for food to pass from the mouth to the cecum, and slowed the rise in postprandial glycemia. On the other hand, guar gum's high viscosity works against it in terms of sensory qualities. Numerous hydrocolloids with nutritional, textural, and health benefits have been found by researchers. These include pectin, alginate, starch, gum Arabic, psyllium, chitosan, carrageenan, and many more. The use of hydrocolloids has been linked to several health advantages, such as better bowel movements, appetite control, and a lower risk of type 2 diabetes, colon cancer, and coronary heart disease.

3.4. FUTURE PROSPECTS

The demand for ready-to-eat meals, consumers' growing health consciousness, and the fast-paced nature of modern life will all contribute to the future growth of hydrocolloid use in the food industry. To enhance thickening and gelling properties, new hydrocolloids could be introduced or existing hydrocolloids could be used synergistically. However, the high cost of obtaining food approval may present several obstacles to this exercise. Food industries could employ synergistic combinations of hydrocolloids that have already been approved for use as food additives to combat this issue. The synergy of stiff linear polymers like galactomannans with agarose, carrageenan, and xanthan has been known since the 1970s; however, in recent years, there has been an increasing interest in pectin-alginate combinations that form gels with improved characteristics through calcium ion crosslinking.

Future studies may concentrate on the solubility of hydrocolloids in food systems as another factor. It is known that certain hydrocolloids can dissolve entirely or partially in water. Their use could be further enhanced if methods or tools were developed to make them more soluble, or instantly soluble. Thus, to create a hydrocolloid combination with better textural qualities and expand the hydrocolloid market, researchers must go above and beyond. Furthermore, these researchers' efforts are hindered by regulatory and social factors. The hydrocolloid market has also been impacted by social issues like vegetarianism, halal, and kosher cuisine, as well as regulatory issues like usage and application levels of different foods. Because consumer perceptions and subsequent actions shape the commercial future of any food ingredient, it is imperative that consumers are made aware of the excellent properties of hydrocolloids when they are used as food additives (Cazón et al., 2017).

REFERENCES

1. Azarakhsh, N., Osman, A., Ghazali, H. M., Tan, C. P., & Mohd Adzahan, N. (2014). Lemongrass essential oil incorporated into alginate-based edible coating for shelf-life extension and quality retention of fresh-cut pineapple. *Postharvest Biology and Technology, 88*, 1–7.
2. Aziznia, S., Khosrowshahi, A., Madadlou, A., & Rahimi, J. (2008). Whey protein concentrate and gum tragacanth as fat replacers in nonfat yogurt: Chemical, physical, and microstructural properties. *Journal of Dairy Science, 91*(7), 2545–2552.
3. Barcenas, M. E., & Rosell, C. M. (2005). Effect of HPMC addition on the microstructure, quality, and aging of wheat bread. *Food Hydrocolloids, 19*(6), 1037–1043.
4. Bedane, A. H., Eić, M., Farmahini-Farahani, M., & Xiao, H. (2015). Water vapor transport properties of regenerated cellulose and nanofibrillated cellulose films. *Journal of Membrane Science, 493*, 46–57.
5. Bourtoom, T. (2008). Edible films and coatings: Characteristics and properties. *International Food Research Journal, 15*(3), 237–248.
6. Cazón, P., Velázquez, G., Ramírez, J. A., & Vázquez, M. (2017). Polysaccharide-based films and coatings for food packaging: A review. *Food Hydrocolloids, 68*, 136–148.
7. Cerqueira, M. A., Bourbon, A. I., Pinheiro, A. C., Martins, J. T., Souza, B. W. S., Teixeira, J. A., & Vicente, A. A. (2011). Galactomannans use in the development of edible films/coatings for food applications. *Trends in Food Science & Technology, 22*(12), 662–671.
8. Chanamai, R., & McClements, D. J. (2002). Comparison of gum Arabic, modified starch, and whey protein isolate as emulsifiers: Influence of pH, $CaCl_2$, and temperature. *Journal of Food Science, 67*(1), 120–125.
9. Chandra, R., & Rustgi, R. (1998). Biodegradable polymers. *Progress in Polymer Science, 23*(7), 1273–1335.
10. Chang, Y., & McClements, D. J. (2016). Influence of emulsifier type on the *in vitro* digestion of fish oil-in-water emulsions in the presence of an anionic marine polysaccharide (fucoidan): Caseinate, whey protein, lecithin, or Tween 80. *Food Hydrocolloids, 61*, 92–101.
11. Cheng, L. H., Karim, A. A., & Seow, C. C. (2008). Characterization of composite films made of konjac glucomannan (KGM), carboxymethyl cellulose (CMC), and lipid. *Food Chemistry, 107*(1), 411–418.
12. Cheng, Y., Chen, J., & Xiong, Y. L. (2014). Interfacial adsorption of peptides in oil-in-water emulsions co-stabilized by Tween 20 and antioxidative

potato peptides. *Journal of Agricultural and Food Chemistry, 62*(50), 11575–11581.

13. Choi, J. H., Choi, W. Y., Cha, D. S., Chinnan, M. J., Park, H. J., Lee, D. S., & Park, J. M. (2005). Diffusivity of potassium sorbate in κ-carrageenan based antimicrobial film. *LWT—Food Science and Technology, 38*(4), 417–423.
14. Chung, C., Rojanasasithara, T., Mutilangi, W., & McClements, D. J. (2016). Enhancement of color stability of anthocyanins in model beverages by gum Arabic addition. *Food Chemistry, 201*, 14–22.
15. Dickinson, E. (2009). Hydrocolloids as emulsifiers and emulsion stabilizers. *Food Hydrocolloids, 23*(6), 1473–1482.
16. Keskin, S. O., Sumnu, G., & Sahin, S. (2007). A study on the effects of different gums on dielectric properties and quality of breads baked in infrared-microwave combination oven. *European Food Research and Technology, 224*(3), 329–334.
17. Kiani, H., Mousavi, M. E., Razavi, H., & Morris, E. R. (2010). Effect of gellan, alone and in combination with high-methoxy pectin, on the structure and stability of dough, a yogurt-based Iranian drink. *Food Hydrocolloids, 24*(8), 744–754.
18. Koocheki, A., Ghandi, A., Razavi, S. M., Mortazavi, S. A., & Vasiljevic, T. (2009). The rheological properties of ketchup as a function of different hydrocolloids and temperature. *International Journal of Food Science and Technology, 44*(3), 596–602.
19. McClements, D. J., Decker, E. A., Park, Y., & Weiss, J. (2008). Designing food structure to control stability, digestion, release, and absorption of lipophilic food components. *Food Biophysics, 3*, 219–228.
20. Mun, S., Decker, E. A., & McClements, D. J. (2007). Influence of emulsifier type on *in vitro* digestibility of lipid droplets by pancreatic lipase. *Food Research International, 40*(6), 770–781.
21. Nguyen, V., Duong, C. T. M., & Vu, V. (2015). Effect of thermal treatment on physical properties and stability of whipping and whipped cream. *Journal of Food Engineering, 163*, 32–36.
22. Shittu, T. A., Aminu, R. A., & Abulude, E. O. (2009). Functional effects of xanthan gum on composite cassava-wheat dough and bread. *Food Hydrocolloids, 23*(8), 2254–2260.
23. Srikaeo, K., & Pradit, M. (2011). Simple techniques to increase the production yield and enhance the quality of organic rice bran oils. *Journal of Oleo Science, 60*(1), 1–5.

24. Tongdeesoontorn, W., Mauer, L. J., Wongruong, S., Sriburi, P., & Rachtanapun, P. (2011). Effect of carboxymethyl cellulose concentration on physical properties of biodegradable cassava starch-based films. *Chemistry Central Journal, 5*(1), 6.
25. Turabi, E., Sumnu, G., & Sahin, S. (2008). Rheological properties and quality of rice cakes formulated with different gums and an emulsifier blend. *Food Hydrocolloids, 22*(2), 305–312.
26. Viebke, C., Al-Assaf, S., & Phillips, G. O. (2014). Food hydrocolloids and health claims. *Bioactive Carbohydrates and Dietary Fibre, 4*(2), 101–114.

CHAPTER 4

LOCAL AND SYSTEMIC EFFECTS OF BIOACTIVE FOOD INGREDIENTS: ROLE FOR FUNCTIONAL FOODS TO PRIME THE GUT FOR RESILIENCE

Contents

INTRODUCTION

Food has been used as medicine for thousands of years since it was common practice to treat illness and restore health with plants, herbs, and foods. The relationship between diet and human physiology and health is now widely recognized. It has been demonstrated that eating a healthy, plant-based diet reduces the risk of developing chronic illnesses. Additionally, it has been demonstrated that certain plant-based diet components affect human physiology by influencing the gut microbiome. The body of knowledge about how different bioactive affect gut microbiota and broader physiology has expanded quickly in recent years. The idea of functional foods is expanded into a new realm by the innovative opportunity to create foods that support a resilient gut microbiota to protect against more widespread pathologies linked to gut dysbiosis. The discovery that the gut microbiota's resilience allows the body to adapt to stress is a crucial component of this investigation. The intricate community of microorganisms known as the gut microbiota, which inhabits the gastrointestinal tract (GIT), is essential for regulating both the systemic and local reactions to stressors. By interacting with the gut microbiota, bioactive food ingredients may enhance microbiota resilience, thereby bolstering the body's ability to adapt.

Knowledgeable consumers are searching for functional foods with active ingredients that promote gut health as a result of growing awareness of the possible advantages between bioactive food ingredients, the gut microbiota, and broader physiology. This is a challenge as well as an opportunity for experts in food innovation and product development. On the one hand, there is a great chance to incorporate combinations of active ingredients into functional food and beverage carriers as an easy and successful intervention for gut health. Conversely, researchers studying a wide range of bioactive and their possible impact on the microbiota and broader physiology, as well as differing degrees of efficacy and evidence, present challenges for product developers. Furthermore, research on the reciprocal relationship between gut microbiota and overall organ health, or systemic health, is emerging, which presents both opportunities for innovation and difficulties in the development of new products. Nonetheless, the deliberate choice of bioactive in the creation of functional foods creates opportunities for the development of tactics to boost the resilience of the gut microbiota and thereby advance general well-being.

4.1. HORMESIS IN THE CONTEXT OF DIET: PRIMING THE GUT TO BECOME MORE RESILIENT

One of the key indicators of health has been identified as having a sufficient reaction to stress. Numerous stressors pose a constant threat to one's health.

To achieve biological stability, organisms use a variety of techniques, such as hormesis, repair, homeostatic resilience, and, when possible, the regeneration of damaged tissues and organs. Strong resistance to stress and protection against a range of dysbiosis-related pathologies, including inflammatory bowel disease, metabolic syndrome, cardiovascular dysfunctions, depression, asthma, rheumatoid arthritis, colon cancer, and autism spectrum disorders, can be attributed to a "healthy" microbiota.

Sommer et al. and others have talked about the potential effects of life course disruptions on the gut microbiota. This is due to the fact that both in health and illness, the gut microbiota ecosystem maintains its own homeostasis. Disturbances usually cause the gut microbiota to temporarily change as part of that homeostasis, and then it returns to its "normal" state. Microbiota resilience has been defined as this ability to self-regenerate, or to return to a stable state. Scenario (a) in Figure 4.1 depicts a return to "normal" health after a disturbance, while scenario; (b) depicts an incapacity to recover, leading to a dysbiosis of the gut microbiota. Frequent consumption of functional foods that are well-designed may help prime the gut to support gut microbiota recovery as in scenario "a," thereby fostering microbiota resilience.

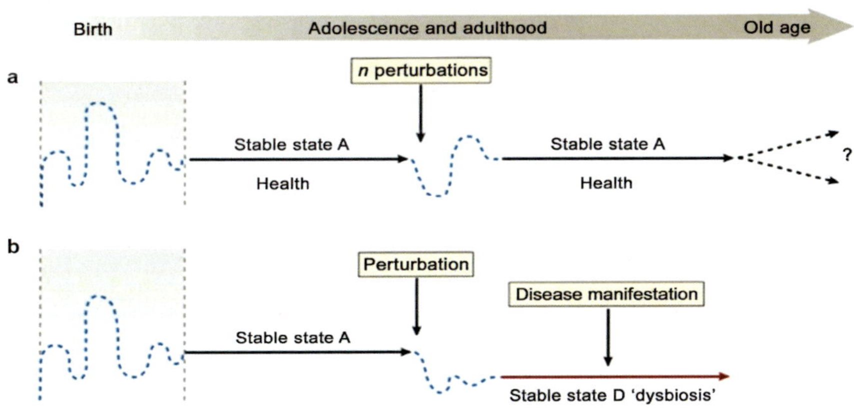

Figure 4.1. *How perturbations (or challenges) may influence the homeostasis of the gut microbiota.*

Source: Jacquier EF, van de Wouw M, Nekrasov E, Contractor N, Kassis A, Marcu D. (2024). Local and Systemic Effects of Bioactive Food Ingredients: Is There a Role for Functional Foods to Prime the Gut for Resilience? Foods, 13(5):739. https://doi.org/10.3390/foods1.

Sommer et al. asserts that based on the gut microbiota's resilience, scenarios a and b are conceivable. In case A, following a disturbance (such as medication, brief dietary changes, or illness), the microbiota returns to its stable state. As a result of the host's inability to "bounce back," dysbiosis and the diseases it causes developed in scenario b. In the context of human physiology, where the gut-organ axis and immune system are in constant communication with the microbiota, scenarios a and b may have consequences.

Therefore, resilience and stress adaptation are essential elements of health, referring to an organism's capacity to react and adjust to stressors in a way that ultimately enhances general well-being. This flexibility is especially noticeable when considering the gut microbiota, which is important for immune response, nutritional metabolism, and general gastrointestinal health. For the microbiota to maintain or restore homeostasis following disruptions, it must engage in intricate interactions with the host's stress response systems, immune system, and gut-organ axes, in addition to genetic and environmental factors. However, the fact that gut microbiota resilience is now part of the health framework emphasizes how important nutrition is as a major modulator.

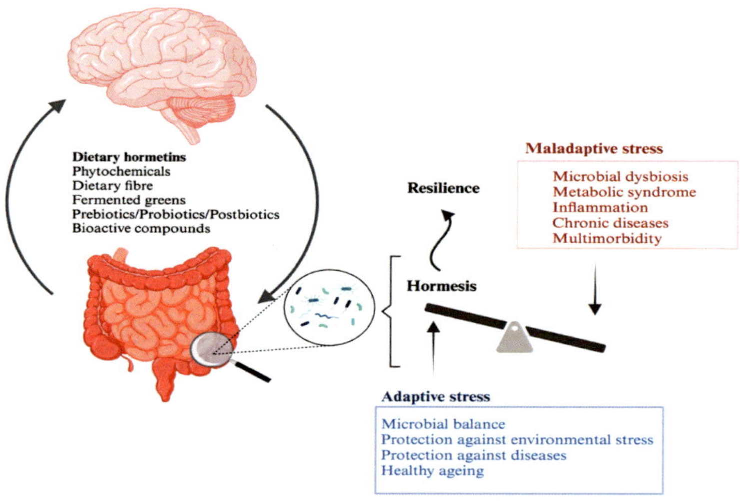

Figure 4.2. *The relationship between dietary hormetins, gut microbiota resilience, and the bidirectional communication with the brain that may shape the body's ability to respond to stress.*

Source: Jacquier EF, van de Wouw M, Nekrasov E, Contractor N, Kassis A, Marcu D. (2024). Local and systemic effects of bioactive food ingredients: Is there a role for functional foods to prime the gut for resilience? Foods; 13(5):739. https://doi.org/10.3390/foods1.

The hormetic responses from dietary interventions may be the key to improving gut resilience. How can we use diet to prime the gut to be more resilient and improve the gut microbiota's response to perturbations/stress? Hormesis is a biphasic dose-response phenomenon in which low stressor exposure triggers adaptive reactions that provide resistance against further challenges. When it comes to nutrition, exposure to phytochemicals, dietary polyphenols, and other bioactive substances found in fruits, vegetables, and some spices can have hormetic effects. Prominent research lends credence to the idea that dietary hormetins—hormetic stressors derived from specific food ingredients—can improve gut health. Research has demonstrated that dietary hormetins can support and fortify cells and tissues' adaptive and repair mechanisms, thereby promoting the preservation of health. For example, it has been demonstrated that polyphenols found in foods like berries and green tea can cause mild oxidative stress, which in turn can trigger cellular adaptations that lead to increased resilience of the gastrointestinal epithelium.

Additionally, dietary hormetins increase microbiota resilience through modifying the composition of the gut microbiota, promoting the growth of beneficial bacteria, preserving microbial equilibrium, boosting metabolic processes and diversity, and improving cellular maintenance and repair. All of these actions improve overall health and decrease the risk of disease. Furthermore, it has been recognized that a balanced and ideal diet rich in hormetic phytochemicals—which include polyphenols, carotenoids, sulforaphane, and other bioactive compounds—can initiate intracellular signaling cascades and alter gene expression, thereby bolstering the resilience of the gut microbiota. The body's reaction to stress is influenced by the gut-brain axis-mediated enhanced resilience of the gut microbiota. The dual potential of this interaction is illustrated in Figure 4.2, which shows scenarios in which the gut microbiota modulation promotes either maladaptive or adaptive stress responses.

Dietary compounds can cause hormetic responses that affect systemic health in addition to their local effects on the gut. Activating cellular stress response pathways, such as the Nrf2 antioxidant pathway, which is essential for preserving cellular homeostasis and reducing inflammation, is one of these adaptive mechanisms. Hormesis may also help people live longer and be more resistant to age-related illnesses. Hormetic responses may help prevent or lessen chronic age-related diseases like metabolic syndromes and neurodegenerative disorders by generating stress resistance at the cellular and molecular levels. In conclusion, knowing how hormesis functions about diet can help us better understand the mechanisms that underlie its beneficial effects on systemic and gut health. Tailoring dietary components to elicit hormetic responses can provide advantages in terms of strengthening gut resilience and enhancing general health.

Foods, dietary patterns, and purposefully created functional foods may, in theory, help prime the body to react to everyday stressors over time:

- The gut microbiota or host physiology is less affected by a future perturbation (or challenge) (e.g., the organism is less affected by daily stressors, or sickness is less intense/impactful).
- Following a disruption (or challenge), recovery is quicker and shorter.
- When a person recovers partially from a disruption (or challenge), the food or supplement can help address some long-term deficiencies (e.g., reducing inflammation) (Jacquier et al., 2024).

4.2. THE IMPACT OF GUT MICROBIOME ON HOST HEALTH

Microbes have always coexisted and evolved alongside humans and animals. The collective genomes of bacteria, viruses, eukaryotic microbes, and archaea that live inside and on top of humans make up the human microbiome. The gut microbiota coevolves with the host from birth and coexists symbiotically in the human body in a number of habitats, including the skin, gastrointestinal tract (GIT), respiratory system, and oral cavity. The largest population of bacteria is found in the GIT, where they are collectively referred to as the gut microbiota. These bacteria are primarily members of the phyla *Bacillota*, *Bacteroidota*, and *Actinomycetota*. Early research showed that as the baby grows and approaches an adult-like composition around the age of three, the gut microbiota exhibits significant compositional shifts in its bacterial communities. Changes in diet, ageing, illness, and the use of antibiotics have all been linked to subsequent changes in the microbial landscape.

4.2.1. The Functional Relevance of Microbiota

A growing body of research indicates that the development and course of metabolic disorders are significantly influenced by the gut microbiota. Transplanting gut microbes from conventionally raised mice into germ-free mice led to an increase in adipose tissue accumulation in the germ-free mice, even though their food intake was kept under control. This shows that gut microbiota transplantation can cause mice to become more obese. Furthermore, it was demonstrated that the intestinal microbial composition differed in lean and obese individuals, with the former having a higher prevalence of *Bacillota* and the latter having a lower prevalence of *Bacteroidota*. Comparing the composition of intestinal microbiota in fecal samples from type 2 diabetes (T2D) patients and healthy individuals has also revealed differences in microbiota composition. Higher quantities of *Bifidobacterium* and fewer Lactobacillus bacteria were found in the T2D samples. These studies highlight the importance of microbiota in mitigating the risk of chronic illness.

Evidence suggests that the gut microbiota plays a significant role in general well-being, especially in later life, and goes beyond specific diseases. Studies conducted on humans and a variety of model organisms show how important the gut microbiota is to health and longevity. In general, as organisms get older, opportunistic bacteria and total microbial counts rise while the diversity of their gut microbiota and health-promoting bacteria decrease. Studies on humans also show that the microbiota's composition varies with age, which may have a causal effect on late-life health. It has been demonstrated that Centenarians are able to withstand these changes by preserving the diversity of their gut microbiota and by raising the frequency of health-associated microorganisms (like *Bifidobacterium* and *Akkermansia*). According to these studies, age and disease are correlated with changes in the composition of the gut microbiota. This suggests that interventions that target the gut microbiota may be able to extend lifespans and improve health.

4.2.2. What is a "Healthy" Gut Microbiota? A Holistic Exploration of Dysbiosis-Induced Pathogenesis

Determining what constitutes a "healthy" gut microbiota and how it relates to host physiological processes are areas of intense research. However, because everyone's gut microbiota is different, the idea of a "healthy" gut microbiota is still subjective and constantly changing in scientific discourse. However, some traits related to the composition of the gut microbiota are frequently thought to be suggestive of a possible state of health. In a state of health, the gut microbiota demonstrates resilience, stability, and a symbiotic association with the host. A balanced and diverse microbial community is typically indicative of a "healthy" gut microbiota and is necessary for immune system modulation, nutrient metabolism, and gut barrier integrity.

The gut microbiota is crucial for maintaining homeostasis because it helps break down complex carbohydrates, extract nutrients from food, and synthesize bioactive molecules like lipids, vitamins, amino acids, and short-chain fatty acids. Furthermore, the production of antimicrobial compounds by the gut microbiota protects the host against external pathogens and plays a major role in the development of the immune system and intestinal mucosa. Dysbiosis can trigger disease processes via a variety of methods. Among the main causes are immune system changes, dysregulation of metabolic pathways, and disruption of the gut barrier.

However, since the gut microbiota can generate and control signals that are absorbed by the bloodstream and alter the physiology of entire organisms, how could gut-confined microbes have such widespread effects on health? These signals may originate from the host or from the microbiota. Microbial signals can be metabolites or structural elements of bacteria. Pattern-recognition receptors on epithelial and immune cells identify structural components like lipopolysaccharide (LPS), peptidoglycan, and flagellin; these components are typically not diffused across the epithelial barrier. According to reports, LPS and chylomicrons may co-transport into the bloodstream, which may have consequences for inflammation and obesity. It has also been demonstrated that a change in the makeup of gut bacteria catabolizes tryptophan into the metabolite indole-3-aldehyde, which has been linked to inflammatory bowel disease. Furthermore, the saccharolytic fermentation of dietary fiber produces SCFAs like butyrate, propionate, and acetate, which can function as signaling molecules at different body sites. For instance, acetyl-CoA can be obtained from dietary fructose, which in turn can initiate hepatic de novo lipogenesis. This is known as acetate. Molecules originating from bacteria can therefore have a systemic impact on host health.

The human gut is the largest endocrine organ, housing specialized hormone-producing enteroendocrine cells in addition to signal-producing bacteria. It has been demonstrated that SCFAs in mice activate receptors on those cells, which in turn causes the secretion of gut peptides like peptide YY (PYY) and glucagon-like peptide (GLP-1). PYY has been demonstrated to enhance the amount of energy retrieved from food, whereas GLP-1 is known to control appetite, insulin release, and pancreatic function. Research on humans has shown that the consumption of fermentable fibers alters the gut microbiome. These changes are thought to increase satiety, decrease appetite, and raise GLP-1 and PYY levels when fructooligosaccharides (FOS) influence the gut microbiome. Therefore, the bacteria that live in our stomachs produce chemicals that can affect the endocrine signaling of the host, which could have an impact on the overall health of the body (Jacquier et al., 2024).

4.3. THE INTERPLAY BETWEEN DIET, MICROBIOTA, AND HUMAN HEALTH

Human health is largely dependent on the gut microbiota, and dysbiosis can lead to the emergence of many diseases. Studies have shown that the variety and relative abundance of microorganisms in the gut microbiota are strongly influenced by the diet of humans. Food's macronutrients, micronutrients, and bioactive substances affect the microbial community differently and can cause dysbiosis or stability. More than half of the variation in the gut microbiota is linked to dietary modifications, making the diet a critical modifiable factor influencing the gut microbial communities. Poor nutritional status—which includes undernutrition, deficiencies in certain micronutrients, and excessive calorie intake—has long been linked to several unfavorable health outcomes, including cancer, heart disease, type II diabetes, gastrointestinal distress, and poor mental health outcomes (such as anxiety and depression).

A healthier body weight, a better gut microbiome profile, and a lower risk of chronic disease are linked to dietary patterns that follow recommended guidelines for a healthy diet. For instance, the Mediterranean diet, which is high in fruits, vegetables, whole grains, and olive oil, encourages the growth of good bacteria and lowers the number of potentially harmful microorganisms. This highlights how the gut microbiota is impacted by holistic dietary patterns. Conversely, Western diets, which are high in processed foods, sugars, and saturated fats and low in fiber, have been widely associated with a higher risk of developing gut dysbiosis and chronic illnesses like obesity, T2D, cardiovascular disease (CVD), and some types of cancer. Furthermore, a diet high in plants and fish has been linked to optimal nutrient metabolism and the synthesis of short-chain fatty acids, while processed foods and foods derived from animals have been linked to the production of endotoxins. These studies also show that a diet high in processed and animal-derived foods was linked to altered gut microbiota, worse gut health, and increased intestinal inflammation.

A more varied plant-forward diet that prioritizes plant foods while keeping animal foods in moderation will offer the gut microbiota a greater variety of nutrients, which may in turn enable a wider variety of different gut microbes to live there. Initiatives like the American Gut Project served as examples of this, as it discovered correlations between microbiota diversity and a greater variety of plant species (participants consuming > 30 different plant species per week) compared to those consuming < 10 different plant species per week. Additionally, a more stable gut microbiota is probably linked to a more varied diet, which is a sign of gut microbiota resilience. The gut microbiome ecosystem can withstand brief individual variations in nutrition and supplementation. For example, short-term dramatic dietary interventions may alter the gut microbiota's composition, but these modifications are only temporary and disappear after a few days. Significantly, a person's distinct gut microbiota composition is more closely associated with their long-term dietary habits. This implies that to induce changes in the gut microbial ecosystem that may potentially confer health benefits, long-term, sustainable dietary interventions are required. You could think of these sustainable interventions—which include foods that have been carefully chosen—as priming the gut for resilience.

It can be difficult to improve dietary diversity for some groups of people because eating a varied diet can be expensive. Compared to nutrient-dense foods like lean meats, dairy products, fruits, and vegetables, energy-dense foods offer energy to the diet at a lower cost. A lack of knowledge about preparing healthy meals and choosing the right foods is linked to lower-quality diets. Additionally, it may be challenging to obtain and eat fresh plant-based foods "on the go." This is where functional foods can provide convenient, targeted nutrition. Food items classified as functional foods have bioactive ingredients that contribute to their overall health benefits in addition to their basic nutritional value. When included in a regular diet, functional ingredients may offer physiological benefits or help lower the risk of chronic diseases. Therefore, it is theoretically feasible to carefully create functional foods that include a range of distinct bioactive ingredients. If regularly consumed, these foods may support a healthy gut microbiota and possibly enhance systemic and gut health (Jacquier et al., 2024).

4.4. A MULTI-PRONGED APPROACH TO THE DESIGN OF FUNCTIONAL FOODS FOR GUT AND SYSTEMIC HEALTH

Therefore, the functional food developers go about choosing ingredients that have the best chance of influencing the functional parameters of gut and systemic health and reshaping the microbiota to a state of health? In recent times, nutrition science has moved from studying the relationship between dietary patterns and health outcomes to studying how intakes of specific nutrients affect disease risk. Crucially, a "holistic approach" to diet is becoming more and more important because (i) many nutrients we eat have synergistic effects on health, for example, vitamin C (ascorbic acid) improves iron absorption; and (ii) many foods that we eat naturally contain a complex matrix of various nutrients. Global dietary guidelines of today also emphasize the need for varied and diverse diets that emphasize whole grains, fruits, and vegetables—consumption of which needs to be sustained throughout life.

It is theoretically possible to create food-based interventions that may act on functional parameters of gut health in addition to shaping the gut microbiota towards a resilient, healthy state, even though nutritional strategies to modulate the gut microbiota towards a resilient state are still a developing field of study. Product developers might be well-equipped to choose ingredients with the necessary technical qualities for incorporation into their product matrix, but it can be difficult to choose the right bioactive that affect the gut microbiota and other areas. The gut has a reciprocal relationship with other organs and systems in the body, so even with the increasing amount of evidence linking certain functional ingredients to gut health, the field is still quite complicated. As a result, various bioactive components may influence distinct gut outcomes directly or indirectly, having either localized or systemic effects. Gaining knowledge about the potential local or systemic effects of specific bioactive ingredients could aid in the innovation of products in the gut health industry. When new ideas are developed for consumer testing, this could be especially helpful for innovation (Jacquier et al., 2024).

4.5. INGREDIENTS FOR FUNCTIONAL FOODS AND BEVERAGES WITH PROMISING GUT-HEALTH-RELATED OUTCOMES

This section covers a variety of possible ingredients that may be beneficial to gut health, from well-researched ingredients to newly-emerging ingredients that are supported by mounting scientific data. Specifically, we differentiate between the possible hormetic, systemic, and local effects of functional ingredients.

4.5.1. Dietary Fibers

Dietary fibers are a broad category of indigestible carbohydrates that come from the cell walls of plants. They are essential to human nutrition and have a significant impact on health. According to the most recent consensus, dietary fiber is made up of polymers of carbohydrates that are not absorbed or digested in the human gut. Rather, they travel to the large bowel, where they may be partially or completely fermented by the colonic microbiota. Conventionally, dietary fiber variations can be divided into groups according to how soluble they are.

Although insoluble fibers were believed to shorten transit times and slow down the absorption of some nutrients, like glucose, early research suggested that the solubility of fibers was crucial to their effects on gastrointestinal health. However, the idea of fiber solubility has become contentious in recent years. This can be attributed to a few factors: (i) inconsistent results from different methods of measuring fiber solubility; (ii) variations in gastrointestinal pH between individuals that affect fiber solubility; (iii) the fact that solubility is not the only indicator of the physiological effects of fiber; and (iv) the fact that dietary fibers are frequently a combination of soluble and insoluble fibers.

However, through a variety of mechanisms, fibers—soluble and insoluble—support both the general health of the system and the health of the local gut. A balanced gut microbiota, optimal intestinal physiology, and reduced risk of various systemic diseases are all dependent on dietary fibers being incorporated into daily nutrition. It is noteworthy that some reviews have already proposed that dietary fibers may help prime the gut for increased gut microbiome resilience.

4.5.1.1. The Local Gut Health Effects of Dietary Fibers

It becomes clear that dietary fibers are crucial for promoting and preserving gut health. Through their effects on immune responses, mucosal integrity, and the composition of the gut microbiota, they can greatly alter the local gut environment. Insoluble fibers, such as cellulose, help to promote regular bowel movements and stave off constipation by giving the stool more volume. Likewise, a variety of health advantages are provided by fiber powders, especially those that contain partially hydrolyzed guar gum (PHGG). As a soluble fiber, PHGG forms a gel-like substance in the intestines by absorbing water. This gives the stool more volume, which encourages peristalsis and makes regular bowel movements easier.

This effect is especially helpful for those who experience constipation. Furthermore, two randomized clinical trials demonstrated that PHGG is a potentially helpful dietary intervention for people with irritable bowel syndrome (IBS) due to its prebiotic qualities, which also help to control bowel movements and lessen bloating.

Soluble fibers can also function as "prebiotics," boosting the resilience of the gut microbiota by providing a substrate for good bacteria and supporting a diverse and well-balanced gut microbiota. Prebiotics are defined as fermentable dietary substances that, when ingested, specifically promote the development and/or activity of advantageous gut bacteria. Bacteria like Bifidobacteria and Lactobacilli ferment prebiotics like fructans (FOS and inulin) and galactans (galactooligosaccharides, or GOS), which in turn promotes the growth of these microorganisms. Short-chain fatty acids (SCFAs) are created during this fermentation process, especially butyrate, which is well-known for its ability to support the integrity of the gut epithelium, improve the function of the mucosal barrier, and lower inflammation in the gut. This supports the general health of the gut and could aid in the prevention of illnesses like inflammatory bowel diseases (IBDs). By lowering the pH, SCFAs change the environment in the intestines and stop the growth of harmful bacteria that are sensitive to pH. Furthermore, they reduce the activity of proteases, which are connected to the production of toxic metabolites such as ammonia, a byproduct of protein fermentation that may cause cancer. In the intestinal defense mechanism against pathogens and harmful substances, SCFAs play a proactive role.

By controlling the expression of mucins, which serve as the main physical barriers against pathogens and increase gut permeability, SCFAs support improved gut barrier function.

Furthermore, new research is highlighting the potential benefits of algal-based polysaccharides like fucoidan as prebiotics. These benefits include immunomodulation, as well as antioxidative, antiviral, and antimicrobial qualities, as well as improvements in gut health and serum dyslipidemia in rats.

4.5.1.2. The Systemic Health Effects of Dietary Fibers

Dietary fibers offer a multitude of health benefits, including improved metabolism, cardiovascular protection, weight management, and immune modulation. As such, including them in one's diet is becoming increasingly important for promoting and maintaining general health. Fiber fermentation results in the production of SCFAs, which have anti-inflammatory properties that affect immune responses and lessen systemic inflammation.

These SCFAs have been linked to better metabolic health, decreased systemic inflammation, and gastrointestinal well-being. Robust evidence indicates that SCFAs may have an effect on other organ systems, including the brain, host energy metabolism, and immune system. Additionally, SCFAs cause the release of GLP-1 and PYY, two satiety hormones. The hypothalamus is influenced by both hormones to encourage satiety. Furthermore, intestinal gluconeogenesis (IGN) can convert SCFAs into glucose. This process triggers the production of leptin by adipocytes, which improves satiety and has an anti-obesity effect. Moreover, enhanced glucose tolerance results from decreased hepatic gluconeogenesis caused by elevated IGN by SCFAs. Furthermore, non-digestible carbohydrates (NDCs) derived from plant sources have been shown to slow the onset and spread of colorectal cancer, according to systematic reviews and meta-analyses of case-control and cohort studies.

Furthermore, the long-term effects of resistant starch supplementation in patients with an increased risk of hereditary cancers (Lynch Syndrome) were examined in a randomized double-blind controlled trial. Moreover, SCFAs have immunomodulatory effects by controlling the production of antimicrobial peptides (AMPs), growing regulatory T cells, and affecting myeloid cell function, all of which help to lower inflammation. Therefore, the cumulative effect of SCFA production induced by NDCs is linked to significant markers associated with metabolic diseases, such as diabetes, obesity, and CVD.

It is still unknown which fibers are most effective at raising SCFA levels, although numerous studies demonstrate that most, if not all, fibers raise SCFA levels. This link between gut health and overall health makes it abundantly evident how important dietary fiber is in preventing long-term inflammatory diseases. In conclusion, dietary fibers offer a variety of properties for ingredients that may support gut health due to their soluble or insoluble qualities as well as their fermentable or poorly fermentable qualities. Additionally, when creating functional foods, choosing ingredients that might encourage the production of SCFAs may play a major role in priming the microbiota for resilience, gut health, and systemic well-being. As such, these ingredients may be crucial choices that could potentially improve overall health.

4.5.2. Local and Systemic Effects of Probiotics and Postbiotics

The availability of foods, drinks, and supplements containing probiotics has skyrocketed globally over the past 20 years. When it comes to adding probiotic bacteria to different dairy matrices, the dairy industry continues to be the largest market segment. "Live microorganisms that, when administered in adequate amounts, confer health benefits to the host" is the definition of probiotics. The most popular probiotics are yeast strains like *S. boulardii* and bacterial strains like *Lactobacillus* and *Bifidobacterium*. Changes in the makeup and functionality of the intestinal microbiome are thought to be the underlying mechanisms for the health benefits of probiotics. It is well recognized that probiotics produce metabolic or antimicrobial compounds that inhibit the growth of other microorganisms. Furthermore, as noted by Collado et al., they compete with other microbes in the intestinal environment for receptors and binding sites on the intestinal mucosa. in 2007. Probiotics have the ability to alter intestinal immunity by affecting the way that immune cells and intestinal epithelial cells (IECs) react to microorganisms in the intestinal lumen. Several research works have exhibited the capability of *L. reuteri* to control the synthesis of cytokines by human immune cells.

Lactobacillus strains improve the integrity of the intestinal barrier, which may help to preserve immunological tolerance and decrease the movement of bacteria across the intestinal mucosa. Thus, phenotypes of diseases like IBS, inflammatory bowel disease (IBD), and gastrointestinal infections may be affected. In a clinical study, patients with IBS who received a 4-week treatment with a rosehip drink containing *L. plantarum* DSM 9843 showed decreased pain and flatulence. Reduced levels of enterococci in fecal specimens were correlated with improved clinical symptoms in patients whose rectal biopsies revealed the presence of *L. plantarum*.

Another investigation focused on patients with diarrhea-dominant IBS (IBS-D) who experienced symptomatic relief with a probiotic mixture (*L. acidophilus*, *L. plantarum*, *L. rhamnosus*, *Bifidobacterium* breve, *B. lactis*, *B. longum*, and *Streptococcus thermophilus*).

Probiotics and their corresponding postbiotics have been demonstrated to influence systemic immunity and health via a sophisticated communication system between the gut and the organs, in addition to their beneficial effects on intestinal integrity, immunity, and general health locally. The microbiota–gut–brain axis (MGBA), for example, is continuously transmitting and interpreting data between the brain and the peripheral. The vagus nerve, which plays a crucial role in MGBA communication, relays environmental cues to the central nervous system (CNS) for the latter to continuously maintain homeostasis. The production of peripheral cytokines sets off the vagal anti-inflammatory reflex, which in turn produces acetylcholine. This averts tissue damage caused by an overabundance of cytokines. Recent studies have shown that patients with anxiety disorders, schizophrenia, and depression all have changes in their gut microbiota and vagal tone. Certain probiotics, like *Bifidobacterium*, have been demonstrated to communicate with the brain through vagal pathways. Thus, probiotics have no effect on the brain or behavior when the vagal nerve is severed (Jacquier et al., 2024).

An alternative to traditional probiotics is heat-killed bacteria. Postbiotics are generally defined as substances that contain inanimate microorganisms and/or their constituents and that, when taken in sufficient quantities, benefit users. As a solution to the drawbacks of probiotics, postbiotics have become more popular. These restrictions include things like unclear molecular mechanisms, short-lived and niche-specific actions, uncertain beneficial effects, the possibility of virulence genes being transferred or antibiotic resistance developing, difficulties maintaining viability and stability during the production process, barriers to commensal gut microflora colonization, and the capacity to cause opportunistic infections and inflammatory reactions. Furthermore, there are practical obstacles to the incorporation of probiotics into complex, multi-component functional foods, which are mostly associated with formulation difficulties and stability problems. Because probiotic viability and functionality can be impacted by a variety of factors, including food matrix interactions, processing conditions, and exposure to environmental elements, it can be difficult to maintain the desired concentration of viable probiotics throughout the product's shelf life. Notably, it has been demonstrated that non-viable microorganisms retain the ability to benefit the host at the intestinal level *in vivo*, which has led to the creation of safer preparations with enhanced medicinal qualities.

4.5.2.1. Heat-Inactivated Lactobacillus

One study by Saito et al. showed that giving rats oral doses of heat-inactivated (HI) preparations of L. brevis SBC8803 increases the concentration of acyl ghrelin, which raises the blood's acyl to des-acyl (inactive) ghrelin ratio. In addition, it has been demonstrated that HI *Lactobacillus casei* subsp. *casei* 327 supplementation enhances the synthesis of serotonin (5-HT) in the colonic mucosa, thereby enhancing GI motility in mice. *L. reuteri* 100-23 that has been HI has been demonstrated to elicit an immunoregulatory response in mice that inhibits immune responses against it in the colon, allowing the bacteria to remain in the gut and continue to have a commensalistic relationship with the host. Fish that receive HI L. planetarium L-137 supplementation have altered intestinal morphology, including increased villus height, villus surface area, and muscularis mucosae. Additionally, the gut microbiota of the fish has higher relative abundances of *Bacteroidota*, *Delftia*, *Brevinema*, and *Elizabethkingia*.

One uncontrolled human study that added HI L. After a 12-month exposure to planetarium L-137, the relative abundance of the gut microbial *Bacteroidota* increased while the abundance of *Bacillota* and Clostridium sub-cluster XIVa decreased in comparison to the baseline.

Systemically, HI L. planetarium L-137 supplementation decreases the gene expression of inflammatory markers in adipose tissue in rats with metabolic syndrome and increases interferon (IFN) ß gene expression in pig spleens and white adipose tissue in mice fed a high-fat diet. Daily management of HK L. In human subjects, exposure to planetarium L-137 improves self-reported quality-of-life scores, influences Th1-related immunity, raises IFNß levels, and shortens the duration and severity of upper respiratory tract infections.

4.5.2.2. Heat-Inactivated Bifidobacterium

The concept that nonviable cells retain probiotic efficacy is supported by a prior study that showed the HI probiotic *Bifidobacterium animalis* subsp. *lactis* CECT 8145 (Ba8145) reduced the fat content in the model organism Caenorhabditis elegans by more than 30%. Furthermore, in obese rats with metabolic syndrome, HI Ba8145 consumption improves insulin sensitivity, increases energy expenditure and lean mass, and reduces mesenteric adiposity and dyslipidemia. Lastly, a comparison of B's HI and living cells. When breve M-16V was given to gnotobiotic mice, it had immune-modulating effects that both affected intestinal metabolism and inhibited the production of pro-inflammatory cytokines in spleen cells. The effects were more pronounced when the cells were alive. In a double-blind, placebo-controlled trial, individuals with IBS were selected at random from 20 study sites in Germany and given capsules containing HI *B. bifidum* HI-MIMBb75 cells or a placebo for eight weeks. Their findings showed that, compared to 22% in the placebo group, slightly more than 30% of subjects in the *B. bifidum* HI-MIMBb75 met the primary endpoint of improvement of abdominal pain and adequate relief of overall IBS symptoms being fulfilled in at least 4 of 8 weeks during treatment. This study showed that certain beneficial bacterial effects are mediated independently of cell viability in a real-world setting.

4.5.2.3. Heat-Inactivated Saccharomyces cerevisiae Fermentate

Because of their high nutritional content and wide range of bioactive compounds, yeasts have been used widely as probiotics. The reason why *S. cerevisiae* is the most studied eukaryotic microorganism is due to its widespread usage in the food, beverage, and pharmaceutical industries. A novel functional ingredient called EpiCor® is made from a dried fermentate that has been HI through a specialized fermentation process using *S. cerevisiae*. It has been discovered that

EpiCor® affects gut health by encouraging the development of advantageous bacteria and preserving a balanced gut microbiota. This ingredient's complex profile of bioactive compounds—which includes vitamins, polyphenols, sterols, yeast metabolites, and phospholipids—has drawn attention for its wide range of health benefits. Enhancement of innate and adaptive immune responses is one way that EpiCor® has been linked to immune system modulation. EpiCor® has been shown in prior *in vitro* investigations involving human cells to activate natural killer (NK) cells, which are essential for recognizing and eliminating malignant or infected cells.

Additionally, over 300 adult participants in five randomized, double-blind, placebo-controlled clinical trials have examined this HI fermentate of *S. cerevisiae*. Comparing these published trials to conventional and nonconventional products, the results point to several different mechanisms of action. Without unduly suppressing or stimulating the immune system as a whole, EpiCor® has demonstrated the ability to positively affect immune responses. Notably, regardless of an individual's influenza vaccination status, EpiCor® has demonstrated significant clinical benefits by reducing the incidence and duration of symptoms similar to colds and the flu during the fall and winter. Daily ingestion of dried yeast ferment resulted in a significant increase in salivary IgA levels and a decrease in pollen-associated allergy symptoms, such as summertime nasal congestion. Additionally, it has been demonstrated that EpiCor® increases the production of anti-inflammatory cytokines, raises erythrocyte hematocrit levels, and strengthens mucosal immune protection while decreasing the secretion of pro-inflammatory cytokines. Overall, this product made of dried yeast fermentate has promise as a nutritional supplement to support the immune system in healthy individuals experiencing acute stress. Patients suffering from specific chronic conditions that affect the innate and adaptive immune systems may also benefit from this product.

This has been demonstrated to help people with gastrointestinal discomfort symptoms and reduced bowel movements have better GI symptoms and stool parameters. A higher quality of life and lower levels of stress were positively connected with the improvement of these symptoms. The comparatively low dosage required (500 mg/day) for EpiCor® administration to support gut health is another benefit, especially in comparison to the high recommended doses for prebiotic fibers. In conclusion, *S. cerevisiae* HI yeast fermentate seems to have a variety of intriguing health benefits, such as immune system modulation, anti-inflammatory effects, improved gut health, antioxidant activity, and possible support for respiratory health.

Additional human trials are essential to strengthen the evidence supporting the advantages of heat-treated yeast fermentations, stemming from *S. cerevisiae*, and postbiotics in totality, in serving as gut stimulants. Furthermore, it is crucial to delve deeper into their influence on both local and systemic immunity and overall wellness. Nevertheless, these substances prove to be a valuable tool for product creators, given their prospective technological advantages and simplicity in being integrated into various product formulations (Jacquier et al., 2024).

4.5.3. Plant-Based Fermented Foods

Food ingredients undergo enzymatic and microbial transformation to produce fermented foods. During the fermentation process, carbohydrates like sugars and starch are broken down into molecules that are easier to digest by microorganisms like yeast, fungi, or bacteria. This improves the vitamins, minerals, and other nutrients' digestibility and absorption, transforming the products into functional foods with possible advantages for gut health. Because microorganisms on these substrates have a transformative effect, fermented plant ingredients have drawn attention for their possible health benefits. Nutrients in plant-based diets can become more bioavailable through fermentation. For example, it has been demonstrated that fermenting grains and legumes lowers antinutritional factors and improves the absorption of minerals like zinc and iron. Furthermore, probiotic microorganisms found in fermented plant foods improve digestion, boost immunity, and modify the microbiota to promote gut health. Fermentation improves the digestibility of plant ingredients by dissolving complex proteins and carbohydrates into simpler forms. For people who are sensitive to specific ingredients in raw plants, this procedure might be especially helpful. Additionally, the fermentation of plants results in the production of bioactive substances. For instance, the fermentation of tea leaves yields polyphenols and catechins, which prime the body to become more resilient to stress through a process called hormesis and are known to induce a multitude of health benefits. Consumer interest in using fermented foods as a source of psychobiotics is growing due to these potential therapeutic benefits for the brain. For instance, in rats given amyloid-β, black carrots fermented with *Lactobacillus plantarum* or *Aspergillus oryzae* improve hippocampal insulin signaling, preventing cognitive dysfunction. Furthermore, in a randomized, double-blind clinical trial, *Saccharina japonica* algae fermented with *Lactobacillus brevis* BJ20 exhibited memory-enhancing effects through the regulation of the SOD antioxidant system. In conclusion, research suggests that adding fermented plant ingredients to functional foods or your diet may provide many health advantages. These benefits include increased digestibility, increased nutrient bioavailability, probiotic effects, the synthesis of bioactive compounds, immunomodulation, and possible mental health benefits.

The potential health benefits of fermented grass products, like grains and oats, have garnered attention, especially when it comes to gut health. Fermented oat drinks have been produced for more than thirty years, coinciding with the growing trend of functional foods. Products made from fermented oats have been linked to improved digestive health by influencing gut motility. According to *in vitro* research, fermentation by advantageous microbes may improve digestion and increase the nutrients' bioavailability. Furthermore, since β-glucan is easily obtained as a byproduct of barley and oats, it is frequently utilized as a functional ingredient in food products. Previous research showed that β-glucan from barley or oats improved gut motility by positively influencing gastrointestinal transit time. Furthermore, β-glucan—extracted specifically from oats—reduces energy intake and body weight and increases satiety-related hormones like PYY and NPY, according to animal studies. These results lend credence to the idea that these dietary components could regulate the microbiota-gut-brain axis. Due to their broad acceptance and the emerging, yet still-to-be-proved benefits to systemic health, they are undoubtedly appealing ingredients to include in functional foods or new food concepts for gut health.

Fermented oat milk has been shown to have anti-diabetic properties in two studies involving rats. These effects include improved lipid metabolism, glucose regulation, liver health, and the composition of the intestinal microbiota. One study, in particular, highlights the beneficial effects of fermented oats and Sidr honey on diabetic rats, showing increased anti-diabetic effects over fermented oats alone. These results highlight the significance of a comprehensive dietary strategy. Studies on humans also highlight the importance of fermented oats in promoting effects on lipid metabolism that are anti-obesity and anti-cholesterol.

Polyphenols are also abundant in fermented oats. According to a remarkable study by Zhang and colleagues, ethyl acetate subfractions (EASs) with a high polyphenol content that are produced from fermented oats using *Rhizopus oryzae* may have anti-cancer potential. They tested this *in vivo* using mouse models and found that the EASs inhibited the growth of human cancer cells without causing toxicity to normal cells. Additionally, they show how the administration of EAS raises intracellular reactive oxygen species (ROS) and triggers the ROS/JNK signaling pathway, both of which aid in the induction of apoptosis. Overall, hormesis may be associated with these potential anti-cancer benefits, indicating that priming the gut through dietary-induced hormetic responses may result in improved stress adaptation. Further investigation is necessary to validate these results, though.

4.5.4. Polyphenols and Phytonutrients

4.5.4.1. Polyphenols, Hormesis Activation, and Adaptation to Stress

Plants have evolved biosynthetic pathways for chemicals that keep insects and microbes from eating them. In fact, it has been possible to identify over 100 of these biopesticides that aid in plant defense. These phytochemicals are usually found in areas of the plant that are exposed and vulnerable, like the fruit skin and developing buds. Subtoxic doses of these phytochemicals have been shown to cause adaptive stress responses in mammals, even though high concentrations of them can be toxic to mammalian cells. Research using a range of cell types has suggested hormetic mechanisms of action for particular phytochemicals, highlighting their importance in biology as cytoprotectants.

By promoting redox homeostasis, these polyphenols activate several vitamins signaling pathways, such as the sirtuins system, HO-1, thioredoxin/thioredoxin reductase, Hsp70, and HO-1. Nrf2, a transcription effector for more than 500 cytoprotective genes, mediates the transcriptional modulation of these cytoprotective genes. In pre-clinical models, Nrf2 activation promotes a healthy physiological state and lengthens lifespan by inducing a mild stress response. On the other hand, persistent Nrf2 stimulation can cause pathophysiological events, which classifies Nrf2 signaling as a pathway similar to hormetic. More and more data points to the activation of Nrf2 signaling by plant polyphenols, which supports redox homeostasis in stressful environments and may have protective effects due to adaptive stress responses.

According to Naviaux and associates, dangers surpass the ability of cells to maintain homeostasis, resulting in perturbing molecular alterations. A growing amount of research indicates that most chronic illnesses are not caused by the original injury or the agent that caused it, but rather by the body's reaction to a stressor. Hormesis, then, reinforces the body's first adaptive stress response, which in turn mediates protection against a range of possible injuries (i.e., diseases) and primes the body to become more resilient. Plant polyphenols are known to cause biphasic dose responses of the hormetic type, according to recent research. Because they activate adaptive responses, these polyphenols are thought to be a "preventive treatment of disease," causing biological effects with therapeutic applications. Plant polyphenols appear to have beneficial effects through hermetic like actions, activating pathways involved in adaptive stress response, and bringing the concept of hormesis fully into the realm of nutrition, according to mounting evidence. This creates creative new opportunities for functional foods in the future.

4.5.4.2. Reciprocal Interaction Between Polyphenols and Gut Microbiota

Research has indicated that plant polyphenols play a role in the gut's heightened microbial diversity. Diets high in polyphenols have been linked to a wider variety and abundance of bacterial species, which supports a stable and well-balanced gut microbiota. Plant polyphenols inhibit the growth of harmful pathogens while specifically promoting the growth of beneficial bacteria like Lactobacillus and *Bifidobacterium*. This prebiotic-like effect helps to keep the microbial community in the gut in a healthy state. Additionally, the metabolism of complex polyphenolic compounds involves the active participation of the gut microbiota. Bioactive metabolites, like urolithins and phenolic acids, are produced as a result of microbial degradation and may have additional health benefits over their parent compounds. Thus, the changes that polyphenols' microbiota induces greatly influence their bioavailability and impact. Several investigations have been carried out to clarify how gut microorganisms convert particular kinds of polyphenols and to identify the microorganisms that are in charge of this process. The *Bacillota* phylum members Clostridium and Eubacterium were found to be the primary genera involved in the metabolism of polyphenols.

In summary, there are important health implications for humans arising from the dynamic interplay between polyphenols and the gut microbiota. Through prebiotic-like effects, commensal microbes' transformation of dietary polyphenols affects not only the bioavailability of these compounds but also the makeup and functioning of the gut microbial community. Future studies will help determine which particular polyphenols to include in functional foods that may improve gut health.

4.5.4.3. Foods Rich in Phytonutrients and Their Associated Gut and Systemic Health Benefits

It has been demonstrated that isothiocyanates, which are abundant in cruciferous vegetables like broccoli, watercress, Brussels sprouts, cabbage, Japanese radish, and cauliflower, stimulate the expression of cytoprotective Nrf-2 proteins in cells of the stomach, liver, and intestinal tract. In rodent models of myocardial infarction and stroke, resveratrol—a compound found in the skin of red grapes and wine—shows the ability to activate stress response pathways and shield cells. It has been demonstrated that polyphenols, such as theobromine and catechin from cocoa and chocolate, reduce blood pressure and arterial stiffness. Green tea is well known for having a high polyphenol content, especially in catechins, which have anti-inflammatory, cardioprotective, and antioxidant properties. Higdon and Frei draw attention to the possibility of green tea lowering the risk of heart problems and specific types of cancer. Berries have anti-inflammatory and neuroprotective properties because they are high in anthocyanins, flavonoids, and other polyphenols. Berries may improve cognitive function and slow down age-related cognitive decline, according to prior research.

Quercetin is an additional polyphenol that can be found in citrus fruits, onions, apples, and berries. It has anti-inflammatory and antioxidant qualities. Quercetin functions as a coagulation, hyperglycemia, inflammatory, and hypertension-lowering agent. Quercetin supplementation is used to lower the risk of CVD, according to several clinical studies. Citrus fruits—which include grapefruits, oranges, lemons, and limes—are well known for having a high concentration of bioactive substances, especially bioflavonoids (which include hesperidin, naringin, and quercetin). These substances not only give citrus fruits their vivid hues, but they also improve endothelial function, lower inflammation, and regulate lipid metabolism—all of which are beneficial to cardiovascular health.

Furthermore, brown seaweeds contain phlorotannins, which have shown promise in modulating the gut microbiota and may have consequences for overall gastrointestinal health. Phlorotannins may help regulate diabetes because of their positive effects on the gut microbiota, which include protecting intestinal cells, improving structure, raising SCFA levels, minimizing changes brought on by obesity, and promoting the growth of commensal bacteria. These benefits are supported by pre-clinical research.

Through their antioxidant and anti-inflammatory qualities, carotenoids, like fucoxanthin and astaxanthin, which are present in a variety of fruits and vegetables, may also have antidepressant and neuroprotective effects in addition to improving the immune system, metabolic health, and gut barrier integrity. They may also help to improve gut health by modulating the gut microbiota. To fully comprehend the possible benefits of carotenoids and phlorotannins in relation to gut health, more study is necessary.

Investigating the biological activities of beets—especially red beets—and their possible benefits for GI health has received more attention in recent years. Consuming beets has been proposed as an adjunctive therapeutic strategy for many diseases associated with inflammation and oxidative stress. The presence of a unique class of phytochemicals called betalains, which is present in beets and has been shown to have chemo-preventive, anti-inflammatory, and antioxidant effects in both *in vitro* and *in vivo* studies, is largely responsible for these potential benefits and uses. Additionally, research indicates that these phytochemicals might affect the gut microbiota's composition and activity.

De Oliveira et al. conducted a systematic review, examining the impact of the bioactive substances present in beets on the microbiome and gastrointestinal health, proposing that these substances have prebiotic properties that encourage the development and activity of beneficial gut bacteria.

Overall, fungi, coffee, turmeric, orange juice, cranberries, blueberries, green tea, grapes, and sorghum bran are some of the foods that have been the subject of the most research on phytonutrient content. Phytochemicals that have been thoroughly studied include procyanidins, epicatechin, resveratrol, quercetin, epicatechin, and epigallocatechin-3-gallate. Polyphenols primarily have a positive impact on health outcomes by lowering inflammation and oxidative stress. Furthermore, there is growing evidence that the gut microbiota metabolizes specific dietary polyphenols, and these by-products are thought to have positive effects on physiology and health. The relationship between phytonutrients, gut microbiota, and hormesis offers an intriguing insight into how these plant-derived compounds aid in adaptive stress responses and may prime the host physiology to become more resilient.

4.5.5. Spices and Aromatic Herbs

Spices and aromatic herbs have been used for preservation, coloring, and flavor enhancement since ancient times. Spices, which have long been used medicinally in many cultures, have attracted interest from the chemical, pharmaceutical, and food industries because of their advantageous physiological effects and potential for use as preventative measures in a range of illnesses. Some herbs and spices are emerging as viable options for functional food inclusion due to the rise in chronic diseases and the demand for safe, affordable interventions in healthy aging. Spices include bioactive substances like phenolic acids, flavonoids, and sterols in addition to being high in fiber, minerals, pigments, and essential oils (EOs). The functional qualities of spices may also be influenced by the EOs, which include terpenes, monoterpenes, and sesquiterpenes. However, the kind, quantity, and presence of phytochemicals are closely related to the functional characteristics of spices.

A multitude of research works have emphasized the functional properties of phenolic compounds, specifically flavonoids, such as their anti-inflammatory, antiviral, antioxidant, and cardioprotective properties. The basic building block of EOs is terpenes. They are all very volatile and have been shown to have several functional qualities, such as antioxidant, antimicrobial, and antiviral capabilities, despite having very different chemical structures. Consequently, because of their high phytochemical content, eating herbs and spices shows the potential to modify host health both locally and systemically.

4.5.5.1. Local Influence of Herbs and Spices on Intestinal and Microbiota Health

The gut health benefits from the addition of spices to meals. They increase the concentration of biliary acids, which are necessary for the digestion and absorption of fatty acids, and they also increase the secretion of saliva and gastric juices. Additionally, they promote the pancreas to secrete digestive enzymes that are essential to digestion, such as lipase, amylase, trypsin, and chymotrypsin.

For example, traditional remedies for digestive health have long included the use of ginger and fennel (*Foeniculum vulgare*). Fennel has been shown in animal studies to increase the pancreatic secretion of digestive enzymes such as lipase, protease, and amylase, which improves nutrient absorption and breakdown.

Herbs and spices, which are high in phytonutrients and polyphenols, have prebiotic-like qualities that increase microbial diversity and encourage the growth and activity of good gut bacteria, which in turn supports a robust gut microbiota. A sizable body of research, the majority of which is derived from animal studies, indicates that capsaicin influences the gut microbiota in many ways through intricate and diverse processes, primarily addressing inflammatory and metabolic disorders. For instance, after giving capsaicin to mice for a week, the experimental group's relative *Faecalibacterium* abundances rose while those of the control group did not. One of the most significant symbiotic elements of the human gut microbiome is the gut bacterium *Faecalibacterium* from the *Bacillota* phylum). It is thought to be a bioindicator of human health because it is inversely correlated with major depressive disorder, IBD, obesity, diabetes, asthma, and colorectal cancer. However, more evidence of the positive effects on people is needed. The *Bacillota/Bacteroidota* ratio and *Faecalibacterium* abundance increased in a single small human trial ($n = 12$) wherein low (5 mg/day) and high (10 mg/day) capsaicin concentrations were given for two weeks to healthy subjects. Additionally, there were increases in the plasma levels of GLP-1 and gastric inhibitory polypeptide, and a decrease in the levels of ghrelin.

It has been proven that curcumin interacts with commensal microbes in the gut. Rats were randomly assigned to three groups in a non-alcoholic fatty liver disease rat model, which was induced by a high-fat diet: standard diet, high-fat diet, and high-fat diet plus curcumin. By supplementing with curcumin, the microbiota of the control rats given a standard diet exhibited a significantly different composition. This was due to the counteraction of the abundance of several genera, including *Ruminococcus*, which had been previously linked to diabetes and inflammation when fed a high-fat diet. Additional research verified that curcumin decreased pathogenic bacteria linked to systemic illnesses, such as *Prevotellaceae*, *Coriobacterales*, *Enterobacteria*, and *Rikenellaceae*, while increasing beneficial strains like *Bifidobacteria*, *Lactobacilli*, and butyrate-producing bacteria.

The ability of ginger to modulate microbes is also well-known. 6-gingerol, a bioactive compound found in ginger, was demonstrated to alter the gut microbiota's composition in an obese mouse model induced by a high-fat diet. This was achieved by decreasing the abundance of bacteria linked to obesity (such as *Lactobacillus reuteri* and Lachnospiraceae) and increasing the abundance of genera associated with weight loss (such as *Muribaculaceae*, *Alloprevotella*, and *Akkermansia*). Furthermore, a double-blind, placebo-controlled trial was conducted to examine the effects of ginger root powder in healthy adults. The results indicated that although taking ginger root powder supplements changed the composition of the gut microbiota, they did not affect microbial diversity, bowel function, gastrointestinal symptoms, mood, or quality of life.

In conclusion, the bioactive substances found in spices have a variety of effects on the gut microbiota, influencing it and encouraging the growth of good bacteria and bacterial metabolites, which may help to maintain the resilience of the gut microbiota as a whole.

4.5.6. Digestive Enzymes

It is valuable to take a quick look at digestive enzymes when developing functional foods for gastrointestinal health. Digestive enzymes are produced and released by the GIT to aid in the breakdown of fats, proteins, and carbohydrates, thereby promoting optimal digestion and subsequent absorption of nutrients. When considered necessary, supplementing with these enzymes can be a dependable adjunct in the treatment of a variety of digestive disorders. The majority of digestive enzymes are sold to consumers as food supplements in the form of pills or capsules. A helpful dietary supplement to maintain and improve digestive health may be added to functional foods in the form of carefully chosen digestive enzymes. Many enzyme supplementation formulations are currently available on the market and are being used in clinical practice to treat a wide range of digestive disorders, including diabetes, cystic fibrosis (CF), pancreatic cancer, exocrine pancreatic insufficiency in chronic pancreatitis, lactose intolerance, celiac disease, and postprandial distress. It is interesting to note that probiotics and lactase supplements have been shown to help patients with lactose malabsorption feel less intolerant. Consequently, customized products that include probiotics and enzyme supplements appear to be advantageous for the therapeutic management of these conditions.

Lastly, new research suggests that the gut microbiota may have a role in pancreatic exocrine insufficiency (PEI), as patients with PEI exhibit changes in this regard. Oral pancreatic enzyme supplementation has been shown to modify the gut microbiota in mice, with both *A. muciniphila* as well as *L. reuteri* raising the proportion of their abundance. According to the authors' theory, pancreatic enzymes might influence the gut microbiota and raise the number of strains that are important for maintaining the intestinal barrier. To find out how giving some digestive enzymes might help prime the gut for resilience through their effects on the gut microbiota, more research is needed (Jacquier et al., 2024).

REFERENCES

1. Ademowo, O. S., Dias, H. K. I., Pararasa, C., & Griffiths, H. R. (2019). Chapter 6—Nutritional hormesis in a modern environment. In S. I. S. Rattan & M. Kyriazis (Eds.), *The Science of Hormesis in Health and Longevity* (pp. 75–86). Academic Press.
2. Ahlman, H., & Nilsson, O. (2001). The gut as the largest endocrine organ in the body. *Annals of Oncology, 12*, S63–S68.
3. Asnicar, F., Berry, S. E., Valdes, A. M., Nguyen, L. H., Piccinno, G., Drew, D. A., Leeming, E., Gibson, R., Le Roy, C., Khatib, H. A., et al. (2021). Microbiome connections with host metabolism and habitual diet from 1,098 deeply phenotyped individuals. *Nature Medicine, 27*, 321–332.
4. Augustin, L. S. A., Aas, A. M., Astrup, A., Atkinson, F. S., Baer-Sinnott, S., Barclay, A. W., Brand-Miller, J. C., Brighenti, F., Bullo, M., Buyken, A. E., et al. (2020). Dietary fibre consensus from the International Carbohydrate Quality Consortium (ICQC). *Nutrients, 12*, 2553.
5. Bäckhed, F., Ding, H., Wang, T., Hooper, L. V., Koh, G. Y., Nagy, A., Semenkovich, C. F., & Gordon, J. I. (2004). The gut microbiota as an environmental factor that regulates fat storage. *Proceedings of the National Academy of Sciences USA, 101*, 15718–15723.
6. Barber, T. M., Kabisch, S., Pfeiffer, A. F. H., & Weickert, M. O. (2020). The health benefits of dietary fibre. *Nutrients, 12*, 3209.
7. Beam, A., Clinger, E., & Hao, L. (2021). Effect of diet and dietary components on the composition of the gut microbiota. *Nutrients, 13*, 2795.
8. Biagi, E., Rampelli, S., Turroni, S., Quercia, S., Candela, M., & Brigidi, P. (2017). The gut microbiota of centenarians: Signatures of longevity in the gut microbiota profile. *Mechanisms of Ageing and Development, 165*, 180–184.
9. Birch, C. S., & Bonwick, G. A. (2019). Ensuring the future of functional foods. *International Journal of Food Science & Technology, 54*, 1467–1485.
10. Blumberg, R., & Powrie, F. (2012). Microbiota, disease, and back to health: A metastable journey. *Science Translational Medicine, 4*, rv137.
11. Bolte, L. A., Vich Vila, A., Imhann, F., Collij, V., Gacesa, R., Peters, V., Wijmenga, C., Kurilshikov, A., Campmans-Kuijpers, M. J. E., Fu, J., et al. (2021). Long-term dietary patterns are associated with pro-inflammatory and anti-inflammatory features of the gut microbiome. *Gut, 70*, 1287–1298.
12. Brooks, L., Viardot, A., Tsakmaki, A., Stolarczyk, E., Howard, J. K., Cani, P. D., Everard, A., Sleeth, M. L., Psichas, A., & Anastasovskaj, J. (2017).

Fermentable carbohydrate stimulates FFAR2-dependent colonic PYY cell expansion to increase satiety. *Molecular Metabolism, 6*, 48–60.

13. Calabrese, V., Cornelius, C., Cuzzocrea, S., Iavicoli, I., Rizzarelli, E., & Calabrese, E. J. (2011). Hormesis, cellular stress response, and vitagenes as critical determinants in aging and longevity. *Molecular Aspects of Medicine, 32*, 279–304.
14. Calabrese, V., Cornelius, C., Dinkova-Kostova, A. T., Calabrese, E. J., & Mattson, M. P. (2010). Cellular stress responses, the hormesis paradigm, and vitagenes: Novel targets for therapeutic intervention in neurodegenerative disorders. *Antioxidants & Redox Signaling, 13*, 1763–1811.
15. Calabrese, V., Cornelius, C., Dinkova-Kostova, A. T., Iavicoli, I., Di Paola, R., Koverech, A., Cuzzocrea, S., Rizzarelli, E., & Calabrese, E. J. (2012). Cellular stress responses, hormetic phytochemicals and vitagenes in aging and longevity. *Biochimica et Biophysica Acta, 1822*, 753–783.
16. Canani, R. B., Costanzo, M. D., Leone, L., Pedata, M., Meli, R., & Calignano, A. (2011). Potential beneficial effects of butyrate in intestinal and extraintestinal diseases. *World Journal of Gastroenterology, 17*, 1519–1528.
17. Cani, P. D., & Jordan, B. F. (2018). Gut microbiota-mediated inflammation in obesity: A link with gastrointestinal cancer. *Nature Reviews Gastroenterology & Hepatology, 15*, 671–682.
18. Cani, P. D., Possemiers, S., Van de Wiele, T., Guiot, Y., Everard, A., Rottier, O., Geurts, L., Naslain, D., Neyrinck, A., Lambert, D. M., et al. (2009). Changes in gut microbiota control inflammation in obese mice through a mechanism involving GLP-2-driven improvement of gut permeability. *Gut, 58*, 1091–1103.
19. Cathomas, F., Murrough, J. W., Nestler, E. J., Han, M. H., & Russo, S. J. (2019). Neurobiology of resilience: Interface between mind and body. *Biological Psychiatry, 86*, 410–420.
20. Conklin, A. I., Monsivais, P., Khaw, K. T., Wareham, N. J., & Forouhi, N. G. (2016). Dietary diversity, diet cost, and incidence of type 2 diabetes in the United Kingdom: A prospective cohort study. *PLoS Medicine, 13*, e1002085.
21. Conlon, M. A., & Bird, A. R. (2014). The impact of diet and lifestyle on gut microbiota and human health. *Nutrients, 7*, 17–44.
22. Consortium, H. M. P. (2012). Structure, function, and diversity of the healthy human microbiome. *Nature, 486*, 207–214.
23. Cook, J. D., & Reddy, M. B. (2001). Effect of ascorbic acid intake on nonheme-iron absorption from a complete diet. *American Journal of Clinical Nutrition, 73*, 93–98.

24. Cummings, J. H., & Stephen, A. M. (2007). Carbohydrate terminology and classification. *European Journal of Clinical Nutrition, 61*(Suppl. S1), S5–S18.

25. Davenport, E. R., Sanders, J. G., Song, S. J., Amato, K. R., Clark, A. G., & Knight, R. (2017). The human microbiome in evolution. *BMC Biology, 15*, 127.

26. David, L. A., Materna, A. C., Friedman, J., Campos-Baptista, M. I., Blackburn, M. C., Perrotta, A., Erdman, S. E., & Alm, E. J. (2014). Host lifestyle affects human microbiota on daily timescales. *Genome Biology, 15*, R89.

27. David, L. A., Maurice, C. F., Carmody, R. N., Gootenberg, D. B., Button, J. E., Wolfe, B. E., Ling, A. V., Devlin, A. S., Varma, Y., Fischbach, M. A., et al. (2014). Diet rapidly and reproducibly alters the human gut microbiome. *Nature, 505*, 559–563.

28. De Filippis, F., Pellegrini, N., Laghi, L., Gobbetti, M., & Ercolini, D. (2016). Unusual sub-genus associations of faecal *Prevotella* and *Bacteroides* with specific dietary patterns. *Microbiome, 4*, 1–6.

29. De Jesus Raposo, M. F., De Morais, A. M. M. B., & De Morais, R. M. S. C. (2016). Emergent sources of prebiotics: Seaweeds and microalgae. *Marine Drugs, 14*, 27.

30. Dhingra, D., Michael, M., Rajput, H., & Patil, R. (2012). Dietary fibre in foods: A review. *Journal of Food Science and Technology, 49*, 255–266.

31. Dogra, S. K., Doré, J., & Damak, S. (2020). Gut microbiota resilience: Definition, link to health, and strategies for intervention. *Frontiers in Microbiology, 11*, 572921.

32. Drewnowski, A., & Barratt-Fornell, A. (2004). Do healthier diets cost more? *Nutrition Today, 39*, 161–168.

33. Elsenbruch, S., & Enck, P. (2017). The stress concept in gastroenterology: From Selye to today. *F1000Research, 6*, 2149.

34. Fassarella, M., Blaak, E. E., Penders, J., Nauta, A., Smidt, H., & Zoetendal, E. G. (2021). Gut microbiome stability and resilience: Elucidating the response to perturbations to modulate gut health. *Gut, 70*, 595–605.

35. Filippou, C. D., Tsioufis, C. P., Thomopoulos, C. G., Mihas, C. C., Dimitriadis, K. S., Sotiropoulou, L. I., Chrysochoou, C. A., Nihoyannopoulos, P. I., & Tousoulis, D. M. (2020). Dietary approaches to stop hypertension (DASH) diet and blood pressure reduction in adults with and without hypertension: A systematic review and meta-analysis of randomized controlled trials. *Advances in Nutrition, 11*, 1150–1160.

36. Gibson, G. R., & Roberfroid, M. B. (1995). Dietary modulation of the human colonic microbiota: Introducing the concept of prebiotics. *Journal of Nutrition, 125*, 1401–1412.
37. Gibson, G. R., Hutkins, R., Sanders, M. E., Prescott, S. L., Reimer, R. A., Salminen, S. J., Scott, K., Stanton, C., Swanson, K. S., Cani, P. D., et al. (2017). Expert consensus document: The International Scientific Association for Probiotics and Prebiotics (ISAPP) consensus statement on the definition and scope of prebiotics. *Nature Reviews Gastroenterology & Hepatology, 14*, 491–502.
38. Gill, S. K., Rossi, M., Bajka, B., & Whelan, K. (2021). Dietary fibre in gastrointestinal health and disease. *Nature Reviews Gastroenterology & Hepatology, 18*, 101–116.
39. Goff, H. D., Repin, N., Fabek, H., El Khoury, D., & Gidley, M. J. (2018). Dietary fibre for glycaemia control: Towards a mechanistic understanding. *Bioactive Carbohydrates and Dietary Fibre, 14*, 39–53.
40. Hasler, C. M. (2002). Functional foods: Benefits, concerns, and challenges—A position paper from the American Council on Science and Health. *Journal of Nutrition, 132*, 3772–3781.
41. Heiman, M. L., & Greenway, F. L. (2016). A healthy gastrointestinal microbiome is dependent on dietary diversity. *Molecular Metabolism, 5*, 317–320.
42. Hersoug, L. G., Møller, P., & Loft, S. (2016). Gut microbiota-derived lipopolysaccharide uptake and trafficking to adipose tissue: Implications for inflammation and obesity. *Obesity Reviews, 17*, 297–312.
43. Hou, K., Wu, Z. X., Chen, X. Y., Wang, J. Q., Zhang, D., Xiao, C., Zhu, D., Koya, J. B., Wei, L., Li, J., et al. (2022). Microbiota in health and diseases. *Signal Transduction and Targeted Therapy, 7*, 135.
44. Jacquier, E. F., van de Wouw, M., Nekrasov, E., Contractor, N., Kassis, A., & Marcu, D. (2024). Local and systemic effects of bioactive food ingredients: Is there a role for functional foods to prime the gut for resilience? *Foods, 13*(5), 739. https://doi.org/10.3390/foods13050739.
45. Jandhyala, S. M., Talukdar, R., Subramanyam, C., Vuyyuru, H., Sasikala, M., & Reddy, D. N. (2015). Role of the normal gut microbiota. *World Journal of Gastroenterology, 21*, 8787.
46. Johnson, A. J., Vangay, P., Al-Ghalith, G. A., Hillmann, B. M., Ward, T. L., Shields-Cutler, R. R., Kim, A. D., Shmagel, A. K., Syed, A. N., Personalized Microbiome Class, S., et al. (2019). Daily sampling reveals personalized diet-microbiome associations in humans. *Cell Host & Microbe, 25*, 789–802.

47. Karl, J. P., Hatch, A. M., Arcidiacono, S. M., Pearce, S. C., Pantoja-Feliciano, I. G., Doherty, L. A., & Soares, J. W. (2018). Effects of psychological, environmental, and physical stressors on the gut microbiota. *Frontiers in Microbiology, 9*, 2013.
48. Kim, H., Caulfield, L. E., Garcia-Larsen, V., Steffen, L. M., Coresh, J., & Rebholz, C. M. (2019). Plant-based diets are associated with a lower risk of incident cardiovascular disease, cardiovascular disease mortality, and all-cause mortality in a general population of middle-aged adults. *Journal of the American Heart Association, 8*, e012865.
49. Kim, S., & Jazwinski, S. M. (2018). The gut microbiota and healthy aging: A mini-review. *Gerontology, 64*, 513–520.
50. Küçükgöz, K., & Trząskowska, M. (2022). Nondairy probiotic products: Functional foods that require more attention. *Nutrients, 14*, 753.
51. Lamas, B., Richard, M. L., Leducq, V., Pham, H. P., Michel, M. L., Da Costa, G., Bridonneau, C., Jegou, S., Hoffmann, T. W., & Natividad, J. M. (2016). CARD9 impacts colitis by altering gut microbiota metabolism of tryptophan into aryl hydrocarbon receptor ligands. *Nature Medicine, 22*, 598–605.
52. Lee, C., Park, G. H., Lee, S. R., & Jang, J. H. (2013). Attenuation of β-amyloid-induced oxidative cell death by sulforaphane via activation of NF-E2-related factor 2. *Oxidative Medicine and Cellular Longevity, 2013*, 313510.
53. Leeming, E. R., Johnson, A. J., Spector, T. D., & Le Roy, C. I. (2019). Effect of diet on the gut microbiota: Rethinking intervention duration. *Nutrients, 11*, 2862.
54. López-Otín, C., & Kroemer, G. (2021). Hallmarks of health. *Cell, 184*, 33–63.
55. Martín-Peláez, S., Fitó, M., & Castañer, O. (2020). Mediterranean diet effects on type 2 diabetes prevention, disease progression, and related mechanisms: A review. *Nutrients, 12*, 2236.
56. Mattson, M. P. (2008). Dietary factors, hormesis, and health. *Ageing Research Reviews, 7*, 43–48.
57. McDonald, D., Hyde, E., Debelius, J. W., Morton, J. T., Gonzalez, A., Ackermann, G., Aksenov, A. A., Behsaz, B., Brennan, C., Chen, Y., et al. (2018). American Gut: An open platform for citizen science microbiome research. *mSystems, 3*, e00198–18.
58. McRorie, J. W., Jr., & McKeown, N. M. (2017). Understanding the physics of functional fibers in the gastrointestinal tract: An evidence-based approach to resolving enduring misconceptions about insoluble

and soluble fiber. *Journal of the Academy of Nutrition and Dietetics, 117*, 251–264.

59. Micha, R., Shulkin, M. L., Penalvo, J. L., Khatibzadeh, S., Singh, G. M., Rao, M., Fahimi, S., Powles, J., & Mozaffarian, D. (2017). Etiologic effects and optimal intakes of foods and nutrients for risk of cardiovascular diseases and diabetes: Systematic reviews and meta-analyses from the Nutrition and Chronic Diseases Expert Group (NutriCoDE). *PLoS ONE, 12*, e0175149.
60. Mötteli, S., & Hotzy, F. (2022). The assessment of cooking skills and food skills and their relationship with nutrition knowledge, attitude toward a healthy diet, and food intake: Results of a German validation study. *Nutrients, 14*, 3157.
61. Mozaffarian, D. (2016). Dietary and policy priorities for cardiovascular disease, diabetes, and obesity: A comprehensive review. *Circulation, 133*, 187–225.
62. Mozaffarian, D., & Wu, J. H. Y. (2018). Flavonoids, dairy foods, and cardiovascular and metabolic health: A review of emerging biologic pathways. *Circulation Research, 122*, 369–384.
63. Nicholson, J. K., Holmes, E., Kinross, J., Burcelin, R., Gibson, G., Jia, W., & Pettersson, S. (2012). Host-gut microbiota metabolic interactions. *Science, 336*, 1262–1267.
64. Nie, Q., Hu, J., Gao, H., Fan, L., Chen, H., & Nie, S. (2019). Polysaccharide from *Plantago asiatica* L. attenuates hyperglycemia, hyperlipidemia and affects colon microbiota in type 2 diabetic rats. *Food Hydrocolloids, 86*, 34–42.
65. Niv, E., Halak, A., Tiommny, E., Yanai, H., Strul, H., Naftali, T., & Vaisman, N. (2016). Randomized clinical study: Partially hydrolyzed guar gum (PHGG) versus placebo in the treatment of patients with irritable bowel syndrome. *Nutrition & Metabolism, 13*, 10.
66. O'Sullivan, L., Murphy, B., McLoughlin, P., Duggan, P., Lawlor, P. G., Hughes, H., & Gardiner, G. E. (2010). Prebiotics from marine macroalgae for human and animal health applications. *Marine Drugs, 8*, 2038–2064.
67. Odamaki, T., Kato, K., Sugahara, H., Hashikura, N., Takahashi, S., Xiao, J. Z., Abe, F., & Osawa, R. (2016). Age-related changes in gut microbiota composition from newborn to centenarian: A cross-sectional study. *BMC Microbiology, 16*, 90.
68. Ohashi, Y., Sumitani, K., Tokunaga, M., Ishihara, N., Okubo, T., & Fujisawa, T. (2015). Consumption of partially hydrolyzed guar gum stimulates Bifidobacteria and butyrate-producing bacteria in the human large intestine. *Beneficial Microbes, 6*, 451–455.

69. Omenetti, S., & Pizarro, T. T. (2015). The Treg/Th17 axis: A dynamic balance regulated by the gut microbiome. *Frontiers in Immunology, 6*, 639.
70. Oriach, C. S., Robertson, R. C., Stanton, C., Cryan, J. F., & Dinan, T. G. (2016). Food for thought: The role of nutrition in the microbiota-gut–brain axis. *Clinical Nutrition Experimental, 6*, 25–38.
71. Palmer, C., Bik, E. M., DiGiulio, D. B., Relman, D. A., & Brown, P. O. (2007). Development of the human infant intestinal microbiota. *PLoS Biology, 5*, e177.
72. Parada Venegas, D., De la Fuente, M. K., Landskron, G., González, M. J., Quera, R., Dijkstra, G., Harmsen, H. J. M., Faber, K. N., & Hermoso, M. A. (2019). Short chain fatty acids (SCFAs)-mediated gut epithelial and immune regulation and its relevance for inflammatory bowel diseases. *Frontiers in Immunology, 10*, 277.
73. Parnell, J. A., & Reimer, R. A. (2009). Weight loss during oligofructose supplementation is associated with decreased ghrelin and increased peptide YY in overweight and obese adults. *American Journal of Clinical Nutrition, 89*, 1751–1759.
74. Petersen, K. S., & Kris-Etherton, P. M. (2021). Diet quality assessment and the relationship between diet quality and cardiovascular disease risk. *Nutrients, 13*, 4305.
75. Rastelli, M., Cani, P. D., & Knauf, C. (2019). The gut microbiome influences host endocrine functions. *Endocrine Reviews, 40*, 1271–1284.
76. Rossnerova, A., Izzotti, A., Pulliero, A., Bast, A., Rattan, S. I. S., & Rossner, P. (2020). The molecular mechanisms of adaptive response related to environmental stress. *International Journal of Molecular Sciences, 21*, 7053.
77. Saarela, M., Lähteenmäki, L., Crittenden, R., Salminen, S., & Mattila-Sandholm, T. (2002). Gut bacteria and health foods: The European perspective. *International Journal of Food Microbiology, 78*, 99–117.
78. Schroeder, B. O., & Bäckhed, F. (2016). Signals from the gut microbiota to distant organs in physiology and disease. *Nature Medicine, 22*, 1079–1089.
79. Sedighi, M., Razavi, S., Navab-Moghadam, F., Khamseh, M. E., Alaei-Shahmiri, F., Mehrtash, A., & Amirmozafari, N. (2017). Comparison of gut microbiota in adult patients with type 2 diabetes and healthy individuals. *Microbial Pathogenesis, 111*, 362–369.
80. Shiby, V. K., & Mishra, H. N. (2013). Fermented milks and milk products as functional foods: A review. *Critical Reviews in Food Science and Nutrition, 53*, 482–496.

81. Shreiner, A. B., Kao, J. Y., & Young, V. B. (2015). The gut microbiome in health and in disease. *Current Opinion in Gastroenterology, 31*, 69.

82. Singh, R. K., Chang, H. W., Yan, D., Lee, K. M., Ucmak, D., Wong, K., Abrouk, M., Farahnik, B., Nakamura, M., & Zhu, T. H. (2017). Influence of diet on the gut microbiome and implications for human health. *Journal of Translational Medicine, 15*, 73.

83. Slavin, J. (2013). Fiber and prebiotics: Mechanisms and health benefits. *Nutrients, 5*, 1417–1435.

84. Sommer, F., Anderson, J. M., Bharti, R., Raes, J., & Rosenstiel, P. (2017). The resilience of the intestinal microbiota influences health and disease. *Nature Reviews Microbiology, 15*, 630–638.

85. Steinert, R. E., Feinle-Bisset, C., Asarian, L., Horowitz, M., Beglinger, C., & Geary, N. (2017). Ghrelin, CCK, GLP-1, and PYY(3–36): Secretory controls and physiological roles in eating and glycemia in health, obesity, and after RYGB. *Physiological Reviews, 97*, 411–463.

86. Sun, Y., & O'Riordan, M. X. (2013). Regulation of bacterial pathogenesis by intestinal short-chain fatty acids. *Advances in Applied Microbiology, 85*, 93–118.

87. Tan, J. K., Macia, L., & Mackay, C. R. (2023). Dietary fiber and SCFAs in the regulation of mucosal immunity. *Journal of Allergy and Clinical Immunology, 151*, 361–370.

88. Tap, J., Mondot, S., Levenez, F., Pelletier, E., Caron, C., Furet, J. P., Ugarte, E., Muñoz-Tamayo, R., Paslier, D. L., & Nalin, R. (2009). Towards the human intestinal microbiota phylogenetic core. *Environmental Microbiology, 11*, 2574–2584.

89. Temple, N. J. (2022). A rational definition for functional foods: A perspective. *Frontiers in Nutrition, 9*, 957516.

90. Timm, M., Offringa, L. C., Van Klinken, B. J., & Slavin, J. (2023). Beyond insoluble dietary fiber: Bioactive compounds in plant foods. *Nutrients, 15*, 4138.

91. Townsend, J. R., Kirby, T. O., Marshall, T. M., Church, D. D., Jajtner, A. R., & Esposito, R. (2023). Foundational nutrition: Implications for human health. *Nutrients, 15*, 2387.

92. Turnbaugh, P. J., Bäckhed, F., Fulton, L., & Gordon, J. I. (2008). Diet-induced obesity is linked to marked but reversible alterations in the mouse distal gut microbiome. *Cell Host & Microbe, 3*, 213–223.

93. Turnbaugh, P. J., Ley, R. E., Mahowald, M. A., Magrini, V., Mardis, E. R., & Gordon, J. I. (2006). An obesity-associated gut microbiome with increased capacity for energy harvest. *Nature, 444*, 1027–1031.

94. Usuda, H., Okamoto, T., & Wada, K. (2021). Leaky gut: Effect of dietary fiber and fats on microbiome and intestinal barrier. *International Journal of Molecular Sciences, 22*, 7613.

95. Van de Wouw, M., Boehme, M., Lyte, J. M., Wiley, N., Strain, C., O'Sullivan, O., Clarke, G., Stanton, C., Dinan, T. G., & Cryan, J. F. (2018). Short-chain fatty acids: Microbial metabolites that alleviate stress-induced brain-gut axis alterations. *Journal of Physiology, 596*, 4923–4944.

96. Wastyk, H. C., Fragiadakis, G. K., Perelman, D., Dahan, D., Merrill, B. D., Yu, F. B., Topf, M., Gonzalez, C. G., Van Treuren, W., Han, S., et al. (2021). Gut-microbiota-targeted diets modulate human immune status. *Cell, 184*, 4137–4153.

97. Willett, W. C., Koplan, J. P., Nugent, R., Dusenbury, C., Puska, P., & Gaziano, T. A. (2006). Prevention of chronic disease by means of diet and lifestyle changes. In *Disease Control Priorities in Developing Countries* (2nd ed.). Oxford University Press.

98. Wu, G. D., Chen, J., Hoffmann, C., Bittinger, K., Chen, Y. Y., Keilbaugh, S. A., Bewtra, M., Knights, D., Walters, W. A., Knight, R., et al. (2011). Linking long-term dietary patterns with gut microbial enterotypes. *Science, 334*, 105–108.

99. Yasukawa, Z., Inoue, R., Ozeki, M., Okubo, T., Takagi, T., Honda, A., & Naito, Y. (2019). Effect of repeated consumption of partially hydrolyzed guar gum on fecal characteristics and gut microbiota: A randomized, double-blind, placebo-controlled, and parallel-group clinical trial. *Nutrients, 11*, 2170.

100. Zhang, C., Zhang, M., Wang, S., Han, R., Cao, Y., Hua, W., Mao, Y., Zhang, X., Pang, X., Wei, C., et al. (2010). Interactions between gut microbiota, host genetics, and diet relevant to development of metabolic syndromes in mice. *ISME Journal, 4*, 232–241.

101. Zhao, S., Jang, C., Liu, J., Uehara, K., Gilbert, M., Izzo, L., Zeng, X., Trefely, S., Fernandez, S., & Carrer, A. (2020). Dietary fructose feeds hepatic lipogenesis via microbiota-derived acetate. *Nature, 579*, 586–591.

CHAPTER 5

FUNCTIONAL FOODS AND CORONARY HEART DISEASE (CHD)

Contents

INTRODUCTION

Cardiovascular disease (CVD) is a leading cause of adult mortality in the Western world, encompassing both coronary heart disease (CHD) and stroke. A concerning rise in CHD has been observed in the Ukraine, Russian Federation, and Kazakhstan, despite a trend over the past 25 years towards a decline in CHD rates in the major industrialized nations (such as the USA, Australia, and the UK) due to improved health screening, drug treatment, smoking cessation, and dietary advice.

The illness known as CHD occurs when the heart's main coronary arteries are unable to provide the myocardium, or heart muscle, with enough blood and oxygen. Atherosclerosis, a disease that mostly affects the artery's intima, is the primary cause of the decreased flow. This is a slowly developing but progressive illness that typically starts in childhood but does not show symptoms until later in life. During the course of CHD, four syndromes may emerge, depending on the rate and final severity of artery narrowing. These consist of myocardial infarction, abrupt death, and unstable angina pectoris. Literally meaning "strangling sensation in the chest," angina pectoris is frequently brought on by exertion or stress. Reduced blood flow in the coronary artery is the cause of this pain, which frequently travels from the chest to the left arm and neck. However, pain goes away very quickly when the patient rests. As the condition worsens, angina pectoris becomes unstable, causing pain to be felt both during physical activity and during rest. It is believed that this kind of angina involves the rupture or fissuring of a fixed lesion, which can cause an acute myocardial infarction if left untreated. The illness, which is largely to blame for CHD deaths, is brought on by a protracted blockage of the coronary artery, which causes part of the heart muscle to die. An irregular heartbeat, known as ventricular fibrillation, or a complete cessation of heart activity can result from disturbances in the electrical contraction of the heart muscle. Within an hour of the attack, a severe myocardial infarction usually results in sudden death and is typically linked to advanced atherosclerosis.

Many risk factors that have been linked to an increased risk of CHD have been divided into two categories: modifiable risk factors (like obesity, diet, cigarette smoking, and lack of exercise) and modifiable risk factors (like age, gender, race, and family history). A positive correlation has been found between an individual's fasting cholesterol level, particularly elevated levels of low-density lipoprotein cholesterol (LDL-C), and the development of CHD in epidemiological studies looking at the risks of the disease in various populations. Due to their inverse correlation with the development of CHD, low levels of high-density lipoprotein cholesterol (HDL-C) have been demonstrated to be an additional risk factor for the disease. Due to its role in reverse cholesterol transport, which involves moving cholesterol from peripheral cells to the liver for excretion, a low level of HDL-C is believed to indicate a malfunction in this process.

Several observational studies have shown a correlation between CHD and elevated levels of triacylglycerol (TAG), a major fat in the blood, during both the fasted and fed (postprandial) states. More recently, it has been noted that elevated postprandial TAG is more discriminatory than fasting TAG and can be considered an independent risk factor for CHD. The metabolic syndrome and the atherogenic lipoprotein phenotype (ALP) represent a group of classical and nonclassical risk factors that raise the chance of CHD in an otherwise healthy person. The identification of the prevalence of ALP in the population has been made possible by the development of methods to characterize the LDL subclass of people with and without CHD. According to heritability studies, environmental factors like diet, smoking, and a sedentary lifestyle account for the remaining 50% of the variability in expression, with genetic factors accounting for 50% of the variability.

Treatment for individuals with congestive heart failure (CHF) and those at high risk of developing the disease involves lowering total cholesterol, LDL-C, and TAG concentrations while increasing HDL-C. Adverse dietary habits have been linked to CHD by numerous lines of evidence. Therefore, diet modification is the first line of treatment for people with moderately elevated cholesterol (5.0–6.5 mmol/l) and/or TAG (1.5–2.2 mmol/l) levels. To achieve this, the proportion of dietary energy derived from fat is decreased to about 30%, and the percentage of dietary energy derived from saturated fatty acids (SFAs) is decreased to 10%. Polyunsaturated fatty acids (PUFAs) of the n -6 (linoleic) series and monounsaturated fatty acids (MUFAs) are candidate fats to replace a significant amount of SFAs. While PUFAs (n-6 series) and MUFAs have both been demonstrated to lower LDL-C and total cholesterol levels, when MUFAs are substituted for SFAs, it has also been demonstrated that MUFAs maintain or raise HDL-C concentrations. PUFAs of the n-3 series, eicosapentaenoic acid (EPA) and docosahexaenoic acid (DHA), have been demonstrated to lower plasma TAG levels in both fed and fasted states as well as to decrease thrombosis when added to the diet; however, they typically do not lower total and LDL-C levels.

Lipid-lowering medications are necessary in addition to dietary modifications for people whose total cholesterol is higher than 6.5 mmol/l and/or whose TAG is higher than 2.2 mmol/l to help manage their condition.

Available drugs are effective in a number of ways, including:

- enhanced LDL clearance (e.g., cholestyramine) and hydroxy-methyl-glutaryl CoA (HMG-CoA) reductase inhibition (e.g., simvastatin and atorvastatin);

- reduced synthesis of very low-density lipoprotein (VLDL) and LDL (e.g., nicotinic acid);
- enhanced VLDL clearance (e.g., benzafi brate).

Fenofibrate has been demonstrated to raise HDL-C levels, lower TAG and cholesterol levels, and change the LDL class profile away from the more atherogenic small dense LDL (LDL-III). Compared to diet modification, these drugs are more aggressive and can result in greater reductions in lipid levels; however, they are frequently linked to unpleasant side effects. This emphasizes the need for efficient dietary approaches that can lower blood lipid levels during fed and fasted states and that have long-term efficacy on par with the majority of successful medication therapies. A dietary approach that has been suggested to improve the lipid profile includes adding probiotics, prebiotics, and synbiotics to the diet. These are mechanisms that work to improve the host's health by fortifying and/or supplementing the host's diet with certain gut bacteria that promote health.

A "probiotic" is a live microbial feed supplement that improves the intestinal microbial balance of the host animal. Probiotics are typically fermented milk products that contain lactic acid bacteria like lactobacilli and/or bifidobacteria. Probiotics may help prevent gastrointestinal infections, lower total cholesterol and TAG levels, boost immunity, and stimulate the body's defenses against disease. Several mechanisms, including the inhibition of HMG-CoA reductase by a "milk factor" and the assimilation of cholesterol by specific bacteria, have been proposed to explain their putative lipid-lowering capacity.

A "prebiotic" is a non-viable food item that enters the colon undigested and is fermented specifically by lactobacilli and bifidobacteria in the colon. Both synthetically and naturally occurring non-digestible oligosaccharides derived from plants are referred to as prebiotics. Prebiotic dietary supplements significantly decreased circulating TAG and, to a lesser extent, cholesterol concentrations, according to animal studies. One theory explaining their lipid-lowering effects through inhibition of de novo lipogenesis-related enzymes is the production of short chain fatty acids (SCFAs) during the prebiotic's fermentation by the gut microbiota. While most human studies that have been done so far have produced positive results, there have been mixed results regarding alterations in lipid levels.

Currently, there is interest in "synbiotics," a combination of a prebiotic and a probiotic, as this combination may increase the probiotic bacteria's ability to survive in the colon. Nevertheless, more carefully monitored feeding studies are needed to investigate the effects and mechanisms of probiotic, prebiotic, and synbiotic action.

Manufacturers in the USA and Europe have been investigating the commercial potential for foods containing probiotics, prebiotics, and synbiotics—ingredients that promote health—for the last 10 years. Safety, consumer education, pricing, and appropriate health claims are among the issues deemed crucial to the market's continued growth. The most widely consumed food products are those that contain probiotics added to dairy products. However, because prebiotics can be added to a wide variety of long-lasting foods, like ice cream and bread, a more varied food market has emerged as a result of the growing interest in prebiotics. Probiotics, prebiotics, and synbiotics are becoming more and more popular as dietary supplements, but it is important to make sure that any claims made for these products are supported by well-executed human trials that make use of the most recent techniques.

5.1. CORONARY HEART DISEASE AND RISK FACTORS

An important fraction of adult global morbidity and mortality is attributed to diseases of the circulatory system. In the UK, cardiovascular disease (CVD) caused 35% of all deaths in 2006. Of these deaths, 48% were from coronary heart disease (CHD) and 28% were stroke-related. The global mortality rate from CVD varies. For instance, the American Heart Association recently reviewed statistics on heart disease and stroke and found that, among men aged 35 to 74, the global CVD death rates (per 100,000 people) ranged from 149.8 in France (2006) to 1555.2 in the Russian Federation (2002) The rates are falling in most countries, the UK saw a later decline in CHD mortality than Austria and Denmark. However, in contrast to many other regions of Europe, rates of CHD are sharply rising in the Ukraine. This downturn can be partially attributed to enhanced surgical techniques, increased utilization of cholesterol-lowering drugs and other medications, decreased smoking rates, and increased health screening programs. Furthermore, dietary modifications, particularly regarding the kind of fat ingested, might have positively impacted the incidence of the disease (Lloyd-Jones, 2010).

5.1.1. Pathology of Coronary Heart Disease Development

Atherosclerosis and thrombosis work together to form the fundamental cause of clinical CVD. The disorder known as atherosclerosis causes the lining of the arteries to thicken in certain areas due to elevated plaques and an excessive buildup of modified lipids. It also causes the migration and proliferation of smooth muscle cells from the arterial wall's deeper layers. Atherosclerotic plaque typically appears where there has been minor arterial wall damage. This site of damage draws tissue macrophages, a subset of white blood cells that take up and accumulate LDL particles from the circulation. Oxidized LDL particles are more likely to be ingested by macrophages and result in the buildup of cholesterol, according to studies. The cholesterol-loaded macrophages change into foam cells that are also loaded with cholesterol and stay within the arterial wall. The phrase "hardening of the arteries" refers to the later stage of the plaque's calcification and sclerose.

One or more arteries, primarily the cerebral and coronary arteries, may become blocked by the formation of an atherosclerotic plaque, leading to a stroke or CHF, respectively. Additionally, the artery may become even more occluded by a superimposed thrombus or clot. The discharge of eicosanoids from platelets and the walls of the vessels contributes to the formation of blood clots. The eicosanoid thromboxane, which is produced when PUFA exits platelet membranes, promotes platelet aggregation and vasoconstriction. In addition, the vessel walls release pro-inflammatory fatty acids (PUFAs) that are subsequently transformed into prostacyclins, which are antagonistic eicosanoids that inhibit platelet aggregation and vasodilation. The extent of blood clot that forms will depend on the ratio of prostacyclin to thromboxane production and the relative potencies of these two eicosanoids (Patsch et al., 1992).

These processes are linked to two major clinical conditions: myocardial infarction, also known as coronary thrombosis, and angina pectoris. Angina pectoris is typified by chest pain and discomfort that is triggered by stress and physical exertion. It rarely lasts longer than 15 minutes and is caused by a decrease in or temporary blockage in the blood flow to the heart via the coronary artery. Prolonged total occlusion of the artery causes a coronary thrombosis, which is accompanied by prolonged and typically excruciating central chest pain and infarction, or death of part of the heart muscle.

Arterial thrombus is caused by a multitude of cells and lipids, such as platelets, monocytes, endothelial cells, fibroblasts, cholesterol, lipoproteins, and smooth muscle cells. By altering one or more of these variables, nutrition may have an impact on the development of CHD.

5.1.2. Risk Factors for the Development of Coronary Heart Disease

There is no one cause for CHD, which is a complex illness. A lot of the time, and often very loosely, the term "risk factor" refers to behavioral and lifestyle factors, physical activity levels, and biochemical characteristics that indicate a person's propensity to contract an illness. The identification of new potential risk factors is a continuous process as the etiology of CHD is studied. The established risk factors for the development of CHF can be divided into four groups: those linked to disease states, those that are changeable, and those that are dependent on geographic distribution.

The development of CHD is more influenced by some risk factors than by others. Total plasma cholesterol and the risk of CHD have been shown to be strongly and consistently correlated. About 70% of the blood's cholesterol is transported by the LDL fraction, which is primarily responsible for the positive correlation. A five-fold variation in CHD mortality was noted in the US population across a range of plasma cholesterol levels in a sizable prospective study that was published in 1986. After conducting a comprehensive meta-analysis on individual data from 61 prospective studies involving 55,000 vascular deaths, it was determined that total cholesterol had a positive correlation with ischemic heart disease across all blood pressure levels and in both middle and old age. A 31% reduction in the five-year incidence of myocardial infarction and CHD death was linked to a 20% and a 26% decrease in total and LDL-C, respectively, in a cholesterol-lowering medication trial conducted in a healthy population. Reductions in total fat and SFA intake are recommended by most dietary guidelines because of this relationship between plasma cholesterol levels and CHD. The Western and developing worlds are currently experiencing an obesity epidemic, and numerous initiatives have been proposed to address this grave global issue. Numerous studies have shown a link between obesity and chronic illnesses. A meta-analysis of 57 prospective studies with 900,000 adult participants found that having a body mass index (BMI) higher than 25 kg/m^2 is highly predictive of developing vascular disease and is most likely mostly causative.

It is not a given that someone with CHD will experience any one of the risk factors alone or in combination. It is difficult to forecast when a myocardial infarction will occur in a particular person. A middle-aged man who is obese, has diabetes, smokes 40 cigarettes a day, follows a high-fat diet, and works in a stressful environment may never develop CHD; in contrast, a physically fit, slender, non-smoker who follows a low-fat diet may experience an early myocardial infarction. Despite this anomaly, the severity and quantity of known risk factors are used to determine whether a person or a population is more likely to develop CHD (Shepherd et al., 1995).

Relative risk scores serve a predictive purpose because they contain unmodifiable factors. However, to effectively intervene, one must consider factors other than those that are used to predict risks, such as blood pressure and cigarette smoking, which are considered, as well as factors not included in the scoring system, such as stress, physical activity, diet, and body weight. Many organizations have released dietary guidelines for the general public with the goal of lowering the prevalence and severity of CHD in the general population.

5.2. RELEVANT LIPID PARTICLES

5.2.1. Plasma Lipoprotein Metabolism

Plasma lipoproteins are large molecules that contain one or more apoproteins, which are particular proteins, and complexes of lipids like TAG, cholesterol, and phospholipids. They play a role in the movement of water-insoluble nutrients from the site of synthesis or absorption to peripheral tissue throughout the bloodstream. Lipoproteins mainly depend on apoproteins connected to their surface coat for proper lipoprotein targeting to metabolic sites.

The two main sites of lipoprotein synthesis are the liver and the intestine, and the two transported lipid components, cholesterol and TAG, have different fates. TAGs are mainly transported to muscle, where the fatty acids are oxidized for energy, and adipose tissue, where they are stored. HDL, on the other hand, constantly transports cholesterol between the extrahepatic tissues, the intestine, and the liver. Depending on their flotation density, human lipoproteins are categorized into five main classes. The ratio of high-density surface protein to low-density, non-polar lipid is reflected in the inverse relationship between the particle sizes and densities. TAG predominates in the core structures of the two biggest lipoproteins. These are VLDL, which is secreted by the hepatocytes (liver cells), and chylomicrons (CMs), which are released by the enterocytes (small intestine cells). The core of intermediate-density lipoprotein (IDL) has significant concentrations of cholesterol esters and TAG. The mature forms of the two smallest classes of particles—LDL and HDL—are created by metabolic processes in the bloodstream rather than being secreted directly from the liver. These particles have cholesterol esters in their core structures. While components of HDL are secreted along with CMs and VLDL, LDL are the byproduct of the metabolism of VLDL. The transport of dietary and hepatically derived lipids is accomplished by the exogenous and endogenous cycles of the lipid metabolic pathways.

5.2.1.1. Exogenous Pathway

The enterocyte absorbs long-chain fatty acids after the small intestine has broken down dietary fat. The developing CM particle is composed of free cholesterol (1–2%), phospholipids (7–9%), and apoproteins (1–2%) on the surface, with a core of TAG (84–89% of the mass) and cholesterol ester (3–5%). After secretion, lipoprotein lipase (LPL), which is attached to the luminal surface of endothelial cells in muscle and adipose tissue, hydrolyzes the TAG component of the CM particle. A lipoprotein particle known as a CM remnant is created when 70–90% of the TAG is eliminated. This particle is rich in cholesterol ester. To keep the CM remnant particle stable as its core shrinks, surface materials, phospholipids, cholesterol, and apoproteins are moved to HDL. Hepatic lipase (HL), which is located on the surface of the liver, is likely necessary for the uptake of these particles. Before the CM remnant is absorbed by the liver through a receptor-mediated process, HL further hydrolyzes the TAG and phospholipid components of the remnant. After being endocytosed and catabolized in lysosomes, the residues allow cholesterol to enter hepatocyte metabolic pathways and be excreted into the bile.

5.2.1.2. Endogenous Pathway

TAG is transported from the liver to the peripheral circulation via a channel made possible by VLDL. The liver secretes two subtypes of VLDL: VLDL 1 (a large lipoprotein rich in TAG) and VLDL 2 (smaller, denser particles). When compared to CMs, LPL in the capillary bed extracts TAG from the secreted VLDL 1 less effectively. A portion of the VLDL 2 particles is endocytosed by the liver's LDL receptors, which are responsible for recognizing the particles. The extended presence of certain VLDL particles in the plasma leads to additional metabolism by LPL and partially by HL, resulting in the formation of higher-density IDL. Hepatocyte surface HL hydrolyzes IDL further, leading to the eventual formation of LDL. LDL is a diverse population made up of smaller, denser LDL particles (LDL-III) and larger LDL species (LDL-I and LDL-II subclass). Hepatocytes and extrahepatic cells both have LDL receptors on their surface that can bind and absorb LDL.

5.2.1.3. Reverse cholesterol transport

One of HDL's roles is to start the flow of cholesterol into the liver for excretion from peripheral cells and membranes that are changing. An enzyme called lecithin cholesterol acyltransferase (LCAT), which is attached to certain HDL particle species, mediates the process of reverse cholesterol transport. It functions by ensnaring cholesterol in the center of the developing HDL after the particle interacts with a protein on the cell surface. HDL receives the cholesterol, which is then transformed into HDL. To get the cholesterol to the liver, this particle can travel either a direct or indirect route. In the direct pathway, the liver selectively absorbs the cholesterol ester from the HDL particle or the LDL receptor mediates endocytosis of the HDL. As part of the indirect pathway, the cholesterol ester transfer protein (CETP) gives cholesterol ester to low-density lipoproteins (VLDL and CMs) in exchange for TAG. The liver then absorbs these lipoproteins (neutral lipid exchange).

The amount of cholesterol in the plasma is determined by the balance between the forward (exogenous and endogenous pathways) and the reverse pathway, which is closely controlled by the lipoprotein-based rates of secretion.

5.2.2. Classical Lipid Risk Factors in Coronary Heart Disease

Numerous epidemiological studies have demonstrated a positive association between LDL-C levels and CHD mortality, as well as fasting total cholesterol levels. When LDL builds up in the plasma, the LDL particles undergo oxidative modification, which causes cholesterol to deposit in the arterial wall. The early stages of atherosclerotic plaque are caused by macrophages that take up the oxidized low-density lipoprotein (LDL), which eventually develops into foam cells. A prospective meta-analysis of data from 90,056 participants in 14 randomized control trials of statins unequivocally supports the role of cholesterol-lowering as a public health strategy in the primary prevention of CVD. Following a year of therapy, there was a 19% decrease in coronary mortality, with LDL-C concentrations dropping by about 1 mmol/l. It has been suggested that a 23% reduction in major coronary events can be achieved by maintaining a 1 mmol/l reduction in LDL-C for 5 years.

The relationship between total cholesterol levels and the risk of CVD may be influenced by HDL-C levels. There is a documented strong negative correlation between the risk of developing CHD and fasting plasma HDL-C. Women typically have higher HDL-C levels than men do. Modest alcohol consumption is one factor that can raise HDL-C levels, whereas smoking, inactivity, and diabetes mellitus are factors that can lower them. The Münster Heart Study examined CVD risk factors, stroke, and mortality in individuals who were employed between 1979 and 1991. When fasting lipid parameters were examined at the start of the study and again six to seven years later, it was found that the group with major coronary events had significantly lower HDL-C concentrations than the group that had no such events. Low HDL-C levels have been linked to a five-fold increase in the risk of CHD when compared to normal HDL-C values. Low HDL-C levels are thought to reflect a compromised pathway for the excretion of cholesterol (reverse cholesterol transport).

As of right now, the total to HDL-C ratio is thought to be the most accurate indicator of CVD risk. Results from 68 long-term prospective studies, mostly carried out in Europe and North America, that followed up on 430 participants (who did not have vascular disease at first) have also hinted that the evaluation of lipids to establish the risk of CVD may be reduced to measuring total or HDL-C.

5.2.3. Non-Classical Risk Factors

5.2.3.1. Elevated Fasting and Postprandial TAG Levels

The majority of the epidemiological data regarding the risk of CHD has been derived from fasting lipoprotein concentrations, which are measured after an overnight 12-hour fast. However, when meals high in fat challenge the lipid transport system, people spend a significant amount of the day in the postprandial state. There is a correlation between the risk of developing CHD and the amount and duration of postprandial TAG concentrations after a fat load. This association has been demonstrated in CHD patients, who exhibit a longer and more marked TAG response after eating in comparison to matched individuals without CHD, although the fasting TAG concentrations were comparable in both groups. A multitude of observational studies have shown how strongly TAG levels are associated with CHD. Elevated non-fasting TAG concentrations have been demonstrated to be a more discriminatory predictor of CVD risk than fasting TAG in two prospective cohort studies conducted in 2007 with 26 509 women in the US and 13 981 adults in Denmark. In the Copenhagen Heart Study, non-fasting TAG levels of ≥ 5, 4–4.99, 3–3.99, 2–2.99, 1–1.99 mmol/l versus < 1 mmol/l predicting a 16.8-, 5.1-, 3.9-, 4.4-, and 2.2-fold, increase in risk of myocardial infarction in females, with equivalent values in males of 4.6-, 3.3-, 3.6-, 2.3-, and 1.6-fold respectively.

5.2.3.2. Elevated Levels of Chylomicrons and Chylomicron Remnants

The ability to distinguish between endogenous and exogenous TAG-rich lipoproteins, CMs and VLDL, in postprandial samples has made it possible to examine the atherogenicity of these lipoproteins. Evidence was presented by Karpe and colleagues indicating that the advancement of atherosclerosis is linked to a delayed uptake of small CM remnants. After a meal high in fat, there is a higher concentration of postprandial TAG-rich lipoproteins (CM and CM remnants) in the blood, which may be directly or indirectly linked to the existence of CHD. Research on animals has demonstrated that CM remnants can enter the artery wall directly. One of two processes is thought to be responsible for the mechanisms by which CM remnants supply the structural elements of arterial lesions. First, the rate of atherogenesis should be proportionate to the concentrations of plasma remnants because the CM remnants may bind and penetrate the arterial surfaces like that of plasma LDL. Second, CMs might be taken up by the artery and subsequently broken down into remnants. In each case, the response that causes smooth muscle cells to endocytose leftovers may occur at locations where endothelium has been removed due to local injury.

The rate of clearance by peripheral tissue and the liver's receptor-mediated absorption of the leftovers determines the extent of postprandial lipaemia. Any of these processes that are not working properly will lead to a buildup of CMs and their byproducts, which can affect naturally occurring lipoproteins that are known to be atherogenic. The atherogenic lipid profile may be elevated, specifically by the transfer of TAG from CMs to LDL and HDL and the reciprocal transfer of cholesterol esters to the CMs via CETP.

Neutral lipid exchange is the term used for this. Due to the transfer of cholesterol ester to CMs and VLDL, these TAG-rich lipoproteins are unable to be metabolized normally by lipase. The macrophages in the artery lesion can then absorb the cholesterol ester-enriched CM remnants. However, LDL and HDL turn into good substrates for LPL and HL, which results in a decrease in the size of these particles. As a result, an atherogenic lipoprotein profile develops, characterized by decreased levels of HDL-C, decreased LDL size, and increased cholesterol ester enrichment in TAG-rich lipoproteins. Because LDL is dense and small, it cannot be recognized by liver LDL receptors. This makes it an excellent candidate for absorption by scavenger receptors on macrophages in the artery lesion.

5.2.3.3. Elevated Levels of Small Dense LDL (LDL-III)

Elevated levels of circulating LDL-C have long been recognized as a significant risk factor for the development of CHD. Density gradient ultracentrifugation methods have been utilized to divide low-density lipoprotein (LDL) into three primary subclasses. They are separated into three categories: small dense LDL (LDL-III), intermediate size LDL (LDL-II), and light large LDL (LDL-I). LDL-III is present in small amounts in healthy normolipidemic males, with a preponderance of LDL-II.

Numerous case control, cross-sectional, and prospective studies have looked at the connection between the risk of CHD and the LDL subclass, with varying degrees of success. It was shown in studies involving men with CHD or those who had survived a myocardial infarction that men with CHD had higher levels of small dense LDL-III than men without it. 1987 saw Austin et al. 35 suggested that a three-fold increase in CHD was linked to a preponderance of LDL-III in young men. Since small dense LDL-III is absorbed slowly by the LDL receptor in the liver and peripheral tissues, it is believed that this contributes to the particles' increased residence in the circulation and increased atherogenicity. This gives the small, dense LDL particles enough time to enter the artery wall's intima, where it is believed that extracellular matrix components will hold the particles in place until the LDL particles oxidize. Following the modified small dense LDL-III's absorption by the macrophage scavenger receptor, foam cells are produced. The association between small dense LDL and CVD risk, according to Sachs and Campos, might not be independent of other lipid risk indicators like TAG and HDL-C. Small dense LDL was found to be elevated in the metabolic syndrome in the Framingham Heart Offspring Study. The authors concluded that small dense LDL was a feature of the atherogenic dyslipidemia typical of the ALP and the metabolic syndrome rather than a factor directly associated with a higher risk of CVD.

5.2.3.4. Atherogenic Lipoprotein Phenotype

The ALP is a group of abnormalities related to lipids that increases the risk of CHD in otherwise healthy, normal people. It has been suggested that 30-35% of middle-aged men in the West might be impacted. When this phenotype is present during fasting, the following characteristics stand out: a slightly elevated TAG concentration (1.5-2.3 mmol/l), low HDL-C (less than 1 mmol/l), and a preponderance of small dense LDL-III (more than 35% of LDL mass) and HDL particles. It is significant to remember that total cholesterol levels are typically either slightly elevated or within the normal range. The impairment of TAG-rich lipoproteins (CMs and VLDL) removal, which results in the conversion of small, atherogenic cholesterol-enriched remnants, LDL and HDL by neutral lipid exchange, maybe the cause of the atherogenic nature of ALP. This group of lipid abnormalities is generally thought to be closely linked to the metabolic syndrome, also known as the insulin resistance syndrome.

5.2.3.5. Metabolic Syndrome

The term "metabolic syndrome" refers to a group of abnormalities related to metabolism, including high blood pressure, low HDL-C, central obesity, insulin resistance, and elevated TAG concentrations. An estimated 20–25% of adults worldwide are estimated to have metabolic syndrome, which carries a two–three-fold increased risk of CVD and a five-fold increased risk of type II diabetes. Many expert panels have published criteria for identifying people who have the metabolic syndrome, and most of them agree that a person must exhibit three or more of the five common features tobe diagnosed with the syndrome. Recent research, however, has called into question whether a particular metabolic syndrome feature or combinations of features, can interact to increase the relative risk of CVD beyond the metabolic syndrome classification alone. It is noteworthy that in the Guize et al. study, the addition of either elevated TAG or blood pressure to both increased abdominal obesity and hyperglycemia was one of the three- and four-component combinations highly associated with all-cause mortality.

5.3. DIET AND CORONARY HEART DISEASE RISK: THE EVIDENCE

A sizable, varied, and largely reliable body of research connects diet to CVD. The connection between dietary fat and CHD has the most supporting data. Diet is thought to affect blood pressure, blood lipids, and other risk factors, as well as potentially through thrombogenic mechanisms, to affect the risk of CHD. More recent research points to the potential protective function of dietary antioxidants like carotenes, vitamins E and C, and olive oil, possibly via preventing LDL particle oxidation.

5.3.1. Dietary Fat Intake

Clinical and epidemiological data unequivocally demonstrate that the level of total cholesterol, and particularly LDL-C, in the blood is positively correlated with the risk of dying from CHD. A substantial risk factor for CHD appears to be an increased postprandial TAG response to meals high in fat. This suggestion is supported by mounting data. Several studies have demonstrated that diet-related fat types and quantities have a major impact on plasma cholesterol levels. The Seven Countries Study found a strong correlation between the mean concentrations of total cholesterol in each population group and the percentage of energy derived from SFA. This relationship was even more pronounced when considering a formula that also considered PUFA intake.

Several studies have confirmed the link between saturates in the diet and elevated plasma lipid concentrations. The hypercholesterolaemic response to a specific SFA varies, based in part on LDL receptor activity and in part on the structure of the TAG. Although the exact mechanism causing the rise in plasma cholesterol brought on by SFAs is unknown at this time, one possibility is a decrease in LDL receptor activity. It is advised that SFA's average contribution to dietary energy be lowered to a maximum of 10%. To achieve this recommendation, there would need to be a significant decrease from the average 13% dietary SFA intake that is currently consumed.

Though dietary SFA should be reduced, no guidance is provided regarding the type of fatty acid that should take its place. The impact on plasma cholesterol levels of substituting different fatty acids for SFA ingestion. LDL-C levels are considerably decreased by substituting MUFA or n-6 PUFA for SFA, though n-6 PUFA are more efficient in this regard. The effectiveness of MUFA in lowering cholesterol has been questioned, as has the possibility that the observed drop in plasma cholesterol was only the result of SFA replacement. Nonetheless, it has long been known that populations in the Mediterranean region have a markedly lower risk of CHD. Traditionally, their diet has been higher in long-chain n-3 PUFA and MUFA-containing olive oil, along with higher amounts of fruits and vegetables, than the UK diet. This connection is supported by data from intervention studies like the Lyon Heart Study. There are currently no specific recommendations for dietary MUFA intake, and it is unclear which aspects of this diet are most important for reducing the risk of CHD.

Regarding CHD, trans fatty acids and SFA seem to have similar effects on blood cholesterol, with the exception that trans fatty acids reduced HDL-C while SFA increased it. Elaidic acid, the main trans fatty acid in diets, was shown to dramatically lower HDL-C and raise LDL-C levels in a study comparing the effects of various fatty acids. These conclusions are also supported by some epidemiological evidence. Vegetable margarines and processed foods contain large amounts of hydrogenated fats, which are a significant dietary source of trans fatty acids. However, the amount of trans fatty acids in many margarines and processed foods has significantly decreased recently due to the documented association between trans fatty acids and CHD risk. The amount of trans fatty acids consumed in the UK diet today does not adversely affect the risk of CVD, despite the undeniable link between high ingestion of trans fatty acids and the disease. However, an increase in trans fatty acid consumption should be avoided.

Although there is a significant inter-individual response, the overwhelming body of research points to the idea that increasing the amount of cholesterol in the diet raises plasma cholesterol, particularly LDL-C. Over the past 35 years, studies conducted on humans have shown that an effect can be seen at a threshold of approximately 95 mg/4,300 kJ, with a ceiling at about 450 mg/4,300 kJ. Either no cholesterol is absorbed or endogenous production is suppressed by excess cholesterol. It is advised that current dietary cholesterol intakes, measured per unit of dietary energy, not increase because daily intake is at the lower end of this range.

The benefits of increased intake of long-chain n -3 PUFA lie in their ability to lower plasma TAG levels and reduce thrombosis, in contrast to dietary SFA, MUFA, n -6 PUFA, and trans fatty acids, whose effects on cardiac health primarily reside in their ability to modify plasma LDL-C and HDL-C levels. The consumption of marine foods high in EPA (20:5, n -3) and DHA (22:6, n -3), which are found in marine foods, has been linked to the low incidence of CHD in Greenland Eskimos and Japanese fishermen, despite their high fat intake (fat accounting for 80% of dietary energy). Additional proof that moderate intakes of 1-2 portions of oily fish per week afford a significant reduction in CHD mortality comes from the long-term intervention trials DART and GISSI.

Additionally, n-3 PUFA series fatty acids are beneficial for thrombogenesis, or the formation of blood clots. When producing eicosanoids, which are substances involved in blood clotting, n -3 PUFA is used instead of n -6 PUFA, which can significantly lower the rate of blood clot formation and lower the risk of myocardial infarction. Furthermore, eating n-3 PUFAs may lessen the incidence of cardiac arrhythmias. In Western nations, sudden cardiac death is a major issue, and no medication has significantly reduced the incidence of this condition to date. Consuming n-3 polyunsaturated fats has been linked to a lower risk of sudden death, according to epidemiological research. There is also some evidence of an anti-arrhythmic effect in animals and humans. To lower the risk of CVD, the Scientific Advisory Committee on Nutrition (SACN) recommends consuming 0.45 g of long-chain n-3 PUFA daily.

5.3.2. Carbohydrate Intake

Dietary carbohydrate intake is significantly impacted by recommendations for reducing total fat in the diet. The amount of energy from dietary protein is relatively constant, so the amounts of dietary fat and carbohydrates that contribute to energy are inversely correlated. The degree of substitution has a major impact on the metabolic consequences of substituting fat for carbohydrates. Diets where 60% of energy from food comes from carbohydrates are linked to higher TAG and lower HDL-C levels, and they may also be linked to an increased risk of CHD even though the LDL-C levels are lower. Increases in carbohydrates, especially complex ones, to make up for dietary fat reductions of 35% of energy have not been linked to harmful alterations in TAG and HDL-C and may even be beneficial to the health of people with CHD. There is growing interest in the quality of dietary carbohydrates as well as their absolute quantity, with a preference for a diet high in low-GI foods (carbohydrate-containing foods that have the ability to raise blood glucose levels). Low GI diets are recommended for lowering the risk of CHD because there is evidence linking them to a lower risk of the disease.

5.3.3. Dietary Fiber Intake

Dietary fiber consumption and CHD have been found to be inversely correlated in several prospective studies. The effective source of fiber has been found to vary across studies. While some researchers discovered that vegetable sources of fiber were linked to a lower risk, Morris et al. reported that cereal fiber was inversely related to CHD cases. In contrast, there was no correlation between higher cereal fiber intake and a lower risk of death from CVD in a two-year intervention study involving men who had previously experienced a heart attack. A meta-analysis of cohort studies found that overall higher dietary fiber intakes were significantly associated with reductions in CHD, despite inconsistent results. Higher fiber intakes have been linked primarily to soluble fiber rather than insoluble fiber. The impact of several soluble fibers, categorized as prebiotics, that are specifically broken down by the colonic gut microbiota on blood cholesterol and, in particular, LDL-C levels. Based on available evidence, the UK suggests consuming 18–24 g of dietary fiber per day. Apart from the health benefits of dietary fiber, there is mounting evidence from epidemiological and prospective population studies that the consumption of wholegrain foods, which are rich in antioxidants, vitamins, and minerals, as well as dietary fiber, is linked to a lower risk of CVD. Because of this, it is now advised in the UK to eat three or more servings of wholegrain each day. Nevertheless, despite this compelling epidemiological evidence, there is a lack of supportive data from detailed mechanistic studies and interventions, and further research is needed to confirm the actual relationship between wholegrain consumption and CVD.

5.3.4. Sodium and Potassium Intake

Consumption of sodium seems to play a significant role in determining blood pressure in the general population, at least when it comes to the age-related increase in blood pressure. It would be expected that a diet higher in potassium and lower in common salt would lower blood pressure and slow its rise with aging. The main source of sodium in the diet is salt, and 65–85% of the salt consumed comes from manufactured foods. One significant risk factor for the onset of CHD and strokes is blood pressure. The UK population has been advised to cut back on salt intake to 6 g per day, with recommendations for babies and kids being introduced in 2003.

5.3.4.1. Alcohol Intake

The benefits of drinking alcohol, and more especially red wine consumption, and the risk of CHD have been the subject of a long-running debate. In many research populations, moderate alcohol use seems to be linked to a comparatively low risk of CHD. Drinking less than 20 grams per day has been linked to a lower risk of CHD while consuming more than 89 grams per day has been linked to an increased risk of stroke. The benefit linked to alcohol consumption is thought to be almost exclusively caused by elevated HDL-C levels. However, other factors have also been proposed, including reduced platelet activity, which lowers the risk of thrombosis, and the antioxidant qualities of some beverages, like red wine. Although there is ongoing debate, it is thought that consuming two units of alcohol per day—one unit is equivalent to 8 g of alcohol—may reduce the risk of CHD.

5.4. THE EFFECTS OF PROBIOTICS INCLUDING FERMENTED MILK PRODUCTS AND LACTIC ACID BACTERIA ON CORONARY HEART DISEASE

There has been an effort to find other dietary components that can lower blood cholesterol levels because consumers do not follow dietary recommendations very well. Among these have been studies on the potential hypocholesterolemic and hypotriacylglycerolemic qualities of milk products, which are typically fermented. Due to the fact that studies on the topic have been conducted with varying degrees of quality, design, and statistical analysis—the majority of which have been severely limited by incomplete documentation—the role of fermented milk products as hypocholesterolemic agents in humans remains unclear. However, there has been a lot of interest in the impact of probiotics on human lipid metabolism since 1974, when Mann and Spoerry 81 demonstrated an 18% decrease in plasma cholesterol after feeding 24 Maasi warriors four to five liters of fermented milk per day for three weeks.

5.4.1. Evidence for the 'Milk Factor'

In a follow-up study to the Maasi trial, Mann 82 fed four liters of yogurt daily to a small group of US volunteers (n = 4) with an unspecified microbiological activity over a 12-day period. The results showed a significant reduction in serum cholesterol values of 37% (though the tabulated data only showed a 16.8% fall). The hypercholesterolemic effect persisted even when the daily yogurt intake was lowered to two liters; however, a daily intake of one liter returned cholesterol levels to baseline. Two hours following a pulse of [14 C] acetate, the specific activity of plasma digitonin-precipitated sterols was measured to track the rate of cholesterol biosynthesis.

16 days after the 12-day trial, which involved consuming four liters of yogurt daily, a 28% decrease in acetate incorporation into serum cholesterol was reported. Mann suggested the possibility of a "milk factor," such as an HMG-CoA reductase inhibitor, to explain the drop in serum cholesterol. Investigating possible candidates for the 'milk factor,' Howard and Marks 83 fed lactose ± Ca/Mg, cheese whey or yogurt, to volunteers over a two-week period. The yogurt, but not the lactose ± Ca/Mg or cheese whey, significantly reduced plasma cholesterol by 5.5%. However, this trial was subject to a lack of dietary control with substantial changes made to the volunteers' habitual diet resulting in many confounding factors.

Due to the subjects' lack of diet control, confounding variables were introduced in the majority of the early studies. Hepner and team carried out a study to account for these. In this crossover study, participants received 750 milliliters of milk and 720 milliliters of yogurt daily for four weeks. Following the first week of both supplementation periods, there were notable decreases in plasma cholesterol. It is commonly known that enrollment in a study can cause a significant drop in cholesterol levels. This is most likely the result of the volunteer changing their diet either consciously or unconsciously as a result of their acute awareness of dietary assessment. Baseline run-in periods are crucial to lessen this. The early published negative studies include those by Massay,

Thompson et al. and McNamara et al. included a time for run-in. McNamara et al.'s study was one of the more meticulously planned interventions. In a study with a three-week run-in and a four-week intervention period, they examined the effects of 480 ml of an unidentified yogurt in comparison to a non-fermented milk concentrate (as a control). Body weight and dietary intake did not alter, and serum total, LDL-C, HDL-C, or TAG levels did not fluctuate. In a different well-controlled, double-blind crossover study, 500 ml of kefir—a fermented milk product traditionally consumed in Soviet countries due to possible health benefits—was used. When compared to unfermented milk, no lipid-lowering effects were noted. Based on the aforementioned studies, it can be inferred that there is insufficient proof that fermented milk products directly impact serum lipid parameters.

5.4.2. The Effect of Viable Lactic Acid Bacteria on Lipid Parameters

The first study to try to determine if live bacteria were necessary for the yoghurt's claimed effects on lipid parameters was Hepner et al. the study compared the effects of 750 milliliters of pasteurized and unpasteurized yoghurt while controlling for milk as a placebo. Following a 12-week intervention, all treatments considerably lowered plasma cholesterol levels; however, the reduction from milk was less pronounced. Regrettably, the nutritional and microbiological contents of the products were not disclosed, which makes it very difficult to compare the results with those of other studies. Thompson et al. assessed a wide range of milk based products including milk supplemented with *Lactobacillus acidophilus* (undefined, strain code not given—1.3×10^7 colony forming units (CFU)/ml), buttermilk (a milk product fermented with Streptococcus cremoris and Streptococcus lactis—6.4×10^8 CFU/ml) and a yoghurt (fermented with *Lactobacillus bulgaricus* and *Streptococcus thermophilus*—1.2×10^9 CFU/ml). 1 liter of supplement was fed for a three-week period, but no significant change was reported in serum total cholesterol, LDL-C or HDL-C. The notion that diversity in yoghurt cultures may be significant motivated Jaspers et al. to evaluate the impact of 681 g/day of three strains of yoghurt fermented in a 1:1 ratio with *Lactobacillus bulgaricus* and *Streptococcus thermophilus*. Two culture strains, CH-I and CH-II (species not given), were taken separately over a period of 14 days, and two batches of a third strain, SH-IIIA and SH-IIIB (species not given), were taken separately over a period of 14 days and seven days, sequentially, separated by a 21-day "washout period" between each of the intervention periods. The subjects' body weight did not change, and the only dietary variations were in the amounts of vitamins and minerals. In the first week with the CH-II strain and the second week with the SH-IIIA strain, there were notable declines in serum total and LDL-C levels. These brief variations may reflect the impact of starting a study, as was previously mentioned, or they may reflect a real distinction in the effective characteristics of various strains. It is clear that the type of probiotic or lactic acid bacteria in the product affects the physiological effect. In normocholesterolaemic and hypercholesterolaemic subjects, respectively, controlled studies using the probiotics *Bifidobacterium lactis* (undefined, strain code not given, 300 g/day) for six weeks and *Bifidobacterium longum* BL1 (300 ml/day, 10^8 CFU/ml) for four weeks produced significant hypocholesterolaemic effects, which likely reflected the effectiveness of the specific probiotic bacteria present.

The physiological impact of lactic acid bacteria is significantly influenced by the study group. In a large study (n = 78), De Roos and colleagues examined the effects of 500 ml/day of yoghurt supplemented with *Lactobacillus acidophilus* L-1 in normocholesterolaemic subjects over the course of six weeks in a parallel study design. It was stated that none of the lipid parameters found showed any discernible change. This stands in contrast to research conducted by Anderson and colleagues (94), who observed a significant reduction of 4.3% in total cholesterol in 40 hypercholesterolaemic subjects over the course of four weeks after consuming a smaller volume (200 ml) of fermented milk (yoghurt-type) containing human *Lactobacillus acidophilus* L-1 or swine *Lactobacillus acidophilus* (ATCC 43121). These findings highlight how crucial it was to know study participants' cholesterol levels at enrollment.

Numerous studies have examined the potential lipid-lowering properties of a specific strain of *Enterococcus faecium* (human species, undefined; strain code not given) and two strains of *Streptococcus thermophilus* found in Gaio® (MD Foods Aarhus, Denmark) probiotic yoghurt fermented by Causido®. The results have been inconsistent. Agerbaek and team conducted a parallel study design to examine the impact of 200 ml of Gaio® per day. The active yoghurt was compared to an identical yoghurt that had undergone chemical acidification using an organic acid (delta-gluco-lactone). For a duration of six weeks, 58 middle-aged men with moderately elevated cholesterol (5.0–6.5 mmol/l) received the intervention.

Throughout the course of the intervention, they saw a 9.8% reduction in LDL-C levels for the live yoghurt group. This study was well-controlled, excluding a wide range of factors including body weight, sex, and age. However, there was an unanticipated skew in the subjects' randomization, which led to a significant difference in the baseline total and LDL-C levels. One possible explanation for the decline in these parameters that was seen in the active yoghurt group is a regression back towards the mean. Here, 87 men and women between the ages of 50 and 70 participated in a longer study that used the same yogurt and a similar design over a period of six months. The group consuming the active yoghurt was found to have significantly lower LDL-C levels after

12 weeks. The fact that there was a decrease in the number of viable bacteria (*Enterococcus faecium*) in the yogurt at 12 weeks helped to explain why these reductions did not continue. Between the two groups, there were no discernible drops in total or LDL-C levels at the conclusion of the investigation. Another study looking into the same product included 32 patients with mild to moderate hypercholesterolemia and used a 200 g/day yoghurt ingestion for eight weeks in a randomized double-blind placebo-controlled trial. The test or control product was administered to the patients for two eight-week periods after they were instructed to adhere to a lipid-lowering diet for eight weeks. The active product significantly reduced LDL-C levels by 6.2% and total cholesterol by 10.3%, according to the results. The authors did, however, raise concerns about the clinical significance of the average 5% reduction for both total and LDL-C.

160 middle-aged men and women with slightly elevated cholesterol participated in a comparable trial using the same Gaio® product that contains *Enterococcus faecium*. The study was the biggest to be conducted to date and was a multi-center, randomized, double-blind, placebo-controlled parallel study. During a 12-week period, volunteers drank 200 ml of either the active or chemically acidified yoghurt daily. To guarantee that the groups were similar in terms of age, sex, BMI, and baseline fasting cholesterol levels, stratified randomization was employed. Dietary assessment was used to verify that participants followed the protocol, underscoring the significance of maintaining their current lifestyle and eating habits during the study. Over the course of the investigation, the viability of *Enterococcus faecium* was assessed because it is crucial that the yogurt contain sufficient amounts of viable bacteria. At all times tested, the levels were found to be no less than 1×10^6 CFU/ml. Blood cholesterol levels in both groups significantly decreased during the two-week run-in period, but after that, neither group's blood cholesterol levels changed further at any point in time. Although these data are consistent with the conclusions made by Rossouw et al., they do not support the previous studies that were discussed. This suggested that blood lipid reductions seen in participants starting an intervention trial may be responsible for the apparent effects of some probiotics on blood cholesterol levels. These decreases are well known, but they are also hard to stop, which emphasizes how crucial it is for these kinds of studies to include a run-in period. In conclusion, six published studies have looked into the possibility of lowering plasma cholesterol by incorporating Gaio® into a regular diet. Four studies found that a 6–10% decrease in LDL–C had a marginally positive short-term impact. However, the largest study did not find any significant effects, and the long-term effects were inconclusive. Nevertheless, despite these conflicting reports, a meta-analysis of these studies revealed that Gaio® had a short-term effect, as evidenced by the reductions in total and LDL-C of 6% and 9%, respectively, when compared to the control yoghurt.

Most of the research on the impact of probiotics and lactic acid bacteria has used fermented cow's milk or cow's milk yogurt. On the other hand, the effects of fermented soy milk 100 and soy milk combined with probiotic yoghurt on lipid concentrations have been found in two studies, both of which produced inconsistent findings. Additionally, one study used fermented oat milk. There is growing evidence that soy consumption is associated with a decrease in plasma lipids, despite disagreements regarding the specific active ingredients in soy.

While isoflavones by themselves typically do not lower lipids, they seem to have positive effects when combined with soy proteins. As a result, research was done to determine how adding a probiotic to soy products affected their ability to lower cholesterol. In middle-aged men with normal cholesterol levels, Rossi and colleagues 100 examined the effects of adding 200 milliliters of fermented soy milk made with *Enterococcus faecium* (undefined, strain not given) and *Lactobacillus jugurti* (undefined, strain not given) over the course of six weeks. Although it contained isoflavones, soy protein, and the same species as Gaio (*Enterococcus faecium*), this fermented soy milk increased HDL-C by 10% but had no discernible effect on total or LDL-C levels. The findings of a study that examined the effects of soy milk and either 100 ml probiotic yogurt (containing 10^8 CFU/serving of each of *Lactobacillus acidophilus* (undefined, strain not given), *Bifidobacterium bifidus* (undefined, strain not given), and *Lactobacillus rhamnosus* GG) or yoghurt containing the standard yogurt starter cultures are in contrast to these data. A substantial decrease in total cholesterol (−4.7%) was observed when soy and probiotic yogurt were consumed together, and possible synergistic hypocholesterolemic effects between soy and probiotic bacteria were proposed. In another study, 18 men with atherosclerotic plaques were given 100 ml of fermented oat milk containing *Lactobacillus plantarum* (DSM 9843, 10^{11} CFU/ml) in a double-blind parallel trial. There was no discernible change in lipid concentrations when compared to a control product, despite reports of a significant increase in colonic bacterial diversity and plasma SCFAs.

Based on the research to date, it can be said that there is some evidence to support a hypocholesterolemic effect of consuming certain products containing lactic acid bacteria on a daily basis. This effect is more noticeable in people who have hypercholesterolemia and may vary depending on gender. The most consistent result is a decrease in total cholesterol, which in some but not all studies is linked to a significant decrease in LDL-C. Nevertheless, no study that has measured plasma TAG has found a significant decrease in this lipid level; in fact, one study found a significant 16% increase in plasma TAG. It follows that lactic acid bacteria's possible hypocholesterolemic effects are what have an impact on CHD lipid risk reduction. Nevertheless, the findings are usually contradictory, necessitating more adequately powered and carefully monitored research before definitive conclusions regarding the effectiveness of lactic acid bacteria use and CHD risk reduction can be made.

5.4.3. Possible Mechanisms of Action of Lactic Acid Bacteria

Before delving into potential mechanisms, it is imperative to emphasize that probiotics must possess not only the qualities essential for eliciting the intended biological effects but also the necessary viability and ability to resist the host's natural defenses against ingested bacteria, as viable and biologically active microorganisms are typically required at the target site in the host. Although they are technologically efficient, the traditional yogurt starter bacteria, *Streptococcus thermophilus* and *Lactobacillus bulgaricus*, do not enter the lower intestine in a viable form. Therefore, when considering probiotic organisms, intrinsic microbiological properties like tolerance to bile, gastric acid, and pancreatic juice are crucial.

Probiotics and lactic acid bacteria may reduce cholesterol through a variety of mechanisms, some of which have been proposed and which may work alone or in concert. These include the assimilation of cholesterol, the deconjugation of bile acids, and the binding of cholesterol to bacterial cell walls, among other physiological actions of the fermentation-derived SCFAs. Acetic, propionic, and butyric acids are the saturated fatty acids (SCFAs) that are generated when bacteria break down carbohydrates anaerobically.

The peculiar effect of microbial bile acid metabolism in the therapeutic role of certain bacteria has been well-documented. The conjugated bile acid hydrolase enzyme, which is only produced by bacteria, catalyzes the deconjugation reaction. Numerous intestinal bacteria, belonging to the genera *Enterococcus*, *Peptostreptococcus*, *Bifidobacterium*, *Fusobacterium*, *Clostridium*, *Bacteroides*, and *Lactobacillus*, are known to possess this deconjugation ability. Through an increase in the excretion of the deconjugated bile acids in the feces, this reaction releases the amino acid moiety and the deconjugated bile acid, which lowers the absorption of cholesterol again. Numerous *in vitro* investigations have looked into the capacity of different bacteria to deconjugate a range of bile acids. Grill and team found that, when tested against six distinct bile salts, *Bifidobacterium longum* strains (BB536 and OC) were the most effective of the four *Bifidobacterium* strains (*Bifidobacterium animalis*, *Bifidobacterium infantis*, *Bifidobacterium brev*, and *Bifidobacterium thermophilum*). According to a different study, different Lactobacillus species have differing capacities for deconjugating taurocholate and glycocholate 110, while *Lactobacillus fermentum* KC5b is said to be able to deconjugate several bile salts *in vitro*. Research on *in vitro* reactions is helpful, but to fully understand how bile acid deconjugation contributes to cholesterol reduction, *in vivo* investigations on humans and animals are necessary. Oral administration of specific bacterial strains has been demonstrated in intervention studies involving animals and ileostomy patients to increase the excretion of free and secondary bile salts.

Additionally, there is some *in vitro* data in favor of the theory that some bacteria are capable of assimilation—or the uptake—of cholesterol. *In vitro* studies revealed that in the presence of bile and under anaerobic conditions, *Lactobacillus acidophilus* (RP32), *Bifidobacterium bifidum* (strains BYU and BPO), and *Lactobacillus fermentum* (KC5b) could all assimilate cholesterol. Notwithstanding these reports, it remains unclear if the cholesterol is being taken up by the bacteria or if it is precipitating alongside the bile salts. Research has been done to try and answer this. Klaver and Meer concluded that bacterial bile salt deconjugase activity, rather than assimilation, was responsible for the removal of cholesterol from the growth medium where *Lactobacillus acidophilus* (RP32) and a *Bifidobacterium* (MUH80) were growing. Tahri et al. investigated the same issue and came to the contradictory conclusions that cholesterol assimilation and bile acid deconjugase activity could happen concurrently and that some of the eliminated cholesterol was present in the cell extracts.

The hypocholesterolemic effects of probiotics have also been linked to the mechanism of cholesterol binding to bacterial cell walls. According to Hosona and Tono-Oka, *Lactococcus lactis* subsp. Biovar. Of the bacteria examined for this study, Diacylactis R-43 exhibited the highest cholesterol-binding capacity. Following an investigation into the *in vitro* cholesterol-lowering capacities of six strains of *Lactobacillus acidophilus*, Lin and Chen reported that the bacterial *in vivo* hypocholesterolemic effects were caused by the cells' direct assimilation of cholesterol and/or cholesterol's attachment to their surface. It was hypothesized that the structural and chemical characteristics of the cell walls caused the binding variations and that dead cells might still be able to bind cholesterol in the colon. Because different bacterial species can have varying effects on lowering cholesterol, the mechanism of action of probiotics on cholesterol reduction may involve one or more of the above mechanisms. To completely understand the impact and mechanism of probiotics and their potential hypocholesterolemic action, more research is necessary.

5.5. THE EFFECTS OF PREBIOTICS ON CORONARY HEART DISEASE

Prebiotics have become more and more popular as functional food ingredients due to their significant nutritional role during the last 10 years. This interest stems from research conducted on animals, which demonstrated significantly lower TAG and total cholesterol levels when diets containing large amounts of the prebiotic oligofructose (OFS) were fed. "A non-digestible food ingredient that selectively stimulates the growth and/or the activity of one or more bacteria in the colon that has the potential to improve health" is the definition of a prebiotic. Prebiotics, also known as non-digestible oligosaccharides, are made from disaccharides (such as transgalacto-oligosaccharides) or extracted from natural sources (such as inulin and OFS). Prebiotics that are most frequently studied include inulin and OFS, which are present in a wide variety of vegetables, such as chicory root, onion, garlic, leek, asparagus, and Jerusalem artichokes. These are made up of 2 to 60 fructose molecules connected by β-2, 1-osidic linkages. Because of the nature of these linkages, they are not broken down in the upper gastrointestinal tract (GIT) but instead undergo selective fermentation by the colonic microbiota. Currently, inulin is used as a food ingredient in yoghurt, ice cream, bread, and baked goods because it has thickening and gelling qualities that enhance the mouthfeel and appearance of lower-energy foods. The recommended daily intake of OFS and inulin in Europe is between 2 and 12 g.

5.5.1. The Effect of Prebiotics on Lipid Metabolism in Humans

The results of several studies looking into how prebiotics affect fasting plasma lipids have been mixed. One study with participants who had elevated blood lipids showed a reduction in total fasting, LDL-C, and TAG; three studies demonstrated significant reductions in total fasting and/or LDL-C, but no significant changes in TAG levels; and two studies only revealed a reduction in TAG. However, no changes in TAG or cholesterol were seen in two studies involving type II diabetics. Only two studies have shown that inulin significantly alters fasting total cholesterol and TAG levels in normolipidemic participants; two more studies have only shown a significant reduction in TAG levels. Seven studies, however, have found no difference in plasma lipid levels between young, healthy, and elderly subjects when OFS or inulin treatment is administered. According to Jackson et al. four weeks after the inulin supplementation period ended, follow-up blood samples were obtained. At that point, the concentrations of TAG had returned to baseline levels, corroborating the theory that inulin feeding may have been the cause. According to a recent meta-analysis of 15 randomized controlled trials, inulin-type fructans can considerably lower fasting TAG concentrations by 0.17 mmol/l. These trials were conducted between 1996 and 2005. These results are consistent with the effects of inulin on lipid levels that have been seen in animals, where TAG is primarily affected rather than cholesterol concentrations.

It has also been established that elevated postprandial TAG concentrations pose a risk for CHD. Rat studies have demonstrated a 40% decrease in postprandial TAG concentrations when fed diets containing 10% (w/v) OFS. Two studies report that inulin treatment had no effect on postprandial TAG levels in middle-aged men and women with moderately elevated lipids, but there is a dearth of information regarding the impact of prebiotics on postprandial lipaemia in human subjects.

There has not been a consistent replication in human subjects of the significant decreases in fasting lipid levels seen in animal studies. Merely three research studies involving normolipidemic participants have demonstrated noteworthy decreases in fasting total absorption graphs (TAG) in response to inulin, with one study demonstrating a noteworthy long-term impact of inulin treatment on TAG levels in comparison to the placebo group. The fructans used in the human studies range from 5 to 34 grams, which is insignificant in comparison to the fructans used in animal studies (50–200 grams per kilogram of OFS rat chow), which corresponds to a daily dose of about 50–80 grams of OFS/inulin in humans. Because OFS and inulin are fibers, their daily dosage in humans is limited to 15–20 g, as higher doses may result in gastrointestinal symptoms like flatulence, diarrhea, and cramping in the stomach. It is unclear if animals would experience meaningful effects at the dosages used in human studies.

Different food vehicles are used to increase the intake of OFS and/or inulin in the diet. The consumption of 40 g of margarine by Pedersen et al. and 100 g of biscuits daily by Luo et al. may have contributed to the unfavorable blood lipid results. In the instance of Davidson et al., for six weeks, significant alterations in total and LDL-C levels were noted when inulin was added to various study products, such as a chocolate bar, a chocolate spread, and fruit or coffee sweeteners, as opposed to the placebo (sugar). The study determined the percentage change in all lipid parameters for each of the six-week treatment periods. Surprisingly, the total cholesterol, LDL-C, and TAG levels increased during the placebo phase. When the net changes in the variables were computed (change during inulin minus change during control treatment), there were significant differences in total and LDL-C between the two treatments. This is because non-significant falls in these variables were observed during inulin treatment. The increased consumption of SFAs in the chocolate products—two of the study's vehicles—was ascribed by the authors to the rise in total and LDL-C levels during the placebo phase. Subsequent research has shown that the use of inulin in powder form allows it to be added to many foods consumed in the participant's regular diet without dietary advice, thereby preventing changes in body weight. Due to its ability to bind water, inulin powder can be added to soup, yogurt, orange juice, tea, and coffee.

Jackson et al. showed a significant correlation between the participants' starting TAG concentration and the percentage change in TAG levels over the course of the eight-week study. It provides credence to the theory that the initial TAG levels may play a significant role in defining the extent of the TAG reaction to inulin. Some people may not respond to inulin because they are less responsive to it, because their background diet is different, or because they are not following the study protocol. Understanding the mechanism of action of inulin on plasma TAG levels would help with conjecturing potential causes of response variability.

One possible explanation for the contradictory results regarding TAG level changes in human subjects supplementing with inulin could be the duration of the supplementation period employed in the studies. Significant effects started to show up as early as 2-4 weeks into the studies, which lasted between 2 and 24 weeks. The results of a 26-week study that examined the effects of inulin and OFS on lipid levels in normolipidemic individuals raise the possibility that the positive outcomes seen in short-term trials might not last over an extended period of time. It has been demonstrated that a wash-out period of four weeks is adequate to allow the TAG concentrations to return to baseline levels. This could explain the noteworthy results in TAG levels found in the studies by Causey et al. and Canzi et al., which employed very brief wash-out periods and sequential and crossover designs lasting three to four weeks.

Even though some of the studies conducted so far have shown that inulin has positive effects on plasma TAG, the results are not always consistent, and further research is needed to show that consuming prebiotics lowers cholesterol.

5.5.2. Mechanism of Lipid Lowering by Prebiotics

It has been demonstrated that prebiotics are the perfect substrate for the colon's health-promoting bacteria, particularly lactobacilli and bifidobacteria. Numerous byproducts are created during the fermentation process, such as lactate, SCFAs (acetate, butyrate, and propionate), gases (H_2S, CO_2, H_2, and CH_4), and lactate. The liver uses the SCFAs, acetate and propionate, once they reach the portal bloodstream. Propionate has been shown to inhibit lipid synthesis, while acetate functions as a lipogenic substrate for de novo lipogenesis and is transformed into acetyl CoA in the liver. On the other hand, butyrate has been demonstrated to guard against the development of intestinal tumors and is absorbed by colonocytes, the large intestinal cells. The microbiota, which the prebiotic can stimulate, determines the type of SCFAs that are produced during the fermentation process. It has been demonstrated that inulin raises butyrate and acetate levels simultaneously, while artificially manufactured prebiotics, such as galacto-oligosaccharides, increase propionate and acetate production while xylo-oligosaccharides only raise acetate.

A lot of research has been done on inulin and OFS to figure out how prebiotics work in animals. Early *in vitro* research employing isolated rat hepatocytes revealed that the hypolipidemic action of OFS was linked to the SCFA propionate-induced inhibition of de novo cholesterol synthesis, which followed the impairment of the liver's acetate utilization for de novo lipogenesis. This is consistent with human studies where propionate inhibited the incorporation of acetate into TAGs released from the liver following rectal infusions of acetate and propionate. Fiordaliso and team showed a significant decrease in plasma TAGs, phospholipids, and cholesterol in rats fed a rat chow diet containing 10% (w/v) OFS who were normolipidemic. After just one week of OFS, the TAG-lowering effect was shown, and it was linked to a decrease in VLDL secretion. Before being available for assembly into VLDL, TAG and phospholipids are synthesized in the liver through the esterification of fatty acids and glycerol-3-phosphate, indicating that the liver may be the site of the hypolipidaemic effect of OFS. The rats' reduced cholesterol levels were only evident following a 16-week duration of OFS feeding. It is interesting to note that eating 10 g of inulin for three weeks decreased fasting TAG levels and hepatic lipogenesis in healthy men and women, but had no discernible impact on plasma cholesterol levels. The reduction in activity of all lipogenic enzymes (acetyl-CoA carboxylase, fatty acid synthase, malic enzyme, ATP citrate lyase, and glucose-6-phosphate dehydrogenase) and, in the case of fatty acid synthase, through modification of lipogenic gene expression, is thought to be the mechanism by which OFS lowers VLDL TAG secretion from the liver.

5.5.3. The Effect of Prebiotics on Glucose and Insulin Levels

The effects of prebiotics on human fasting insulin and glucose levels are largely unknown. Of the human supplementation studies, type II diabetic subjects showed a significant reduction in glucose, and hyperlipidaemic subjects with OFS showed a trend toward a reduction in glucose. Two studies, however, have not found that OFS has an impact on type II diabetics' blood glucose levels. Inulin and OFS-treated obese slightly dyslipidemic women and healthy middle-aged subjects showed significant reductions in insulin levels; however, there was no effect on plasma glucose levels in either study. A lower glycaemic response and a peak in insulin levels were observed after consuming acute test meals enriched with OFS, according to an analysis of the effects on blood glucose, insulin, and C-peptide levels in healthy adults.

Prebiotics have been shown to work by lowering insulin and glucose levels through a mechanism linked to SCFAs, particularly propionate. A substantial decrease in postprandial glucose concentrations was noted after consuming propionate-enriched bread for both short-term and long-term periods. There was no research done on how propionate consumption affected postprandial insulin levels. After four weeks of feeding with OFS, a recent study on animals revealed a reduction in both postprandial insulin and glucose levels. The actions of OFS on the secretion of the gut hormones, glucagon-like peptide 1 (GLP-1) and glucose-dependent insulinotropic polypeptide (GIP), were responsible for these effects. These hormones help secrete insulin after a meal in the presence of elevated glucose levels. They are secreted from the small intestine (GIP), the terminal ileum, and the proximal colon (GLP-1). According to a recent review, GLP-1 in particular is a key site where inulin-type fructans modulate it. When mice and rats are fed OFS, the serum concentration of this incretin hormone doubles, improving glucose homeostasis. Similar results were seen in a human study wherein feeding 20 g/day of OFS for a week caused an increase in plasma GLP-1 levels, which were then increased upon consumption of a mixed meal.

In conclusion, research on animals has played a major role in determining the mechanisms of action of prebiotics, particularly inulin and OFS. Based on current research, prebiotics appear to primarily act by downregulating the involved enzymes, which inhibit de novo lipogenesis and reduce cholesterol. If this is the case, one could anticipate more subdued or erratic effects in humans, as de novo lipogenesis in them is incredibly low, or variable depending on their diet history. Rats used in animal experiments are fed a diet high in carbohydrates and low in fat; as a result, the synthesis of fatty acids in these animals is facilitated by an upregulated pathway called de novo lipogenesis. Interestingly, TAG levels are believed to be lowered by a different mechanism involving the enhanced clearance of TAG-rich lipoproteins when rats are fed OFS in addition to a high-fat diet typical of the Western diet. Kok et al. found that rats receiving OFS secreted more GIP and it has been demonstrated that this gut hormone increases LPL activity.

TAG-containing lipoproteins after a fat meal. The systemic impact of prebiotics like inulin and OFS on blood lipid, insulin, and glucose levels may be mediated by the release of GIP and GLP-1 in the colon and intestine. To identify the metabolic pathways that prebiotics affect, more research is necessary. It has recently been suggested that their impact on gastrointestinal kinetics, specifically gastric emptying, and its alteration of TAG-rich lipoproteins (CMs) and glucose levels in the bloodstream, may act as a modulator of systemic effects. Because these factors may affect the study's outcome, future research on the effects of prebiotics in humans should take into account the type of vehicle used to increase prebiotic intake in the diet, the duration of the supplementation period, and the subjects' selection for the study.

5.6. THE EFFECTS OF SYNBIOTICS INCLUDING COMBINATIONS OF LACTIC ACID BACTERIA AND PREBIOTIC FIBERS ON CORONARY HEART DISEASE

5.6.1. Synbiotics

By enhancing the survival and implantation of live microbial dietary supplements in the GIT, as well as by selectively stimulating the growth and/or activating the metabolism of one or a limited number of health-promoting bacteria, a synbiotic is characterized as a combination of prebiotic and probiotic that has positive effects on the host. Few human studies have examined the impact of synbiotics, or other combinations of lactic acid bacteria and fibers, on CHD risk factors. This is a developing field in the functional food ingredient space. Schaafsma and colleagues conducted a study that found that after three weeks of supplementing with a fermented milk product (375 ml) containing 2.5% FOS (~9 g), there was a significant decrease in both total and LDL-C when compared to control milk (375 ml) fermented by conventional yogurt starters. The LDL-C to HDL-C ratio improved while HDL-C, TAG, and glucose levels remained constant. Consuming a fermented milk product with 1% OFS (~3 g) for six weeks did not affect serum total, LDL-C, or TAG concentrations, according to research by Kießling and colleagues.

The consumption of 300 g of yoghurt per day resulted in a relatively high intake of milk fat (10.5 g/day), which is thought to be the reason for the significant improvement in HDL-C and the LDL-C to HDL-C ratio. Most of the synbiotic research done to date has focused on how the gut microbiota is impacted by the supplements. In a research involving healthy participants, a *Bifidobacterium* sp. For 12 days, a daily dose of a fermented milk product (undefined, strain undefined) with or without 18 g of inulin was administered. The amount of bifidobacteria in the gut was significantly increased by the fermentation milk product (probiotic) administration, but this increase was not furthered by the addition of 18 g of inulin. Two weeks after the supplementation period ended, the gut microbiota composition was evaluated, and it was discovered that subjects who received both the fermented milk product and inulin preserved their bifidobacterial population in the gut in comparison to those who only received the fermented milk product. While the synbiotic did not appear to have a synergistic effect on gut bifidobacteria, these findings imply that either the probiotic was more effectively implanted or that native bifidobacteria had a prebiotic effect. Feeding healthy individuals, a synbiotic food containing *Bifidobacterium longum* (strain not defined), *L. acidophilus* (strain not defined), and OFS (0.5 g) for 30 days had a significant impact on the metabolic activity of the intestinal microbiota, according to a recent study using 1 H Nuclear Magnetic Resonance Spectroscopy. Specifically, there was a positive correlation and a negative correlation between the number of bifidobacteria in the feces after supplementation and an increase in particular metabolites, such as the SCFA butyrate. It is still unknown how maintaining a high concentration of bifidobacteria in the gut will affect blood lipid levels, but it may be advantageous for gut health. When given over seven weeks to healthy human subjects, a lower dose of prebiotic (2.75 g) added to lactobacillus fermented milk has been shown to significantly increase the number of bifidobacteria. Lower doses of prebiotics that can be used in synbiotic preparations should help to reduce gastrointestinal complaints observed when a prebiotic is used alone and improve the acceptability of these types of products by the general public if this effect was caused by the synbiotic product used in the study (Lei et al., 2023).

REFERENCES

1. Alberti, K. G., Zimmet, P., & Shaw, J. (2006). Metabolic syndrome – A new worldwide definition: A consensus statement from the International Diabetes Federation. *Diabetic Medicine, 23*, 469–480.
2. Austin, M. A. (1991). Plasma triglyceride and coronary artery disease. *Arteriosclerosis and Thrombosis, 11*, 2–14.
3. Austin, M. A., Breslow, J. L., Hennekens, C. H., Burling, J. E., Willett, W. C., & Krauss, R. M. (1988). Low-density lipoprotein subclass patterns and risk of myocardial infarction. *JAMA, 260*, 1917–1921.
4. Austin, M. A., Hokanson, J. E., & Brunzell, J. D. (1994). Characterization of low-density lipoprotein subclasses: Methodological approaches and clinical relevance. *Current Opinion in Lipidology, 5*, 395–403.
5. Balkau, B., & Charles, M. A. (1999). Comment on the provisional report from the WHO consultation: European Group for the study of insulin resistance (EGIR). *Diabetic Medicine, 16*, 442–443.
6. Bansal, S., Buring, J. E., Rifai, N., Sacks, F. M., & Ridker, P. M. (2007). Fasting compared with non-fasting triglycerides and risk of cardiovascular events in women. *JAMA, 298*, 309–316.
7. Brasaemle, D. L., Cornley-Moss, K., & Bensadoun, A. (1993). Hepatic lipase treatment of chylomicron remnants increases exposure of apolipoprotein E. *Journal of Lipid Research, 34*, 455–463.
8. Cholesterol Treatment Trialists' Collaboration. (2005). Efficacy and safety of cholesterol-lowering treatment: Prospective meta-analysis of data from 90,056 participants in 14 randomized trials of statins. *Lancet, 366*, 1267–1278.
9. Einhorn, D., Reaven, G. M., & Cobin, R. H. (2003). American College of Endocrinology position statement on the insulin resistance syndrome. *Endocrine Practice, 9*, 237–242.
10. Executive Summary of the third report of the National Cholesterol Education Program (NCEP) Expert Panel on detection, evaluation, and treatment of high blood cholesterol in adults (Adult Treatment Panel III). (2001). *JAMA, 285*, 2486–2497.
11. Gami, A. S., Witt, B. J., Howard, D. E., Erwin, P. J., Gami, L. A., Somers, V. K., & Montori, V. M. (2007). Metabolic syndrome and risk of incident cardiovascular events and death: A systematic review and meta-analysis of longitudinal studies. *Journal of the American College of Cardiology, 49*, 403–414.
12. Griffin, B. A., & Packard, C. J. (1994). Metabolism of VLDL and LDL subclasses. *Current Opinion in Lipidology, 5*, 200–206.

13. Griffin, B. A., & Zampelas, A. (1995). Influence of dietary fatty acids on the atherogenic lipoprotein phenotype. *Nutrition Research Reviews, 8*, 1–26.
14. Guize, L., Tomas, F., Pannier, B., Bean, K., Jego, B., & Benetos, A. (2007). All-cause mortality associated with specific combinations of the metabolic syndrome according to recent definitions. *Diabetes Care, 30*, 2381–2387.
15. Ingelsson, E., Schaefer, E. J., Contois, J. H., McNamara, J. R., Sullivan, L., Keyes, M. J., Pencina, M. J., Schoonmaker, C., Wilson, P. W., D'Agostino, R. B., & Vasan, R. S. (2007). Clinical utility of different lipid measures for prediction of coronary heart disease in men and women. *JAMA, 298*, 776–785.
16. Karpe, F., & Hultin, M. (1995). Endogenous triglyceride-rich lipoproteins accumulate in rat plasma when competing with a triglyceride emulsion for a common lipolytic pathway. *Journal of Lipid Research, 36*, 1557–1566.
17. Kathiresan, S., Otvos, J. D., Sullivan, L. M., Keyes, M. J., Schaefer, E. J., Wilson, P. W. F., D'Agostino, R. B., Vasan, R. S., & Robins, S. J. (2006). Increased small low-density lipoprotein particle number: A prominent feature of the metabolic syndrome in the Framingham Heart Study. *Circulation, 113*, 20–29.
18. Keys, A. (1970). Coronary artery disease in seven countries. *Circulation, 41*, 1–211.
19. Lei, Z., Xu, M., Li, Y., Chen, L., & Li, H. (2023). Prebiotics, probiotics, and nutrients in cardiovascular and kidney disease. *Nutrients, 15*(19), 4284. https://doi.org/10.3390/nu15194284.
20. Lloyd-Jones, D., Adams, R. J., Brown, T. M., et al. (2010). Heart disease and stroke statistics 2010 update: A report from the American Heart Association. *Circulation, 121*, 948–954.
21. Lorenzo, C., Hunt, K. J., Williams, K., & Haffner, S. M. (2007). The National Cholesterol Education Program–Adult Treatment Panel III, International Diabetes Federation, and World Health Organization definitions of the metabolic syndrome as predictors of incident cardiovascular disease and diabetes. *Diabetes Care, 30*, 8–13.
22. Martin, M. J., Browner, W. S., Wentworth, J., Hulley, S. B., & Kuller, L. H. (1986). Serum cholesterol, blood pressure, and mortality: Implications from a cohort of 361,662 men. *Lancet, 2*, 933–936.
23. Meisenböck, G., & Patsch, J. R. (1991). Coronary artery disease: Synergy of triglyceride-rich lipoproteins and HDL. *Cardiovascular Risk Factors, 1*, 293–299.

24. Nordestgaard, B. G., Benn, M., Schnohr, P., & Tybjörg-Hansen, A. (2007). Nonfasting triglycerides and risk of myocardial infarction, ischemic heart disease, and death in men and women. *JAMA, 298*, 299–308.
25. Patsch, J. R., Meisenböck, G., Hopferweiser, T., Mulhberger, V., Knapp, E., Dunn, J. K., Gotto, A. M., & Patsch, W. (1992). Relation of triglyceride metabolism and coronary artery disease: Studies in the postprandial state. *Arteriosclerosis and Thrombosis, 12*, 1336–1345.
26. Pencina, M. J., D'Agostino, R. B., Larson, M. G., Massaro, J. M., & Vasan, R. S. (2009). Predicting the 30-year risk of cardiovascular disease: The Framingham Heart Study. *Circulation, 119*, 3078–3084.
27. Proctor, S. D., Vine, D. F., & Mamo, J. C. L. (2004). Arterial permeability and efflux of apolipoprotein B containing lipoproteins assessed by *in situ* perfusion and three-dimensional quantitative confocal microscopy. *Arteriosclerosis, Thrombosis, and Vascular Biology, 24*, 2162–2167.
28. Reaven, G. M., Chen, Y. D. I., Jeppesen, J., Maheux, P., & Krauss, R. M. (1993). Insulin resistance and hyperinsulinemia in individuals with small, dense, low-density lipoprotein particles. *Journal of Clinical Investigation, 92*, 141–146.
29. Sacks, F. M., & Campos, H. (2003). Low-density lipoprotein size and cardiovascular disease: A reappraisal. *Journal of Clinical Endocrinology & Metabolism, 88*, 4525–4532.
30. Shaper, A. G., Pocock, S. J., Philips, A. N., & Walker, M. (1986). Identifying men at high risk of heart attacks: Strategy for use in general practice. *BMJ, 293*, 474–479.
31. Shepherd, J., Cobbe, S. M., Ford, I., Isles, C. G., Lorimer, A. R., Macfarlane, P. W., McKillop, J. H., & Packard, C. J. (1995). Prevention of coronary heart disease with pravastatin in men with hypercholesterolemia. *New England Journal of Medicine, 333*, 1301–1307.
32. Sone, H., Tanaka, S., Limuro, S., Oida, K., Yamasaki, Y., Oikawa, S., Ishibashi, S., Katayama, S., Ito, H., Ohashi, Y., Akanuma, Y., & Yamada, N. (2009). Components of the metabolic syndrome and their components as predictors of cardiovascular disease in Japanese patients with type 2 diabetes: Implications for improved definition. Analysis from Japan Diabetes Complications Study (JDCS). *Journal of Atherosclerosis and Thrombosis, 16*, 380–387.
33. Sunström, J., Vallhagen, E., Risérus, U., Byberg, L., Zethelius, B., Berne, C., Lind, L., & Ingelsson, E. (2006). Risk associated with the metabolic syndrome versus the sum of its individual components. *Diabetes Care, 29*, 1673–1674.

34. The Prospective Studies Collaboration, Lewington, S., Whitlock, G., Clarke, R., Sherliker, P., Emberson, J., Halsey, J., Qizilbash, N., Peto, R., & Collins, R. (2007). Blood cholesterol and vascular mortality by age, sex, and blood pressure: A meta-analysis of individual data from 61 prospective studies with 55,000 vascular deaths. *Lancet, 370*, 1829–1839.

CHAPTER 6

FUNCTIONAL FOODS AND THEIR ROLE IN CANCER PREVENTION AND HEALTH PROMOTION

Contents

INTRODUCTION

During regular aerobic metabolism, a variety of reactive oxygen species (ROS) are produced, including the hydroxyl radical, hydrogen peroxide, and superoxide. Furthermore, photochemical processes (like those in the skin and eyes) can produce singlet oxygen, and lipid peroxidation can result in the formation of peroxyl radicals. Through the oxidation of DNA, proteins, and lipids, these oxidants completely contribute to aging and degenerative diseases like cancer and atherosclerosis. Inflammation is another factor in the carcinogenesis of cancer, rather than just oxidation. Numerous processes, including the generation of free radicals by inflammatory cells, contribute to inflammation's ability to cause cancer. One of the main causes of mortality for adults is cancer. One of the biggest health issues, cancer accounts for 1 in 8 deaths globally. It is projected that approximately 25% of Americans will experience cancer at some point in their lives. Surgery, radiation, and costly and frequently traumatic drug use are the usual forms of treatment. Based on estimates from the World Economic Forum in 2011, the estimated cost of the 13.3 million new cases of cancer in 2010 was estimated to be US$ 290 billion; however, by 2030, total costs were projected to rise to US$ 458 billion. These studies suggest that most cancer cases and deaths globally can be prevented. Around 30% of cancer cases have some connection to diet and food. Several studies have demonstrated the link between functional foods and a lower risk of cancer. Scientists studying cancer are interested in using natural products to increase cancer patients' chances of survival. To prevent or treat cancer, Americans, Japanese, and Europeans are using dietary vegetables, medicinal herbs, and their extracts or components. Food manufacturers have recently started developing "functional foods," which are defined as food items with an additional positive health benefit, based on health criteria. Foods and food ingredients that offer health advantages over and above basic nourishment are known as functional foods. These foods resemble conventional foods in appearance; they are functional foods that are meant to be a part of a regular diet. The body receives the necessary amounts of vitamins, fats, proteins, carbs, and other nutrients from functional food. necessary to ensure its healthy survival. All things considered, functional foods are a range of products that contain natural ingredients or components found in regular, enhanced, fortified, and enriched foods. Many naturally occurring substances found in food, especially antioxidant-containing plants, and their extracts as well as essential oils (EOs), have demonstrated promise as possible chemopreventive agents. Antioxidants appear to be able to lessen DNA damage caused by free radicals, which is thought to be the main cause of the majority of cancers. Antioxidant compounds are thought to have the ability to decrease oxidative damage to DNA and oxidant-stimulated cell division, which in turn can reduce mutagenesis and carcinogenesis. It appears that astaxanthin is useful in lessening the intensity of several inflammatory diseases in both humans and rodents. The carotenoid pigments in yellow, orange, and red that have been studied recently are among these phytonutrients. The various categories of phytochemicals include astaxanthin, xanthine, lycopene, carotenoids, and carotenoid pigments. Carotenoids are categorized into various groups, such as functional groups that are alcohols, hydrocarbons, ethers, epoxides, ketones, or acids. Studies have investigated thc relationships between xanthine, other phytochemicals, and the yellow, orange, and red carotenoid pigments. Naturally, functional foods contain more than just phytochemicals; other substances also function similarly to those found in functional foods. Growth factors and conditionally essential nutrients, for instance, are examples. Prebiotic, probiotic, and synbiotic ingredients, as well as amino acids and polyunsaturated fatty acids (PUFAs), may be advantageous in functional foods (Table 6.1) (Ames et al., 1993).

Table 6.1. Bioactive Compounds in Red Food, and their Target in Various Cancer Types

Bioactive Compounds	**Red Foods**	**Target Regulation**		**Cancer Types & Ref.**
Lycopene	Carrots, watermelon, grapefruit, apricot, pink guava, and papaya	↑	Bax, PPARγ, and Caspase-3	Gastric (Conaway et al., 2002; Kurahashi et al., 2007; Plochmann et al., 2007), pancreatic (Kumar et al., 2010), prostate (Limer & Spiers, 2004), esophageal (Wakai et al., 2007), ovarian (Bidoli et al., 2005), oral (Shannon et al., 2007), breast (Naguib, 2000), and lung (Yuan et al., 2011) cancer
		↓	EGFR, Ras ERK, p38, NFκB, Cox2, clAP1, clAP2, and Bcl2	
Anthocyanin	Grapes, berries, black bean, black rice, porunn, potatos, and red onion	↑	Bax and p53	Glioblastoma (Kim et al., 2005), breast (Nishikawa et al., 2005), colon (Gajalakshmi et al., 2012; Chew et al., 1999), oral squamous (Tanaka et al., 1995), and prostate (Jyonouchi et al., 2000) cancer
		↓	STAT3/VEGF, caspase-1, caspase-3, and IL-1β	

β-carotene	Carrots, apricots, paprika, and chili powder	↓	IL-6, STAT3, DNMT, DLK1, Oct3, Oct4, AKT, NFκB, SOD2, Nrf2, Ku70/80, MMP-10, AP-1, Cav-1, and mTOR	Neuroblastoma, colon esophageal squamous cell carcinoma, breast, and gastric cancer
Pectin	Apple and citrus	↑	p53, AMPK, and Nrf2	Colorectal and breast cancer
		↓	β-glucuronidase, β-glucosidase, tryptophanase, PTK2B, PDE4B, and TCF4	
Betaine	Sugar cane, Guji berries, and beets	↑	GSH, SOD, CAT, TAS, p53, Bax, caspase-3, cytochrome c oxidase, SIRT1, and p16,	Breast , prostate , cervical, colon, and liver cancer, and oral squamous cell carcinoma
		↓	PolIII, Brf1, TNFα, iNOS, Cox2, MMP-2, MMP-9, Akt, NFκB, and c-myc	
Rutin	Oranges, wine, and grapes	↑	Caspase-3, Caspase-9, Bax, TNF-α, GSK3-β, and p53	Neuroblastoma, cervical, breast, colorectal, lung, and prostate cancer
		↓	Notch-1, Hes-1, Bcl2, HPV-E6, HPV-E7, Cox2, and IL-6	

Ursolic acid	Apple peels	↑	sFRP4, sp1, Caveolin-1, Bax, p21, p53, Cyclin B1, Caspase-3, Caspase-8, Caspase-9, fibronectin-1, E-Cad, and LC3-II	Gliolastoma, breast, intestinal, lung, colorectal, esophageal, and pancreatic cancer, and papillary thyroid carcinoma
		↓	FN1, CDH2, CTNNB1, TWSIT, Bcl2, Servivin, NFκB, VEGF, MMP, PD-L1, RAGE, N-Cad, p62, and p-AKT	
Kaempferol	Apple, blueberry, cherry, cranberry, black tea, and red wine	↑	DR4, DR5, CHOP, JNK, ERK1/2, p38, and Bax	Renal, breast, prostate, ovarian, colorectal, lung, gastric, and endometrial cancer
		↓	MMP-2, FAK, CDK1, PSA, TMPRSS2, TMEPA1, PI3K, Akt, Bcl-2, TS, RhoA, Rac1, JAK/STAT3, MAPK, and NFκB	
Quercetin	Apple, chili pepper, blueberry, cherry, cranberry, black tea, and red wine	↑	p53	Liver, prostate, colon, leukemia, lymphoma, breast, pancreatic, lung, and ovarian cancer
		↓	Cyclin D1, JAK2, STAT3, MALAT1, PI3K, Akt, hnRNPA1, Bcl-2, PI3K, MAPK, and Wnt	

Myricetin	Tomatoes, oranges, berries, and grapes	↑	Bax, Caspase-3, Caspase-8, Caspase-9, p53, BRCA1, GADD45, and E-Cad	Prostate, ovarian, skin, breast, liver, colorectal, and thyroid cancer
		↓	PIM1, CXCR4, MDR-1, Bcl-2, VEGFR2, p38, p-STAT3, COX2, IL-6, TNF-α, NFκB, and N-Cad	

Source: Chung M-Y, Hwang J-T, Lee J, Choi H-K. (2022). The anti-cancer effects of red-pigmented foods: Biomarker modulation and mechanisms underlying cancer progression. Applied Sciences; 12(5):2584. https://doi.org/10.3390/app12052584.

6.1. ASTAXANTHIN

Astaxanthinis is a type of keto-carotenoid that belongs to the class of chemicals called terpenes or carotenones. A metabolite of canthaxanthin and zeaxanthin, astaxanthin has both ketone and hydroxyl functional groups.

Because of the extended chain of conjugated (alternating double and single) double bonds in the compound's center, it is a lipid-soluble pigment with red coloring characteristics. The molecule is amphiphilic due to the presence of hydrophobic hydrocarbons and hydroxyl functional groups.

The bacteria *Paracoccus carotinifaciens*, the yeast fungus *Xanthophyllomyces dendrorhous* (also known as *Phaffia rhodozyma*), and the freshwater microalgae *Haematococcus pluvialis* naturally produce astaxanthin. The production of astaxanthin occurs when algae are under stress due to a shortage of nutrients, high salinity, or excessive sunshine. The red-orange astaxanthin pigmentation is then reflected in the animals that consume the algae, including flamingos, salmon, red trout, red sea bream, and crustaceans (shrimp, krill, crab, lobster, and crayfish).

Salmon, crabs, shrimp, chickens, and egg production all use astaxanthin as a food colorant in their diets and feed supplements.

6.1.1. Structure

Alpha-hydroxyl-keto-carotenoid is astaxanthin. It is a member of the xanthophyll or oxygenated carotenoids family, which also includes lutein and zeaxanthin.

In 1938, a team of researchers using a lobster extract made the discovery and identification of astaxanthin, which has more hydroxyl groups than the other xanthophylls. Figure 6.1 shows the molecular structure.

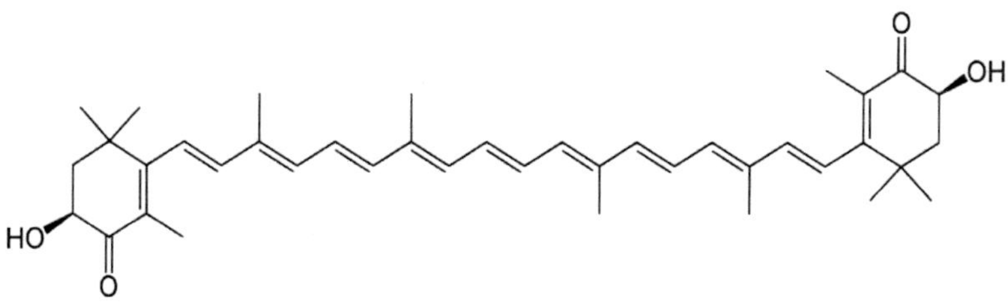

Figure 6.1. *The molecular structure of astaxanthin.*

Source: By Yikrazuul Own work, Public Domain, https://commons.wikimedia.org/w/index.php?curid=3967136.

6.1.2. Function

Astaxanthin is special because it scavenges free radicals while also preventing oxidation and reducing the amount of free radicals produced. Studies that have been published indicate that astaxanthin also improves the activity of other antioxidants, such as vitamins C and E. It has been suggested that astaxanthin provides protection from the damaging effects of ultraviolet (UV) light exposure. There is little evidence that astaxanthin improves immune function, prevents skin ulcers brought on by a variety of conditions, lowers the level of oxidized low-density lipoprotein (LDL) cholesterol, and lessens the risk of inflammation-induced cardiovascular disease (CVD). Due to its ability to scavenge free radicals in the brain and cross the blood-brain barrier, astaxanthin offers neuroprotection. Antioxidant qualities were correlated with anti-inflammatory activity. Studies have demonstrated astaxanthin's anti-inflammatory properties. Anti-inflammatory properties linked to NO synthase inhibition, prostaglandin E2 synthesis, and tumor necrosis factor-alpha (TNF-α) production (Lockwook and Gross, 2005).

6.1.3. Anti-Cancer Activity

Studies have demonstrated its effectiveness against cancer; however, the exact mechanism underlying this relationship is unknown. Due to the importance of gap junction communications for cell development, homeostasis, and growth control, some researchers hypothesized that altering these connections would have an anti-cancer effect. These studies indicate that astaxanthin affects channel functions by altering the phosphorylation pattern of gap junction protein, connexin, and that gap junctional intercellular communication (GJIC) is impaired in cancer cells. Functional connexin proteins in the membrane can be phosphorylated or dephosphorylated, which can change channel gating and set channel action. When compared to other carotenoids, astaxanthin has two keto groups on each ring structure, which gives it potent antioxidant qualities. According to a review study by Tanaka et al., (2012) and Pashkow et al. (2008) verified the claim that astaxanthin is a super-antioxidant. Furthermore, astaxanthin proved to be the most successful of the eight carotenoids examined in inhibiting the invasion of rat ascites hepatoma cells in culture. These studies demonstrate the antioxidant activity of astaxanthin, as well as its ability to scavenge free radicals, effectively quench reactive oxygen species (ROS) and nitrogen oxygen species, and function as a chain-breaking antioxidant. Other studies looked at the roles of astaxanthin and canthaxanthin in rat colon carcinogenesis induced by azoxymethane (AOM), mouse urinary bladder carcinogenesis induced by N-butyl-N (4-hydroxybutyl) nitrosamine (OH-BBN), and rat oral carcinogenesis induced by 4-NQO. The findings of these investigations demonstrated that astaxanthin, in particular, inhibited the growth of cells, thereby playing an inhibitory role in the development of cancer in the tongue, colorectum, and bladder.

As was already mentioned, inflammation plays a role in cancer. According to a study, lycopene or astaxanthin prevented human prostate cancer from proliferating. Yasui and associates. Because astaxanthin modulates the expression of several inflammatory cytokines involved in inflammation-associated carcinogenesis, it has been shown to have anti-inflammatory and anti-carcinogenic properties in inflamed colon tissue. It appears that astaxanthin aids in the downregulation of COX-2. According to a different study, astaxanthin functions by influencing extracellular signal-regulated kinase (ERK)-2, protein kinase B (Akt), matrix metalloproteinases (MMP) 2/9, COX-2, and nuclear factor kappa B (NF-κB). Research indicates that astaxanthin may prevent rats' preneoplastic liver cell lesions from developing by rerouting AFB1 metabolism away from metabolism and toward detoxification pathways. Tetrasodium diphosphate astaxanthin has also been demonstrated to fully inhibit the neoplastic transformation of $C_3H/10T1/2$ cells induced by methylcholanthrene through the up-regulation of connexin43 and gap junctional intercellular communication. Furthermore, astaxanthin inhibited the proliferation of murine mammary tumor cells by 40% in a dose-dependent manner when added to the culture medium in an animal study. In other investigations, astaxanthin supplementation to the diet inhibited the growth of transplanted Meth-A tumor cells in BALB/c mice in a dose-dependent manner. Astaxanthin-containing egg yolks were found to be protective against benzo (a) pyrene-induced mouse fore stomach neoplasia and sarcoma-180 cell-induced mouse ascites cancer in a different study. Conversely, the main cause of skin tumor pathogenesis is UVA radiation. According to Lyons and O'Brien, intestinal CaCo-2 cells, melanocytes, and human skin fibroblasts in culture were significantly protected against UVA-induced DNA damage by astaxanthin and astaxanthin-rich algal extract. Additionally, astaxanthin supplementation prevented vitamin A-deficient hairless mice's skin from accumulating potentially tumor-promoting polyamines following exposure to UVA and UVB irradiation.

6.2. LYCOPENE

Lycopene is an organic compound that belongs to the carotene and tetraterpene classes. Tomatoes and other red fruits and vegetables contain lycopene, a bright red carotenoid hydrocarbon derived from the Neo-Latin Lycopersicon, the name of a former tomato genus.

In addition to tomatoes, red carrots, grapefruits, watermelons, and papayas also contain it. Neither strawberries nor cherries contain it. It does not function as vitamin A.

Lycopene is an intermediate in the biosynthesis of many carotenoids, including beta-carotene, which is in charge of pigmentation in yellow, orange, or red, photosynthesis, and photoprotection in plants, algae, and other photosynthetic organisms. Lycopene is a tetraterpene, just like all other carotenoids. It does not dissolve in water. The deep red color of lycopene is attributed to 11 conjugated double bonds. Because of its vibrant color, lycopene can be used as a food coloring and is legal in the US, Australia, New Zealand, and the EU.

6.2.1. Structure

Lycopene is an open-chain hydrocarbon with 40 carbon atoms that has two non-conjugated and 11 conjugated double bonds arranged in a linear array. Eight isoprene units are assembled to form the tetraterpene-like structure. Through photo or chemical reactions, the structure's bonds can undergo isomerization from the trans configuration to mono or polycis isomers. The structure of lycopene lacks the P-ionone ring, which is responsible for its provitamin-A activity (Figure 6.2). Lycopene's abundance of conjugated double bonds is thought to be the cause of its many protective qualities, particularly its capacity to trap peroxyl radicals and to quench singlet oxygen.

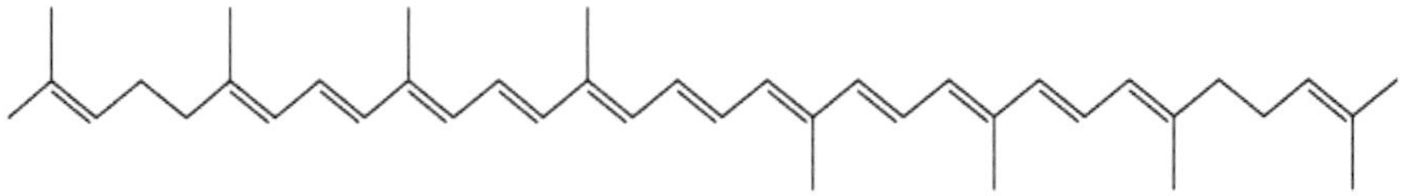

Figure 6.2. *Molecular structure of lycopene.*

Source: https://en.m.wikipedia.org/wiki/File:Lycopene.svg.

6.2.2. Function

Ripe tomatoes and tomato-based products have a distinctive deep red color that is attributed to lycopene. Lycopene has beneficial effects on various diseases and biological properties. Over 85% of lycopene in most diets comes from foods like tomatoes, watermelon, grapefruit, apricots, pink guava, pawpaw, and tomato-based products. Lycopene functions as a potent antioxidant and free radical scavenger due to its numerous conjugated double bonds. It has been demonstrated that lycopene contributes to lower cholesterol levels by inhibiting the synthesis of cholesterol, increasing the breakdown of low-density lipoproteins, and blocking the hydroxyl-methylglutaryl-coenzyme A reductase enzyme. According to Agarwal and Rao's research, lycopene has a singlet quenching capacity that is 10 times greater than alpha-tocopherol's and twice as potent as beta-carotene. Previous research in this area by Fuhrman et al. revealed that six healthy male participants who received a daily dietary supplement of 60 mg of lycopene for three months showed a significant 14% reduction in plasma LDL cholesterol levels, while no effects on HDL cholesterol contents were noted. Additional research has demonstrated the protective effects of lycopene consumption against the risk of atherosclerosis, myocardial infarction, and stroke, among other CVD. Furthermore, lycopene's anti-inflammatory properties have been demonstrated in models of inflammation that are both acute and chronic (Yaping et al., 2003).

6.2.3. Anti-Cancer Activity

Lycopene lacks provitamin A action, but it does have antioxidant activity (singlet oxygen quenching and peroxyl radical scavenging), induction of cell-cell communication, and growth control. Lycopene has numerous conjugated double bonds, which could make them extremely potent antioxidants. It appears that carotenoids, like lycopene, can react with oxygen free radicals through one of three ways: the carotenoid "bleaches," releases excess energy as heat, or transfers the unpaired electron from the carotenoid to an excited triplet state. These studies indicate that lycopene is more effective than α and ß-carotene at stopping the growth of different human cancer cell lines. Lycopene has demonstrated anti-carcinogenic effects in the skin, lungs, liver, and mammary gland in mouse models. It has also been shown to stop the formation of aberrant crypt foci in the rat colon. When α and β-carotenes were combined with lycopene (at a moderate dosage), gap-junctional intercellular communication was enhanced. A study revealed that the combination of lycopene and β-carotene decreased the incidence and quantity of cancer cases. Research has shown that the combination of lycopene from tomatoes and S-allylcysteine from garlic inhibited the growth of chemically induced gastric cancer by modulating apoptosis-associated proteins (up-regulating Bim and caspases 8 and 3) at significantly lower intakes than when these substances were given separately. Without any indication of harmful effects or cell death, lycopene has been shown to inhibit the growth of human cancer cells by interfering with growth factor receptor signaling and cell cycle progression, particularly in prostate cancer cells.

According to research by Sharoni et al. carotenoids and their oxidized derivatives interact with a network of transcription systems that are selectively and affinitively activated by various ligands, synergistically preventing cell growth. In one study, 15 men received 30 mg of lycopene daily, and the other 11 men in the control group were advised to consume at least five servings of fruits and vegetables each day, per the National Cancer Institute's recommendations. These results refuted the theory that lycopene inhibited prostate cancer growth. The lycopene group-treated prostate tissue had a 47% higher lycopene content. Prostate-specific antigen (PSA) levels had dropped by 17% in subjects taking lycopene for three weeks. Additionally, there were fewer tumors, less involvement of the surgical margins, and less diffuse involvement of the prostate by pre-cancerous high-grade prostatic intraepithelial neoplasia. Connexins, including connexin, are significantly less expressed in human tumors than in normal tissue. Kucuk et al. reported that after three weeks of lycopene supplementation, there was a slight, albeit not statistically significant, increase in connexin43 levels among prostate cancer patients. As was previously mentioned, lycopene was present in tomatoes. These studies suggest that lycopene inhibits IGF-1's ability to proliferate human cancer cells. Lycopene treatment significantly reduced insulin growth factor-1 (IGF-1) stimulation of insulin receptor substrate-1 tyrosine phosphorylation and the activator 1 (AP-1) transcription role's DNA binding ability in mammary cancer cells. Additionally, the researchers demonstrated that lycopene treatment of MCF-7 mammary cancer cells inhibited the advancement of the IGF-1-stimulated cell cycle without causing necrotic or apoptotic cell death. Other human cancer cell lines (including leukemia and cancers of the endometrium, lung, and prostate) have also shown evidence of lycopene-induced development delays by the G1 and

S phases. As was previously mentioned, inflammation plays a significant role in cancer. These findings suggest that lycopene inhibits the pro-inflammatory cytokine interleukin (IL)-8 produced in response to cigarette smoke.

According to Yang et al. lycopene inhibits the growth of human prostate cancer cells (LNCaP) by activating the PPARγ-LXRα-ATP-binding cassette transporter 1 (ABCA1) pathway. In mouse models, lycopene demonstrated anticarcinogenic properties in the skin, liver, lungs, and mammary glands. It also inhibited the formation of aberrant crypt foci in the rat colon. In a rat colon carcinogenesis model, tomato juice high in lycopene (17 ppm) was found to have an inhibitory effect. In a rat spontaneous liver carcinogenesis model, long-term administration (6 to 76 weeks of age) of a diet containing 0.005% lycopene did not reduce the risk of hepato-carcinogenesis. According to Faezizadeh et al. lycopene may lessen the risk of cancer caused by K562 cells.

6.3. LUTEIN

One of the 600 known naturally occurring carotenoids is lutein, a xanthophyll. Like other xanthophylls, lutein is exclusively synthesized by plants and is present in large amounts in green leafy vegetables like kale, spinach, and yellow carrots. In green plants, triplet chlorophyll—an excited form of chlorophyll that is overproduced at very high light levels during photosynthesis—is dealt with by xanthophylls, which also act as non-photochemical quenching agents.

Naturally occurring in orange-yellow fruits, flowers, and leafy vegetables, lutein is a component of human diets. The average daily intake of lutein and zeaxanthin by adults in the United States is 1.7 mg, according to the NHANES 2013-2014 survey. As of right now, there is no suggested dietary allowance for lutein. It has been observed that dietary intake levels of 6–10 mg/day have some positive health effects. The only definite adverse consequence of consuming too much lutein is skin bronzing or carotenodermia. Lutein, an ingredient used in food, is extracted from the petals of African marigolds (*Tagetes erecta*) and has the INS number 161b (E number E161b). Australia, New Zealand, and the EU have all approved its use. In the US, lutein is permitted as a dietary supplement for humans and can be added to animal feed, frequently in conjunction with zeaxanthin. However, it is not permitted to be used as a food coloring for foods meant for human consumption.

6.3.1. Structure

The Latin term for yellow (xanthophyll) is the source of the name lutein. The formula for lutein is $C_{40}H_{56}O_2$. The family xanthophyll includes lutein. The two hydroxyl groups that lutein and zeaxanthin have—one on each side of the molecule—set them apart from other carotenoids. The lutein molecular structure is shown in Figure 6.3.

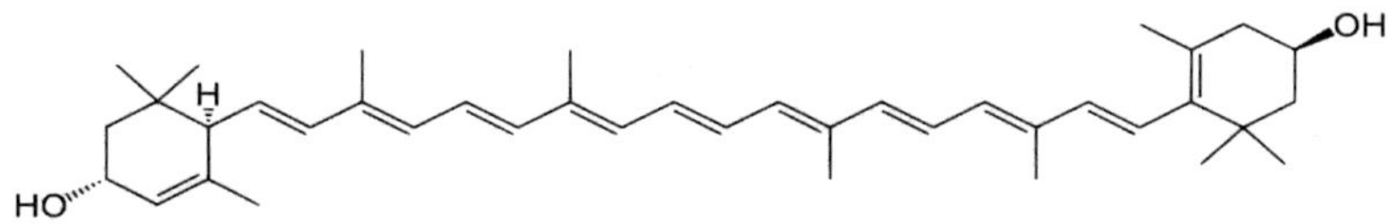

Figure 6.3. *Molecular structure of lutein.*

Source: https://en.m.wikipedia.org/wiki/File:Luteine_-_Lutein.svg.

6.3.2. Function

Leafy green vegetables, numerous fruits, and colorful veggies like sweet peppers, sweet corn, peas, and egg yolks are particularly high in lutein. The vegetable containing the most lutein (60% of the total) was maize; other foods high in lutein include grapes, spinach, orange juice, kiwi fruit, zucchini (or vegetable marrow), and various types of squash. In lenses, lutein is an excellent antioxidant. These studies suggest that antioxidants can prevent cell damage and even some types of cancer by scavenging free radicals. Studies show that when melatonin, glutathione, alpha-tocopherol, and ascorbate are present, lutein is more effective in preventing lipid peroxidation and is also better shielded against secondary oxidative breakdown. Subsequent epidemiological research on lutein revealed a significant inverse relationship between lutein intake and the risk of stroke, and that subjects with the highest serum lutein levels had a significantly lower risk of coronary heart disease (CHD). It has been shown that lutein and zeaxanthin consumption provide protection against skin swelling (edema) and hyperplasia brought on by UV exposure.

6.3.3. Anti-Cancer Activity

It has been demonstrated that zeaxanthin and lutein can cut the risk of breast cancer by 53%. As was previously mentioned, lutein and other antioxidant-rich functional foods prevent cancer. *In vitro*, lutein exhibits antioxidant properties that protect against oxidative damage and squelches peroxy radicals. Analyzed plasma lutein from 37 women was inversely correlated with measured oxidative indices. The combination of lutein and lycopene had the greatest synergistic effect. Lutein may also have anticarcinogenic properties. Aflatoxin B1 (AFB1) and 1-nitropyrene are two mutagens that lutein can interact with. It can also have an anticarcinogenic effect by activating certain genes that are involved in T-cell transformations that are triggered by mitogens, cytokines, and antigens. The Pim-1 gene controls both apoptosis and cell differentiation. Based on these findings, mice fed lutein had increased pim-1 gene expression. Narisawa and associates demonstrated how lutein protects preneoplastic colorectal adenocarcinoma lesions. Researchers found a negative correlation between lutein and lung cancer in a study conducted among Fijians, who consume 200 g of dark green vegetables (or 25 mg of lutein) on average each day. Dietary lutein consumption and colon cancer in both men and women are inversely correlated, according to research by Slattery et al. only those patients who received a colon cancer diagnosis earlier in life showed a statistically significant reduction in risk. Based on these studies, mammary tumor incidence, growth, and latency were all prevented by low dietary lutein levels, specifically at 0.002 and 0.02% of the diet. It has been shown that lutein can cause apoptosis in transformed human mammary cells but not in normal cells. It can also shield normal cells from apoptosis induced in cell culture. Freudenheim and team have shown a strong correlation between lutein and foods high in carotenoids, particularly vegetables, and a decreased risk of premenopausal breast cancer. Raising lutein levels in the blood was associated, albeit weakly, with a decreased risk of breast cancer in a case-control study. When comparing women with benign breast biopsies to those with breast cancer, there was a slight but non-significant correlation found between rising levels of lutein content in their breast adipose tissue and a lower risk of cancer. In a different study, the highest quintile of lutein and zeaxanthin consumers (9 mg/day) showed a significant 21% reduction in breast cancer risk compared to the lowest quintile (2 mg/day). In a related study, Gunasekera and colleagues concluded that while having no effect on the growth of the parental cell line of the benign tumor, lycopene, lutein, or their combination selectively inhibits the growth of a highly malignant line of prostate tumor cells (AT3) (DTE). This inhibitory effect is not indicative of a general cytotoxic response; rather, it depends on the content. As previously indicated, lutein is effective in preventing skin damage; however, we have not been able to locate any research demonstrating lutein's chemoprotective properties in the context of skin cancer. Using lutein at a dose of 20 μg/ml, Chethan Kumar and Veerabasappa Gowda found that it effectively prevented lipid peroxidation, hydroxyl radical production, and 1, 1-diphenyl-2-picrylhydrazyl (DPPH) radical production by 86, 92, and 90%, respectively. At a dose of approximately 12 times higher (400 μM) than lutein, however, curcumin, butylated hydroxy anisole, and curcumin showed 75–95% inhibition of lipid peroxidation and scavenging of hydroxyl and DPPH radicals. Previous research indicates that while higher dietary levels (0.2 or 0.4%) of dietary lutein were less effective, lower levels (0.002 and 0.02%) reduced mammary tumor incidence, tumor growth and lipid peroxidation, and elevated tumor latency.

6.4. A-CAROTENE

A type of beta-carotene that has an α-ionone ring at one end and a β-ionone ring at the other is known as α-carotene (alpha-carotene). It is the second most prevalent type of vitamin A.

The following vegetables are rich in alpha-carotene:

- Yellow-orange vegetables: Carrots, sweet potatoes, pumpkin, winter squash
- Dark-green vegetables: Broccoli, green beans, green peas, spinach, turnip greens, collards, leaf lettuce, avocado

According to a 2018 meta-analysis, there is a correlation between circulating and dietary α-carotene and a decreased risk of death from all causes. A reduction of 32% in the risk of all-cause mortality was associated with the highest circulating α-carotene category when compared to the lowest, and a 21% decrease in the risk of all-cause mortality was linked to increased dietary α-carotene intake.

6.4.1. Structure

β-carotene 15, 15'-monooxygenase (CMO1), formerly known as β-carotene 15, 15'-dioxygenase, cleaves α-carotene symmetrically at their central double bond. By identifying the excentric cleavage enzyme β-carotene 9,' 10'-monooxygenase (CMO2) in mice, humans, and zebrafish through molecular means, an alternative excentric cleavage pathway was also demonstrated and verified. Figure 6.4 shows the α-carotene molecular structure.

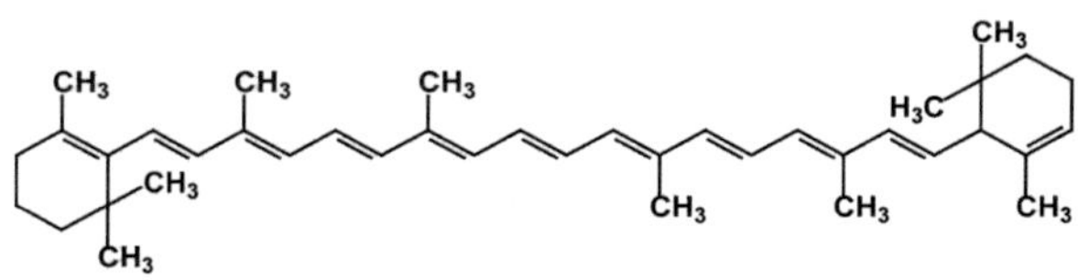

Figure 6.4. *Molecular structure of α-carotene.*

Source: https://en.m.wikipedia.org/wiki/File:Alpha-carotene.png.

6.4.2. Function

Carrots and their juice are the richest sources of α-carotene; the second densest sources are pumpkins and winter squash. Carrots contain about 1 μg of α-carotene for every 2 μg of β-carotene. Lipoproteins carry carotenoids throughout the plasma. The majority of α-carotene is typically found in low-density lipoproteins (LDL). It has been demonstrated that α-carotene offers greater protection than its well-known isomer, β-carotene. In an investigation, Narisawa et al. demonstrated how α-carotene shields preneoplastic colorectal adenocarcinoma lesions.

6.4.3. Anti-Cancer Activity

Research on animals has demonstrated that α-carotene exhibits greater efficacy than β-carotene in preventing the growth of tumors in the skin, lungs, liver, and colon. In a research on skin tumorigenesis, Murakoshi et al. in the positive control group, the percentage of tumor-bearing mice was 69%, while the percentage in the α-carotene-treated groups was 25%. The group treated with α-carotene had 0.13 tumors/mouse, whereas the positive control group had an average multiplicity of 3.73 tumors/mouse. In the identical investigation, which used a lung carcinogenesis model stimulated by glycerol and started by 4-nitroquinoline 1-oxide (4-NQO), the mean number of lung tumors per mouse in the positive control group was 4.06/mouse, while the group treated with α-carotene had 1.33/mouse.

6.5. B-CAROTENE

β-Carotene, also known as beta-carotene, is a red-orange pigment that is abundant in fruits, plants, and fungi. It is an organic substance. It is a terpene (isoprenoid) belonging to the carotenes family. It is biochemically synthesized from eight isoprene units, giving it 40 carbons. Dietary β-carotene is a provitamin A compound that the body converts to vitamin A or retinol. Carrots, pumpkin, spinach, and sweet potatoes are foods high in it. It is a dietary supplement that is sometimes prescribed to treat erythropoietic protoporphyria, a genetic condition that causes extreme sensitivity to sunlight. The most prevalent carotenoid in plants is β-carotene. Its E number, E160a, indicates that it can be used as a food coloring agent. In 1930, the structure was deduced.

6.5.1. Structure

As previously mentioned, β-carotene 15, 15'-monooxygenase (CMO1), formerly known as β-carotene 15, 15'-dioxygenase, cleaves β-carotene symmetrically at their central double bond. By identifying the excentric cleavage enzyme β-carotene 9,' 10'-monooxygenase (CMO2) in mice, humans, and zebrafish through molecular means, an alternative excentric cleavage pathway was also demonstrated and verified. Figure 6.5 shows the α-carotene molecular structure.

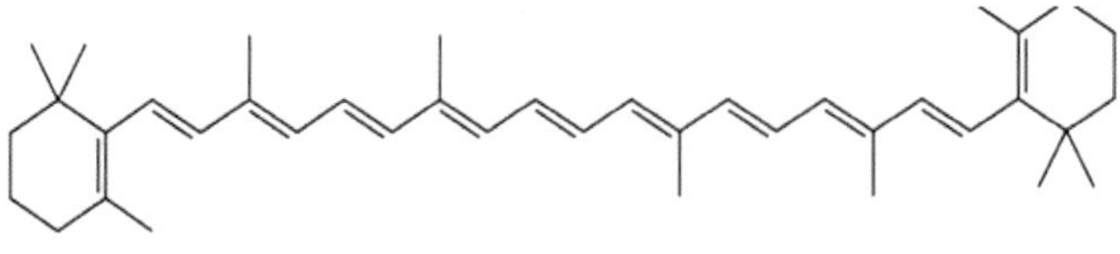

Figure 6.5. *Molecular structure of β-carotene.*

Source: https://upload.wikimedia.org/wikipedia/commons/thumb/2/27/Beta-Carotin.svg/1920px-Beta-Carotin.svg.png.

6.5.2. Function

The β-carotene present in orange and yellow fruits and vegetables, as well as green leafy vegetables. In humans, β-carotene can be transformed into vitamin A. It has been demonstrated that in normal bronchial epithelial cells, β-carotene and its oxidative metabolite, apo-14'-carotenoic acid, reverse the down-regulation of RARβ caused by smoke-borne carcinogens. Furthermore, it has been demonstrated that the metabolism of β-apo-14'-carotenoic acid to all-trans-retinoic acid is responsible for the transactivation of the RARβ enhancer. Thus, RA may mediate the molecular mode of β-carotene's role through transcriptional activation of a number of genes with unique anti-proliferative or pro-apoptotic activity, which open the door for the deletion of irreversibly altered neoplastic and preneoplastic cells. In rats, AFB1-induced preneoplastic hepatocellular lesions were inhibited by β-carotene. In normal human fibroblasts, it has also been demonstrated that β-carotene causes a cell-cycle delay in the G1 phase.

6.5.3. Anti-Cancer Activity

As previously mentioned, β-carotene functions as an antioxidant; however, at high content and particularly at high oxygen tension, it exhibits prooxidant effects. These effects could potentially account for the unexpected rise in lung cancer deaths among smokers treated with β-carotene. *In vitro*, β-carotene and lutein exhibit antioxidant properties that squelch peroxy radicals and prevent oxidative damage. Research indicates that β-carotene has anticancer properties. A study by Toniolo et al. found a negative correlation between serum β-carotene concentration and breast cancer. The USA, China, Turkey, and India have all reported similar findings. In research on skin tumorigenesis, Murakoshi et al. 69% of the positive control group's mice bore tumors, compared to 13% of the groups receiving α-carotene treatment. In an additional investigation, the combination of β-carotene, vitamin E, and selenium decreased the risk of cancer by 13% and stomach cancer by 21%. Despite some research indicating that giving smokers dietary supplements containing β-carotene or vitamins A and E significantly increased their risk of developing lung cancer. Smokers may be more likely to develop lung cancer due to an imbalance of other carotenoids or antioxidants, β-carotene's pro-oxidant role at the high oxygen tensions in their lungs, induction of P450 enzymes, and components of cigarette smoke producing harmful β-carotene oxidation products. In an additional investigation, Rautalahti et al. demonstrated that adding β-carotene and α-tocopherol to the diet did not significantly affect the incidence or mortality rates of pancreatic cancer. According to Jayappriyan et al. the β-carotene found in *D. salina* microalga has anti-cancer properties.

6.6. B-CRYPTOXANTHIN

One naturally occurring carotenoid pigment is β-cryptoxanthin. Numerous sources have been used to isolate it, such as the fruit of Physalis plants, orange rinds, papayas, egg yolks, butter, apples, and bovine blood serum.

6.6.1. Structure

The β-cryptoxanthin is a cyclized hydrocarbon carotenoid. Figure 6.6 shows the α-carotene molecular structure.

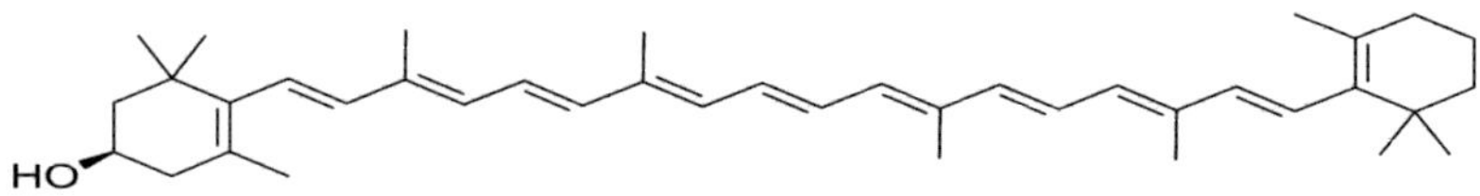

Figure 6.6. *Molecular structure of β-Cryptoxanthin.*

Source: By Oks -Own work using: BKChem, Public Domain, https://commons.wikimedia.org/w/index.php?curid=37097522.

6.6.2. Function

Cryptoxanthin, which have varied degrees of biologic activity, are found in human milk. The bran/germ fraction of whole wheat flour contributed the total amount of β-cryptoxanthin. Oranges and their juice, peaches, papayas, mangoes, watermelon, nectarines, black olives, plums, grapefruit, and fruit cocktail all contain β-cryptoxanthin. Papayas, tangerines, and red bell peppers are also rich in β-cryptoxanthin. It is the body's precursor to vitamin A and is transformed into retinol. A single study demonstrates how β-Cryptoxanthin affects bone calcification by raising calcium concentration and alkaline phosphatase activity in rat femoral tissue, which in turn promotes bone formation and inhibits bone resorption. Numerous studies demonstrate a negative correlation between the levels of β-cryptoxanthin and the morbidity of certain diseases, including liver disorders.

6.6.3. Anti-Cancer Activity

β-cryptoxanthin participates in multiple biological processes, including the elimination of free radicals, augmentation of gap junctions, immune system control, and modulation of the activity of enzymes implicated in the development of cancer. In a research study, Narisawa et al. demonstrated that feeding rats 25 ppm of β-cryptoxanthin for 30 weeks in their diet dramatically stopped the development of colon cancer caused by N-methylnitrosourea. Citrus unshiu segment membrane (CUSM), which contains fiber and β-cryptoxanthin, has been shown in animal studies to prevent colon tumorigenesis associated with obesity and colitis. Human cervical cancer incidence has been found to be inversely correlated with serum levels of β-cryptoxanthin. Furthermore, there is a correlation between a lower risk of lung cancer and the dietary consumption of β-cryptoxanthin. In an additional investigation, Toniolo and colleagues documented a rise in breast cancer cases among subjects with reduced β-cryptoxanthin levels. Furthermore, there exist additional reports demonstrating the effectiveness of β-cryptoxanthin in impeding the development of skin tumors in mice.

6.7. CANTHAXANTHIN

One common keto-carotenoid pigment found in nature is canthaxanthin. Terpenoids are a broader class of phytochemicals that includes carotenoids. $C_{40}H_{52}O_2$ is the chemical formula for canthaxanthin. It was initially identified in edible fungi. Additionally, it has been detected in bacteria, crustaceans, green algae, and fish that bioaccumulate it, including carp, golden grey mullet, seabream, and thrush wrasse. Canthaxanthin, which goes by the E number E161g, is permitted to be used as a food coloring agent in a number of nations, including the EU and the US, but not in Australia or New Zealand. It is generally approved for use in feed in the US, Canada, and EU, at the very least. It is legal in the EU for canthaxanthin to be added to poultry, salmon, and trout feed. The limit for feed products in the European Union is 80 mg/kg for feedstuffs, 8 mg/kg for egg-laying hens' feed, and 25 mg/kg for feed intended for other poultry and salmonids. One strong lipid-soluble antioxidant is canthaxanthin. The biological activities of canthaxanthin are associated, at least partially, with its capacity to act in animal tissues as an antioxidant (scavenging free radicals and sparing vitamin E).

6.7.1. Structure

Canthaxanthin is hydrocarbon carotenoid that is cyclized. The molecular structure of canthaxanthin is presented in Figure 6.7.

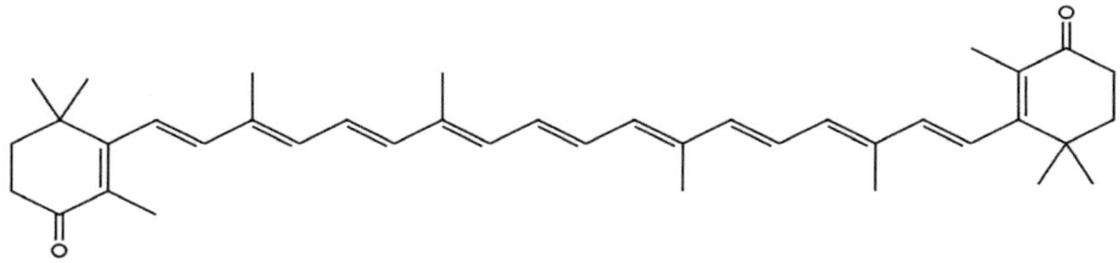

Figure 6.7. *Molecular structure of canthaxanthin.*

Source: https://en.m.wikipedia.org/wiki/File:Canthaxanthin.svg.

6.7.2. Function

Cantharellus cinnabarinus, an edible mushroom, was the source of the initial isolation of this molecule. In addition to primary carotenoids, it is also present in various green and blue-green algae at the end of their growth phase. It is present in a variety of fish species, such as carp (*Cyprinus carpio*), golden mullet (*Mugil auratus*), annular seabream (*Diplodus annularis*), and trush wrasse (*Crenilabrus tinca*), as well as bacteria and crustacea. Although canthaxanthin is widely used as a coloring agent in foods and animal feed, it is not generally regarded as a dietary carotenoid and may therefore be ingested by humans. Canthaxanthin also functions as a chain-breaking antioxidant, a strong quencher of ROS and nitrogen oxygen species, and a free radical quencher. When it comes to scavenging free radicals and acting as an antioxidant, canthaxanthin outperforms carotenoids like β-carotene. Among the carotenoids lacking provitamin activity is canthaxanthin. It has been shown that canthaxanthin can inhibit the growth of preneoplastic liver cell lesions brought on by AFB1 in rats by diverting AFB1 metabolism into detoxification pathways.

6.7.3. Anti-Cancer Activity

Canthaxanthin has the potential to inhibit the growth of human colon cancer cells, shield mouse embryo fibroblasts from metamorphosis, and protect mice against the development of skin and mammary tumors. Additionally, canthaxanthin has been shown to be beneficial in preventing colon and oral carcinogenesis in rats. Despite being a strong antioxidant, canthaxanthin may also have chemopreventive effects due to its capacity to up-regulate gene expression, which enhances gap junctional cell-cell communication. Even at high oxygen tension and carotenoid content, canthaxanthin exhibits little to no prooxidative behavior, providing evidence that it is a pure antioxidant. The capacity of canthaxanthin to induce xenobiotic metabolizing enzymes in rat liver, lungs, and kidneys may also be linked to its chemopreventive effects. The overuse of canthaxanthin as a sunless tanning product has resulted in the development of crystalline deposits in the retina of humans. Other research has demonstrated that canthaxanthin induces the activity of certain enzymes. A research team demonstrated that canthaxanthin contained CYP1A1 and 1A2 as well as P4501A1 and 1A2, which are involved in the metabolism of substances that may cause cancer, including aflatoxin, aromatic amines, and polycyclic aromatic hydrocarbons. Certain P450 enzymes were also induced by this xanthophyll in the lung and kidney tissues of rats, but not in the small intestine. The availability of these retinal administrations has increased caution when it comes to consuming this carotenoid, even though they are reversible and appear to have no negative effects. Based on these investigations, canthaxanthin inhibits the growth of cells, thereby preventing cancer from developing in the bladder, tongue, and colon. Additionally, canthaxanthin has demonstrated chemopreventive effects against cancer in UV-B-induced skin tumorigenesis in mice as well as chemically-induced gastric and breast carcinogenesis. It is possible that canthaxanthin prevents the growth of human colon cancer cells, prevents the transformation of mouse embryo fibroblasts, and prevents mice from developing mammary and skin tumors. In rats, canthaxanthin effectively inhibits the development of colon and oral cancer.

6.8. FUCOXANTHIN

Fucoxanthin, formula $C_{42}H_{58}O_6$, is a xanthophyll. Brown algae and the majority of other heterokonts contain it as an accessory pigment in their chloroplasts, which gives them an olive-green or brown hue. Fucoxanthin is a light-absorbing substance that mainly absorbs light in the blue-green to yellow-green region of the visible spectrum. It peaks, according to different estimates, at 510–525 nm and absorbs heavily in the 450–540 nm range.

6.8.1. Structure

Fucoxanthin is a cyclized hydrocarbon carotenoid. Figure 6.8 shows the fucoxanthin molecular structure.

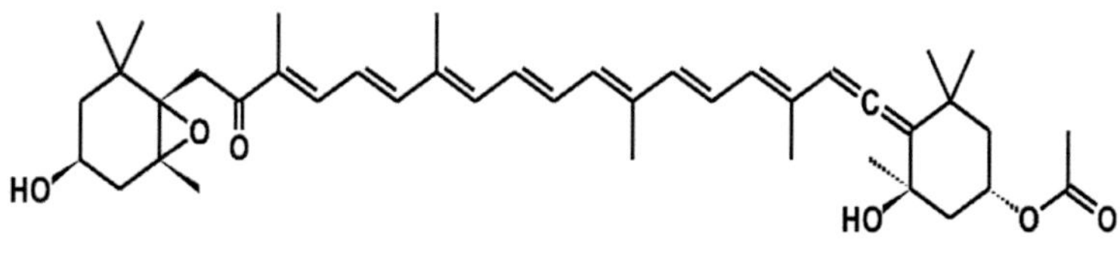

Figure 6.8. *Molecular structure of fucoxanthin.*

Source: https://en.wikipedia.org/wiki/File:Fucoxanthin.png.

6.8.2. Function

Fucoxanthin is a naturally occurring pigment that belongs to the non-provitamin A carotenoids class and has a brown or orange color. In anoxic environments, fucoxanthin functions as an antioxidant. Fucoxanthin is present in diatoms (Bacillariophyta) and brown seaweeds (*Phaeophyceae*) in the Chromophyta (Heterokontophyta or Ochrophyta) order. According to a study, fucoxanthin affects a number of enzymes involved in fat metabolism, increasing the amount of energy that can be produced from fat. One common feature of fucoxanthin in rodents may be an increase in circulating cholesterol levels.

6.8.3. Anti-Cancer Activity

Fucoxanthin has been shown to inhibit the growth of cell lines in the bladder (DLD-1), colon (HT-29, Caco-2, and HepG2), and liver (HepG2). These findings are based on *in vitro* research. Additionally, it exhibits some inhibitory effects on cyclin D levels and the induction of apoptosis, both of which are essential for the growth of cancer cells. Kim and colleagues conducted a study, demonstrated the inhibition of DMH-induced colon cancer in mice by fucoxanthin. Additionally, in a two-stage carcinogenesis experiment involving the skin of ICR mice, it demonstrated antitumor-promoting activity and may inhibit spontaneous liver tumorigenesis in C3H/He male mice. The experiment was initiated with 7,12-dimethylbenz[a] anthracene and increased with 12-O-teradecanoylphorbol-13-acetate and mezerein. Fucoxanthin has also been demonstrated to shield mice against N-ethyl-N'-nitro-N-nitrosoguanidine-induced duodenal carcinogenesis. Fucoxanthin's anticancer activity is thought to be linked to biomolecules that are involved in the cell cycle, apoptosis, and antioxidant activity due to its pro-oxidant function. Another study demonstrates the anti-neoplastic role of fucoxanthin by selectively inhibiting the activities of mammalian DNA polymerase, particularly replicative DNA polymerases (pol α, δ, and ε). Fucoxanthin may alter the course of the cell cycle. According to research by Hosokawa et al. fucoxanthin can cause apoptosis, boost the anti-proliferative effects of troglitazone, a PPARγ ligand, and stop the spread of human colon cancer cells. Fucoxanthin is useful in the cosmeceutical and antioxidant industries because it can protect against oxidative stress brought on by UV-B radiation. Sangeetha et al. demonstrated that fucoxanthin is more effective than beta-carotene at regulating glutathione transferase, catalase, and lipid peroxidation in the plasma and liver of rats deficient in retinol. In UVB-irradiated hairless mice, fucoxanthin prevents skin photoaging, possibly through effects on topical treatment that are antioxidant and antiangiogenic. Tyrosinase activity, UVB-induced skin pigmentation, and melanogenesis in melanoma were all inhibited by fucoxanthin. The Bcl-xL signaling pathway was activated in human leukemia cell HL-60 cells by fucoxanthin-induced apoptosis. Based on these investigations, fucoxanthin inhibited LNCap prostate cancer cell growth in a dose-dependent manner. Fucoxanthin induced G(1) arrest and GADD45A expression, but it also inhibited SAPK/JNK. On the other hand, fucoxanthin activated c-Jun N-terminal kinase (SAPK/JNK). In PC-3 human prostate cancer cells, fucoxanthin treatments were found to induce apoptosis through caspase-3 activation.

6.9. ISOTHIOCYANATES

A functional group that can be found in compounds with the formula R−N=C=S is isothiocyanate. The more prevalent isomers of thiocyanates, with the formula R−S−C≡N, are called isothiocyanates.

Plants produce a large number of isothiocyanates through the enzymatic conversion of glucosinolate metabolites. Mustard oils, or allyl isothiocyanate, are a well-known naturally occurring isothiocyanate.

Broccoli, cabbage, cauliflower, kale, Bok choy, and other cruciferous vegetables are excellent providers of glucosinolate precursors, which are the building blocks of isothiocyanates.

6.9.1. Structure

A class of sulfur-containing phytochemicals known as isothiocyanates arises spontaneously from conjugated glucosinolates. As demonstrated in Figure 6.9, glucosinolates like glucoraphanin are broken down into isothiocyanate sulforaphane by the enzyme myrosinase, which is found in plant tissues or intestinal flora.

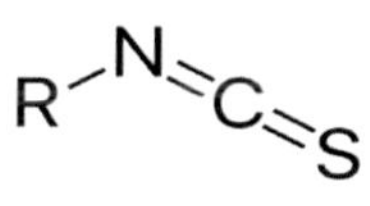

Figure 6.9. *Molecular structure of isothiocyanate.*

Source: By Fvasconcellos-Own work, Public Domain, https://commons.wikimedia.org/w/index.php?curid=1819322.

6.9.2. Function

Among other cruciferous vegetables, broccoli, cauliflower, kale, Brussels sprouts, and cabbage contain isothiocyanates. Cruciferous vegetables are high in glucosinolates, but sadly, food processing techniques like boiling or high-power microwaving greatly reduce their bioavailability. In broccoli (sprouts), isothiocyanates were present as sulforaphane, and in watercress, they were present as phenethyl isothiocyanate.

6.9.3. Anti-Cancer Activity

Isothiocyanates and their metabolites have been shown in numerous studies to help reduce the risk of developing cancers of the lungs, breast, liver, esophagus, stomach, small intestine, and colon. The mechanism of action of isothiocyanates is thought to be through modulation of cytoprotective biotransformation enzymes by the Kelch-like erythroid-cell-obtained protein with CNC homology (KEAP1)/Nuclear factor erythroid 2-related factor 2 (NRFF2)/ antioxidant response element (ARE) pathway; anti-inflammatory activity by blocking NFκB; prevention of proliferation by inducing cell cycle arrest and programmed cell death (apoptosis); induction of hormone receptor expression; potential antiangiogenic and anti-metastasis; and induction of autophagy. There is evidence linking immunoprecipitation to sulforaphanes; consequently, after stimulating human lymphoblastoid cells with sulforaphanes, more than 240 genomic regions bound to nuclear factor erythroid 2-related factor 2 were recently discovered through chromatin-immunoprecipitation using an antibody against the transcription factor nuclear factor erythroid 2-related factor 2 in conjunction with sequencing of the chromatin-bound DNA (ChIP-seq). To suppress oxidative or electrophilic stress and stop the development of cancer, NRF2 activates anti-stress signaling with a protective response. Because of proteasomal degradation brought about by the negative regulator Kelch-like ECH associated protein 1 (KEAP1), NRF2 is dormant during the resting state. Moreover, Wagner and colleagues demonstrated that in mouse macrophages, lipopolysaccharide-induced NF-κB-mediated transcription of proinflammatory proteins was inhibited by sulforaphane and allylisothiocyanate. Phase II enzymes were induced by sulforaphane in a time—and dose-dependent manner, according to *in vitro* investigations. The activity showing the beneficial effects of glutathione reductase (GR), quinone oxidoreductase 1 (NQO1), Aldo-keto reductase (AKR), and GST in a number of mammalian cancer cell lines, including HepG2, MCF7, MDA-MB-231, LNCaP, HeLa, and HT-29.

It appears that sulforaphane prevents cancer through the action of histone deacetylases (HDACs), which records its chemopreventive effects up to the post-initiation phases. HDAC inhibitors have been shown to induce mitotic cell death, growth arrest, apoptosis, and ROS-mediated cell death in cancer models. Seligson et al. provided evidence to support this theory by demonstrating that a decrease in histone acetylation state is correlated with an elevated cancer grade and an increased risk of prostate cancer recurrence. Based on these studies, sulforaphane at 75 μM may induce apoptosis and cause G1/G2 cell cycle arrest by upregulating apoptosis-inducing Bax expression and downregulating anti-apoptotic bcl-2 expression in colon cancer Caco-2 cells. It has been shown that, after 48 hours of treatment, sulforaphane can more successfully stop the growth of MCF-7 human breast cancer cells than MCF-12A normal human breast epithelial cells (IC50 40.5 μM). It is commonly established that Helicobacter pylori increase the risk of stomach cancer. Based on these investigations, sulforaphane was demonstrated to eradicate or stop the growth of several strains, and in certain situations, it is even the first choice. It has been demonstrated that isothiocyanates all preferentially cause apoptosis in cancer cells compared to normal cells. In an intriguing study, Mi et al. used radioactivity-labelled ^{14}C-sulforaphane and ^{14}C-phenethylisothiocyanate to treat human lung cancer cells. Based on previous research, giving rats the phenyl ethyl isothiocyanate 7, 12-dimethylbenz[a] anthracene inhibited the development of mammary cancer. In research involving animals, giving 10 μmol sulforaphane to mice with APC gene mutation-causing multiple intestinal neoplasia prevented HDAC activity in the colonic mucosa and inhibited the growth of tumors. It is interesting to note that the combination of sulforaphane and green tea polyphenols dramatically reduced cellular proliferation in MDA-MB-231 cells. This effect was probably caused by both the strong influence of histone modifications and DNA demethylation-mediated ERα activation. In another study, using a murine UMUC3 invasive bladder cell xenograft model, the tumor weight was reduced by 42%, 33%, and 58%, respectively, when the diet was supplemented with broccoli sprouts, 2% broccoli sprout isothiocyanate extract, or gavage of pure sulforaphane. Sulforaphane inhibits the growth of the epithelial ovarian cancer cell (EOC) line SkOV-3 in a different study by lowering AKT activity. Sulforaphane's effectiveness against human brain malignant glioma GBM 8401 cells and human lung adenocarcinoma LTEP-A2 cells with growth prevention is supported by data in *in vivo* models.

6.10. PHYTOESTROGENS (GENISTEIN AND DAIDZEIN)

A phytoestrogen is a xenoestrogen (a form of estrogen produced by organisms other than humans) derived from plants that are consumed through the consumption of manufactured foods or plants rather than being produced by the endocrine system. Also referred to as dietary estrogen, this is a broad category of naturally occurring nonsteroidal plant compounds that can have either estrogenic or antiestrogenic effects due to their structural resemblance to 17-β-estradiol. Since phytoestrogens are neither known to cause disease nor to be involved in any regular biological function, they are not considered essential nutrients. Foods like soy protein, beans, oats, barley, rice, coffee, apples, and carrots are common sources of phytoestrogens.

6.10.1. Structure

The majority of phytoestrogens are members of a broad class of naturally occurring phenolic compounds that have been substituted; three of this class's most potent estrogenic compounds are coumestans, isoflavones, and prenylflavonoids. Isoflavones, which are frequently found in red clover and soy, have gotten the most research. Despite not being flavonoids, lignans have also been identified as phytoestrogens. Mycoestrogens are mold metabolites of Fusarium, which are primarily found on cereal grains but can also be found on other forages. They share similar structures and effects with plants, but they are not plant components. Mycoestrogens are the compounds that first sparked interest in phytoestrogens, despite the fact that they are rarely discussed in relation to phytoestrogens.

The three main groups of phytoestrogens are coumestans (coumestrol), lignans (enterolactone or enterodiol), and isoflavones (genistein, daidzein, glycitein, or equol).

6.10.2. Function

According to published research, phytoestrogens function similarly to estrogens. They may have either estrogenic or anti-estrogenic effects on the body. Studies have shown that phytoestrogens have a beneficial effect on the cardiovascular system, the skeleton, the incidence of osteoporosis, and the severity of menopausal symptoms.

6.10.3. Anti-Cancer Activity

It has been demonstrated that certain flavonoids, such as genistein, stilbenes, resveratrol, curcumin, and polyphenols, preferentially induce apoptosis in cancer cells compared to normal cells. (Phytoestrogens have the potential to prevent colorectal, thyroid, endometrial, breast, prostate, and skin cancers. In a test, the co-administration of vitamin D3 and genistein inhibited prostate cancer cell growth at significantly lower concentrations than either compound alone. The possible explanation for this correlation could be the co-existence of genistein and vitamin D, as genistein inhibits the expression and activity of the cytochrome P-450 isoenzyme CYP24. Quercetin and genistein work together to inhibit the growth of ovarian carcinoma cells by altering various cell cycle stages and signal transduction pathways, which is somewhat similar to earlier research. Yeh et al. determined that genistein is an effective isoflavonoid that sequentially triggers apoptotic signaling in Hep3B cells. The process results in endoplasmic reticulum stress, which is identified by increased calcium mobilization, m-calpain cleavage, up-regulation of GRP78 and GADD153 expression, and caspase-12 activation. Additionally, Genistein induces caspase-3 and caspase-7 executor activation. Furthermore, the interaction with mitochondrial stress to generate truncated Bad and down-regulate the level of Mcl-1 may help Hep3B cells undergo genistein-mediated apoptosis. These findings suggest that genistein and daidzein may be important in lowering the risk of cancer. Epidemiologic studies have shown that populations with high intakes of soy and isoflavones have lower cancer rates. Interest in these substances has increased further with the discovery of a novel estrogen receptor b (ERb), which is primarily found in the brain, heart, bones, and urogenital system and binds phytoestrogens with a relatively high affinity.

6.11. FIBER

Roughage, also known as dietary fiber, is the part of food derived from plants that human digestive enzymes are unable to fully break down. The chemical makeup of dietary fibers varies, but they can be broadly categorized based on factors like solubility, viscosity, and fermentability that influence the body's processing of the fibers. Soluble fiber and insoluble fiber, which are found in plant-based foods like legumes, whole grains, cereals, fruits, vegetables, nuts, and seeds, make up the two main types of dietary fiber. Regularly consuming a high-fiber diet is generally linked to improving health and reducing the risk of a number of diseases. Dietary fiber is made up of plant components like cellulose, resistant starch, resistant dextrins, inulin, lignins, chitins, pectins, beta-glucans, and oligosaccharides, as well as non-starch polysaccharides.

Dietary fiber sources have historically been classified as either soluble or insoluble fiber-containing foods. Both kinds of fiber are present in plant foods in different proportions depending on the viscosity and fermentability of the fiber. The benefits of eating fiber vary depending on the kind of fiber taken and the potential effects on the gastrointestinal tract (GIT). Bulking fibers, like cellulose and hemicellulose (which includes psyllium), retain and absorb water, which helps to maintain regular bowel movements. Beta-glucan and psyllium are examples of viscous fibers that thicken the fecal mass. Fermentable fibers, like inulin, xanthan gum, and resistant starch, nourish the microbiota and bacteria of the large intestine and are broken down to produce short-chain fatty acids, which play a variety of functions in gastrointestinal health.

The water-soluble fiber known as soluble fiber (also called fermentable fiber or prebiotic fiber) is typically fermented in the colon to produce gases and physiologically active byproducts like short-chain fatty acids, which are made there by gut bacteria. Two such are raw guar gum and beta-glucans, which are found in oats, barley, and mushrooms. Psyllium is a bulking fiber that is soluble, viscous, and non-fermented. It helps with defecation by holding onto water while passing through the digestive system. Since soluble fiber often has a viscous consistency and slows down stomach emptying, it can give people a prolonged feeling of fullness. Soluble non-viscous fibers include inulin (found in chicory root), wheat dextrin, oligosaccharides, and resistant starches (found in legumes and bananas). Consuming soluble fibers on a regular basis, such as beta-glucans from barley or oats, has been shown to reduce LDL cholesterol levels in the blood. Supplements containing soluble fiber also dramatically reduce LDL cholesterol.

Upper GIT digesting enzymes are inactive against insoluble fiber, which is a material that does not dissolve in water. Lignin, cellulose, and wheat bran are a few examples. Insoluble fiber that has been coarsely ground causes the large intestine to secrete mucus, which adds bulk. In fact, finely ground insoluble fiber can have the opposite effect, making constipation more likely. Resistant starches are one type of insoluble fiber that can ferment in the colon.

6.11.1. Structure

Dietary fibers are categorized as either insoluble or soluble. Recently, there has been a proposal to replace the terms soluble and insoluble when describing the benefits of dietary fiber for health by using the terms viscous and fermentability.

6.11.2. Function

Consuming dietary and functional fibers offers numerous potential health benefits, including the potential to lower cholesterol, lessen the incidence of coronary and cardiovascular heart disease, inhibit obesity and diabetes, and lessen the frequency of constipation and irritable bowel syndrome (IBS).

6.11.3. Anti-Cancer Activity

Relatively recent research revealed a negative correlation between dietary fiber and the emergence of several cancers, including those of the breast, small intestine, oral, larynx, and colon. In an investigation by Rafter et al. demonstrated that inulin decreased the biological components linked to colonic cancer, including the release of interleukin-2, genotoxic exposure, colorectal cell proliferation, and water-induced necrosis. Both Asian and Western cultures have found that a high-fiber diet stops prostate cancer from progressing in its early stages.

6.12. OMEGA 3 FATTY ACIDS

The chemical structure of omega-3 fatty acids, which are polyunsaturated fatty acids (PUFAs), is distinguished by the presence of a double bond located three atoms away from the terminal methyl group. They are found in many parts of nature, are crucial to animal lipid metabolism, and have a significant impact on both human physiology and diet. The human physiology involves three types of omega-3 fatty acids: docosahexaenoic acid (DHA), eicosapentaenoic acid (EPA), and α-linolenic acid (ALA). Fish and algae contain DHA and EPA, whereas plants contain ALA. Protozoa and marine algae are the main providers of omega-3 fatty acids. The fish that consume these algae accumulate DHA and EPA. Algae oil and fish oils are common sources of DHA and EPA, while walnuts, edible seeds, flaxseeds, and hempseed oil are common sources of plant oils containing ALA.

6.12.1. Structure

Linolenic acid is the source of omega-3 fatty acids. The number that comes after "omega-" indicates the location of the molecule's first double bond, counting backward from the terminal methyl group. ALA, the most common type of omega-3 fatty acid, EPA, and DHA are the three main types of omega-3 fatty acids.

6.12.2. Function

The two types of omega-3 fatty acids present in high-oil fatty fish are DHA and EPA. Additionally, certain leafy vegetables, nuts, and oils contain omega-3 fatty acids in the form of a-linolenic acid. As demonstrated by the diets of today's Western populations, high levels of omega-6 PUFAs and a very high omega-6/omega-3 ratio promote the pathogenesis of numerous diseases, including cancer, heart disease, autoimmune diseases, and inflammation. Conversely, high levels of omega-3 PUFAs, which have a lower omega-6/omega-3 ratio, have inhibitory effects. According to a different study, taking omega-3 dietary supplements had the opposite effect on rheumatoid arthritis, psoriasis, inflammatory bowel disease, and cancer.

6.12.3. Anti-Cancer Activity

Some studies have attempted to assess the effects of omega-3 fatty acids on cancer treatment through omega-3 supplementation to the diet, either as omega-3 fatty acid-rich foods or as dietary supplements. Epidemiological studies suggest that individuals who consume diets high in omega-3 fatty acids may have a lower prevalence of some types of cancer. Gerber examined prospective and case-control studies assessing the potential preventive effects of omega-3 fatty acid consumption on the development of cancer. Fish supplementation had a reverse association with cell carcinoma of the oral cavity or pharynx, esophagus, or larynx, according to a pertinent study conducted on the Hawaiian island of Oahu. Chyou and team demonstrated that eating fish had no appreciable effect on bladder cancer.

6.13. FLAVONOIDS

Human diets typically include flavonoids, a class of secondary metabolites that are found in plants and are polyphenolic.

Chemically speaking, flavonoids are made up of a heterocyclic ring (C), which contains an embedded oxygen ring, and two phenyl rings (A and B). This makes up the general structure of a 15-carbon skeleton. The carbon structure is represented by the acronym C_6-C_3-C_6.

According to the IUPAC nomenclature, they can be classified into:

- flavonoids or bioflavonoids
- isoflavonoids, derived from 3-phenylchromen-4-one (3-phenyl-1,4-benzopyrone) structure
- neoflavonoids, derived from 4-phenylcoumarin (4-phenyl-1,2-benzopyrone) structure

All three of the aforementioned flavonoid classes are anthoxanthins (flavones and flavonols), which are compounds that contain ketone bodies. This group was initially referred to as bioflavonoids. More precisely, non-ketone polyhydroxy polyphenol compounds—also referred to as flavanoids—have also been more loosely categorized as flavonoids and bioflavonoids. Generally speaking, the flavonoid backbone's three cycles, or heterocycles, are referred to as rings A, B, and C. A phloroglucinol substitution pattern is typically seen in Ring A.

6.13.1. Structure

Plants produce a broad family of polyphenolic compounds called flavonoids. They can be divided into numerous subclasses, each of which is further subdivided into distinct compounds: isoflavones, chalcones, flavonols, flavones, flavanones, and flavonols.

6.13.2. Function

As was previously mentioned, plants synthesize a large family of polyphenolics called flavonoids. The body benefits from flavonoids in a number of ways, including antioxidant, antiviral, antitoxic, antifungal, antibacterial, anti-allergic, and anti-inflammatory properties. According to a few additional studies, flavonoids can help prevent heart diseases.

6.13.3. Anti-Cancer Activity

Studies have demonstrated that flavonoids are beneficial for the treatment and/or prevention of a wide range of cancers, including pancreatic, esophageal, ovarian colon, lung, prostate, hepatocellular, leukemia, and renal cell carcinoma. These studies suggest that phenolic flavonoids, like epigallocatechin-3-gallate and green tea polyphenols, function as anticancer agents by stimulating the transcription system. Additional evidence indicated that the mechanisms underlying the flavonoid-induced anticancer effects include the regulation of cell cycle progression, inhibition of kinase and protease activities, termination of matrix metalloproteinase secretion, and prevention of activator protein-1 function induction. Plant extracts have the ability to inhibit the production of inflammatory mediators, including leukotrienes, cytokines, and PGs mediated by COX-2. Furthermore, it is known that some plant-based products cause cancerous cells to undergo apoptosis. To support this assertion, Hostanska et al. demonstrated that, regardless of the COX-selectivity of the bark extract BNO 1455, its fractions stimulate apoptosis and inhibit cell growth in human colon and lung cancer cell lines. Additionally, in a number of cancer cell lines, apigenin—an isoconformer of genistein—has demonstrated more effective growth prevention.

It has been shown that apigenin has anti-inflammatory, free radical-scavenging, and anti-carcinogenic properties. It has been documented to have growth-inhibiting properties in a number of cancer cell lines, including leukemia, breast, colon, skin, and thyroid. This affiliation is that of Ujiki et al. demonstrated that apigenin inhibits the growth of pancreatic cancer cells by suppressing G2/M arrest and cyclin B-associated Cdc2 activity. Apigenin may therefore be a useful medication for the treatment or inhibition of pancreatic cancer. Rossi and associates. Consuming flavanones has been shown to have a negative correlation with the risk of esophageal cancer and may be responsible, along with vitamin C, for fruit's protective effect against the disease, particularly citrus fruit. Based on these relationships, Fink et al. showed that among Long Island women, consumption of flavonols, flavones, flavan-3-ols, and lignans is associated with a decreased risk of incident postmenopausal breast cancer.

Numerous significant biological functions of flavonoids may be linked to an increased risk of cancer. They have been shown to affect signal transduction pathways, trigger apoptosis, and inhibit inflammation and proliferation in human cancer cell lines using *in vitro* and animal model systems. Certain flavonoids may also increase the transcription of phase II detoxifying enzymes, which aid in the removal of compounds known to cause cancer. Because isoflavones have weak estrogenic activity, inhibit angiogenesis and tyrosine protein kinases, and lower serum testosterone levels, they protect against prostate cancer. Moreover, Numerous significant biological functions of flavonoids may be linked to an increased risk of cancer. They have been shown to affect signal transduction pathways, trigger apoptosis, and inhibit these reasons. Furthermore, studies conducted on rats revealed that while a soy diet is beneficial, it has no effect on invasive prostate cancer. Instead, it is beneficial in the early stages of tumor development. Regarding isoflavonoids and colorectal cancer, Theodoratou et al. indicated a clear, linear relationship between the risk of colorectal cancer and the consumption of flavonoids. Nonetheless, a few substances (folate and fiber) present in plant-based diets offer protection against colorectal cancer. Additional research revealed a 40% decrease in the risk of colorectal cancer in those who ate fiber and a 30% reduction in those who ate folate.

A call to action was issued by Farmer et al. regarding the expansion of cancer care and control in low-and middle-income countries. To effectively prevent and control cancer in low-and middle-income countries, this policy paper highlights the need for implementation science research in these areas. In a different study, Hunter and Reddy stress the importance of international research initiatives to support the diagnosis, prevention, and treatment of non-communicable diseases. They also delineate the fundamental components of successful cancer research and suggest areas of focus for implementation science, which will direct the fight against cancer. Summarize the review's recommendations, including research priorities, from representatives of institutions and organizations that fund and conduct cancer research to address the disease's growing global challenge. The WHO action offers recommendations on public health priorities and, consequently, important implementation science matters concerning the prevention and control of cancer (George and Chandran, 2021).

REFERENCES

1. Abrams, J. S., Mooney, M. M., Zwiebel, J. A., Korn, E. L., Friedman, S. H., Finnigan, S. R., Schettino, P. R., Denicoff, A. M., Kruhm, M. G., Montello, M., Misra, R. R., Ansher, S. S., DiPiazza, K. J., Souhan, E. M., Wickerham, D. L., Giantonio, B. J., O'Donnell, R. T., Sullivan, D. M., Soto, N. I., Fleming, G. F., Prindiville, S. A., Petryshyn, R. A., Hautala, J. A., Grad, O., Zuckerman, B. L., Meyer, R. M., Yao, J. C., Baker, L. A., Buckner, J. C., Hortobagyi, G. N., & Doroshow, J. H. (2013). Implementation of timeline reforms speeds initiation of national cancer institute-sponsored trials. *Journal of the National Cancer Institute, 105*(12), 954–959.
2. Ames, B. N., & Shigenaga, M. K. (1992). Oxidants are a major contributor to aging. *Annals of the New York Academy of Sciences, 663*, 85–93.
3. Ames, B. N., Shigenaga, M. K., & Hagen, T. M. (1993). Oxidants, antioxidants, and the degenerative diseases of aging. *Proceedings of the National Academy of Sciences of the United States of America, 90*(18), 7915–7932.
4. Arbyn, M., Ronco, G., Anttila, A., Meijer, C. J., Poljak, M., Ogilvie, G., Koliopoulos, G., Naucler, P., Sankaranarayanan, R., & Peto, J. (2012). Evidence regarding human papillomavirus testing in secondary prevention of cervical cancer. *Vaccine, 30*(Suppl 5), F88–F99.
5. Bidoli, E., Talamini, R., Bosetti, C., Negri, E., Maruzzi, D., Montella, M., Franceschi, S., & La Vecchia, C. (2005). Macronutrients, fatty acids, cholesterol, and prostate cancer risk. *Annals of Oncology, 16*(1), 152–157.
6. Bragg, F. L., Smith, M., Guo, Y., Chen, Y., Millwood, I., Bian, Z., Walters, R., Chen, J., Yang, L., Collins, R., Peto, R., Lu, Y., Yu, B., Xie, X., Lei, Y., Luo, G., Chen, Z., & China Kadoorie Biobank Collaborative Group. (2014). Associations of blood glucose and prevalent diabetes with risk of cardiovascular disease in 500,000 adult Chinese: The China Kadoorie Biobank. *Diabetic Medicine, 31*(5), 540–551.
7. Center for Science in the Public Interest (CSPI) Reports. (1998). Public health boon or 21st-century quackery? *International Functional Foods, Center for Science in the Public Interest*. Available online: http://www.cspinet.org/reports/functional_foods/introduction.html.
8. Chen, Y., Li, L., Zhang, Q., Clarke, R., Chen, J., Guo, Y., Bian, Z., Pan, X., Peto, R., Tao, R., Shi, K., Collins, R., Ma, L., Sun, H., Chen, Z., & China Kadoorie Biobank Study. (2014). Use of drug treatment for secondary prevention of cardiovascular diseases in urban and rural communities of China: China Kadoorie Biobank Study of 0.5 million people. *International Journal of Cardiology, 172*(1), 88–95.

9. Chen, Z., Chen, J., Collins, R., Guo, Y., Peto, R., Wu, F., Li, L., & China Kadoorie Biobank (CKB) Collaborative Group. (2011). China Kadoorie Biobank of 0.5 million people: Survey methods, baseline characteristics, and long-term follow-up. *International Journal of Epidemiology, 40*(5), 1652–1666.

10. Chen, Z., Yang, G., Offer, A., Zhou, M., Smith, M., Peto, R., Ge, H., Yang, L., & Whitlock, G. (2012). Body mass and mortality in China: A 15-year prospective study of 220,000 men. *International Journal of Epidemiology, 41*(2), 472–481.

11. Chew, B., Park, J., Wong, M., & Wong, T. A. (1999). Comparison of the anticancer activities of dietary β-carotene, canthaxanthin, and astaxanthin in mice *in vivo*. *Anticancer Research, 19*(3B), 1849–1853.

12. Chiller, J. T., & Lowy, D. R. (2014). Virus infection and human cancer: An overview. *Recent Results in Cancer Research, 193*, 1–10.

13. Conaway, C. C., Yang, Y. M., & Chung, F. L. (2002). Isothiocyanates as cancer chemopreventive agents: Their biological activities and metabolism in rodents and humans. *Current Drug Metabolism, 3*(3), 233–255.

14. De Martel, C., Ferlay, J., Franceschi, S., Vignat, J., & Bray, F. (2012). Global burden of cancers attributable to infections in 2008: A review and synthetic analysis. *Lancet Oncology, 13*(6), 607–615.

15. Dikshit, R., Gupta, P. C., Ramsundarahettige, C., Gajalakshmi, V., Aleksandrowicz, L., Badwe, R., Kumar, R., Roy, S., Suraweera, W., Bray, F., Mallath, M., Singh, P. K., Sinha, D. N., Shet, A. S., Gelband, H., & Jha, P. Million Death Study Collaborators. (2012). Cancer mortality in India: A nationally representative survey. *Lancet, 379*(9828), 1807–1816.

16. Du, H., Bennett, D., Li, L., Whitlock, G., & Guo, Y. (2013). Physical activity and sedentary leisure time and their associations with BMI, waist circumference, and percentage body fat in 0.5 million adults: The China Kadoorie Biobank Study. *American Journal of Clinical Nutrition, 97*(3), 487–496.

17. Food and Agriculture Organization of the United Nations (FAO). (2007). *Report on Functional Foods, Food Quality and Standards Service (AGNS)*. Available online: http://www.fao.org/ag/agn/agns/files/Functional_Foods_Report_Nov2007.pdf.

18. Gajalakshmi, V., Whitlock, G., & Peto, R. (2012). Social inequalities, tobacco chewing, and cancer mortality in South India: A case-control analysis of 2,580 cancer deaths among non-smoking non-drinkers. *Cancer Causes & Control, 23*(Suppl 1), 91–98.

19. Galagan, S. R., Paul, P., Menezes, L., & LaMontagne, D. S. (2013).

Influences on parental acceptance of HPV vaccination in demonstration projects in Uganda and Vietnam. *Vaccine, 31*(21), 3072–3078.

20. George, B. P., Chandran, R., & Abrahamse, H. (2021). Role of phytochemicals in cancer chemoprevention: Insights. *Antioxidants, 10*(9), 1455. https://doi.org/10.3390/antiox10091455.
21. Hecht, S. S., Kelloff, G. J., Hawk, E. T., & Sigman, C. C. (2004). *Promising Cancer Chemopreventive Agents, Volume 1: Cancer Chemopreventive Agents*. Humana Press.
22. Hyang-Sook, K., Bowen, P., Longwen, C., Duncan, C., & Ghosh, L. (2003). Effects of tomato sauce consumption on apoptotic cell death in prostate benign hyperplasia and carcinoma. *Nutrition and Cancer, 47*(1), 40–47.
23. Jyonouchi, H., Sun, S., Iijima, K., & Gross, M. (2000). Antitumor activity of astaxanthin and its mode of action. *Nutrition and Cancer, 36*(1), 59–65.
24. Karppi, J., Rissanen, T. H., Nyyssönen, K., Olsson, A. G., Voutilainen, S., & Salonen, J. T. (2007). Effects of astaxanthin supplementation on lipid peroxidation. *International Journal for Vitamin and Nutrition Research, 77*(1), 3–11.
25. Kim, J. H., Kim, Y. S., Song, G. G., Park, J. J., & Chang, H. I. (2005). Protective effect of astaxanthin on naproxen-induced gastric antral ulceration in rats. *European Journal of Pharmacology, 514*(1), 53–59.
26. Kumar, M., Kumar, A., Nagpal, R., Mohania, D., Behare, P., Verma, V., Kumar, P., Poddar, D., Aggarwal, P. K., Henry, C. J., Jain, S., & Yadav, H. (2010). Cancer-preventing attributes of probiotics: An update. *International Journal of Food Science & Nutrition, 61*(5), 473–496.
27. Kuno, T., Tsukamoto, T., & Hara, A. (2012). Cancer chemoprevention through the induction of apoptosis by natural compounds. *Biophysical Chemistry, 3*(2), 156–173.
28. Kurahashi, N., Iwasaki, M., Sasazuki, S., Otani, T., Inoue, M., & Tsugane, S. Japan Public Health Center-Based Prospective Study Group. (2007). Soy product and isoflavones consumption in relation to prostate cancer in Japanese men. *Cancer Epidemiology, Biomarkers & Prevention, 16*(3), 538–545.
29. Kurihara, H., Koda, H., Asami, S., Kiso, Y., & Tanaka, T. (2002). Contribution of the antioxidative property of astaxanthin to its protective effect on the promotion of cancer metastasis in mice treated with restraint stress. *Life Sciences, 70*(22), 2509–2519.
30. Lee, B. M., & Park, K. K. (2003). Beneficial and adverse effects of chemopreventive agents. *Mutation Research, 23*(6), 265–278.

31. Levy, J., Bosin, E., Feldman, B., Giat, Y., Miinster, A., Danilenko, M., & Sharoni, Y. (1995). Lycopene is a more potent inhibitor of human cancer cell proliferation than either α or ß-carotene. *Nutrition and Cancer, 24*(3), 257–266.

32. Limer, J. L., & Spiers, V. (2004). Phyto-estrogens and breast cancer chemoprevention. *Breast Cancer Research, 6*(2), 119–127.

33. Lockwood, S. F., & Gross, G. J. (2005). Disodium disuccinate astaxanthin (Cardax): Antioxidant and anti-inflammatory cardio protection. *Cardiovascular Drug Reviews, 23*(3), 199–216.

34. Naguib, Y. M. (2000). Antioxidant activities of astaxanthin and related carotenoids. *Journal of Agricultural and Food Chemistry, 48*(4), 1150–1154.

35. Nakagawa, K., Kiko, T., Miyazawa, T., Carpentero Burdeos, G., Kimura, F., Satoh, A., & Miyazawa, T. (2011). Antioxidant effect of astaxanthin on phospholipids peroxidation in human erythrocytes. *British Journal of Nutrition, 105*(10), 1563–1571.

36. Nishikawa, Y., Minenaka, Y., Ichimura, M., Tatsumi, K., Nadamoto, T., & Urabe, K. (2005). Effects of astaxanthin and vitamin C on the prevention of gastric ulcerations in stressed rats. *Journal of Nutritional Science and Vitaminology (Tokyo), 51*(2), 135–141.

37. Okada, F. (2002). Inflammation and free radicals in tumor development and progression. *Redox Report, 7*(6), 357–366.

38. Pashkow, F. J., Watumull, D. G., & Campbell, C. L. (2008). Astaxanthin: A novel potential treatment for oxidative stress and inflammation in cardiovascular disease. *American Journal of Cardiology, 101*(10A), 58D–68D.

39. Plochmann, K., Korte, G., Koutsilieri, E., Richling, E., Riederer, P., Rethwilm, A., Schreier, P., & Scheller, C. (2007). Structure-activity relationships of flavonoid-induced cytotoxicity on human leukemia cells. *Archives of Biochemistry and Biophysics, 460*(1), 1–9.

40. Rutovskikh, V., Asamoto, M., Takasuka, N., Murakoshi, M., Nishino, H., & Tsuda, H. (1997). Differential dose-dependent effects of alpha-, beta-carotenes, and lycopene on gap-junctional intercellular communication in rat liver *in vivo*. *Japanese Journal of Cancer Research, 88*(11), 1121–1124.

41. Shannon, J., King, I. B., Moshofsky, R., Lampe, J. W., Gao, D. L., Ray, R. M., & Thomas, D. B. (2007). Erythrocyte fatty acids and breast cancer risk: A case-control study in Shanghai, China. *American Journal of Clinical Nutrition, 85*(4), 1090–1097.

42. Shin, J. A., Kim, J. S., Hong, I. S., & Cho, S. D. (2012). Bak is a key molecule in apoptosis induced by methanol extracts of *Codonopsis lanceolata* and *Tricholoma matsutake* in HSC-2 human oral cancer cells. *Oncology Letters, 4*(6), 1379–1383.

43. Sporn, M. B., & Suh, N. (2002). Chemoprevention: An essential approach to controlling cancer. *Nature Reviews Cancer, 2*(7), 537–543.

44. Tanaka, T., Makita, H., Ohnishi, M., Mori, H., Satoh, K., & Hara, A. (1995). Chemoprevention of rat oral carcinogenesis by naturally occurring xanthophylls, astaxanthin and canthaxanthin. *Cancer Research, 55*(18), 4059–4064.

45. Tanaka, T., Shimizu, M., & Moriwaki, H. (2012). Cancer chemoprevention by carotenoids. *Molecules, 17*(3), 3202–3242.

46. Tantamango-Bartley, Y., Jaceldo-Siegl, K., Fan, J., & Fraser, G. (2013). Vegetarian diets and the incidence of cancer in a low-risk population. *Cancer Epidemiology, Biomarkers & Prevention, 22*(2), 286–294.

47. Vel Szic, K. S., Palagani, A., & Hassannia, B. (2011). Phytochemicals and cancer chemoprevention: Epigenetic friends or foes? In I. Rasooli (Ed.), *Phytochemicals—Bioactivities and Impact on Health* (pp. 3–25). InTech.

48. Wakai, K., Date, C., Fukui, M., Tamakoshi, K., Watanabe, Y., Hayakawa, N., Kojima, M., Kawado, M., Suzuki, K. M., Hashimoto, S., Tokudome, S., Ozasa, K., Suzuki, S., Toyoshima, H., Ito, Y., & Tamakoshi, A. (2007). Dietary fiber and risk of colorectal cancer in the Japan Collaborative Cohort Study. *Cancer Epidemiology, Biomarkers & Prevention, 16*(4), 668–675.

49. Wu, Q. J., Yang, Y., Vogtmann, E., Wang, J., Han, L. H., Li, H. L., & Xiang, Y. B. (2013). Cruciferous vegetables intake and the risk of colorectal cancer: A meta-analysis of observational studies. *Annals of Oncology, 24*(4), 1079–1087.

50. Yaping, Z., Wenlia, Y., Weile, K., & Ying, Y. (2003). Antiinflammatory and anticoagulant activities of lycopene in mice. *Nutrition Research, 23*(10), 1591–1595.

51. Yoshida, H., Yanai, H., Ito, K., Tomono, Y., Koikeda, T., Tsukahara, H., & Tada, N. (2010). Administration of natural astaxanthin increases serum HDL-cholesterol and adiponectin in subjects with mild hyperlipidemia. *Atherosclerosis, 209*(2), 520–523.

52. Yuan, J. P., Peng, J., Yin, K., & Wang, J. H. (2011). Potential health-promoting effects of astaxanthin: A high value carotenoid mostly from microalgae. *Molecular Nutrition & Food Research, 55*(2), 150–165

CHAPTER 7

TARGETED APPLICATION OF FUNCTIONAL FOODS AS IMMUNE FITNESS BOOSTERS IN THE DEFENSE AGAINST VIRAL INFECTION

Contents

INTRODUCTION

Consumer preferences have significantly shifted in favor of foods that strengthen immune systems in recent years as a result of growing awareness of the effects of diet on health. A good, disease-free life is greatly influenced by what one eats. Food is now more important than ever for longevity, with global dietary risk factors estimated to cause 11 million deaths and 255 million disability-adjusted life years annually. It is well known that our bodies are influenced by the food we eat, and that certain nutrients are necessary for immune system support, tissue development, cellular activity, growth, and energy production. There is mounting evidence that food can function as an immunomodulator, and specific foods and nutrients have been shown to enhance immune tolerance while boosting infection resistance and immune defense. In addition, a lack of some nutrients causes malnutrition, which in turn causes diseases. These illnesses are frequently brought on by inadequate nutrition, which compromises immune system performance. A poor diet can weaken the immune system, making the host more vulnerable to viral and other pathogenic infections. Increasing immune function is crucial in light of the recent viral outbreaks caused by the SARS-CoV-2 virus, the monkeypox virus, and, most recently, the Langya virus. Viruses can spread quickly from person to person and cause devastating harm to humanity because of their high transmissibility, as demonstrated by the COVID-19 pandemic. Thus, it is imperative to investigate novel approaches for augmenting viral immunity. Since food has long been known to offer antiviral protection, it can be used as a first line of defense to strengthen the immune system. Functional foods, for example, can improve defense against both well-known viruses like the influenza virus and newly emerging viral infections.

When combined with good personal hygiene and an active lifestyle, functional foods could offer extra natural antiviral support. In this section, we address the health and wellness trends of the present day in light of the growing popularity of functional foods, particularly in Europe. We introduce the idea of immune fitness enhancement, concentrating on those who are susceptible to viral infections. We also go over how functional foods and nutrients from the diet can strengthen immunity in people who are at a higher risk of illness, as well as how diet-based approaches can play a significant role in enhancing immune function in general. Lastly, we discuss potential directions for the future development of functional foods, particularly fermented foods, to improve immune fitness.

7.1. BOOSTING IMMUNE FITNESS IN HIGH-RISK INDIVIDUALS

The public's awareness of health and wellness has shifted in recent years, and many people are changing their diets significantly to boost their immunity and general well-being. The use of functional foods is growing globally as a result of increased health consciousness.

Here, we emphasize how critical it is to strengthen high-risk individuals' immune systems. After introducing the idea of immune fitness, we go over the salient characteristics of the innate and adaptive immune systems of three extremely susceptible cohorts, emphasizing gut immunity. Comprehending the alterations in immune responses of these groups is essential for advancing the creation of functional foods aimed at enhancing immunity against viruses.

7.1.1. Immune Fitness

The term "immune fitness" refers to the body's ability to react to health issues, like infection, by triggering the proper immune response. This helps to ensure quality of life by promoting health, preventing pathogen infection, and speeding up the resolution of diseases. A robust immune system with the "built-in" ability to adjust to obstacles by mounting, sustaining, and controlling a suitable immune response is referred to as immune fitness.

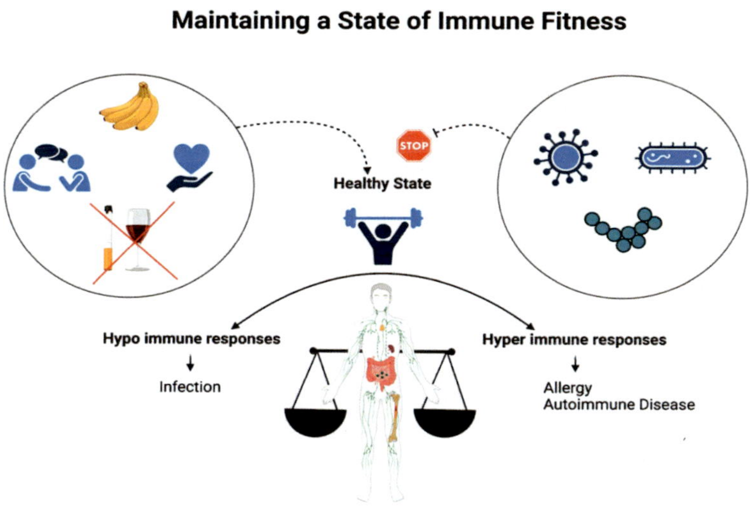

Figure 7.1. *Maintaining immune fitness.*

Source: Finnegan D, Tocmo R, Loscher C. (2023). Targeted application of functional foods as immune fitness boosters in the defense against viral infection. Nutrients; 15(15):3371. https://doi.org/10.3390/nu15153371.

This indicates that the person's immune system is strong enough to get rid of dangerous pathogens like bacteria and viruses while still being able to withstand benign ones like food antigens. By doing this, the body is kept from going into a hyperresponsive state that can cause allergies and autoimmune diseases, or a hyporesponsive state that can weaken immunity and increase infection. The development of fundamental lifestyle practices that can strengthen your immune system—such as healthy eating, social interactions with others, quitting smoking, consuming less alcohol, and managing stress—can be seen as immune fitness (Figure 7.1). These practices can also help to slow down the aging process of the immune system (Franceschi et al., 2018).

The immune system alterations with age lead to the development of "inflammaging," or persistent low-grade inflammation, which subsequently plays a role in the etiology of age-related diseases. The immune system reacts more slowly and ineffectively, which raises the risk of infection because immune defenses are weaker. Immunosenescence is the term for this progressive immune system deterioration brought on by advanced age. Because the immune system's proper operation and nutrition are closely related, our diets have a significant impact on our immune response and determine our propensity to contract infections or inflammatory diseases. Long recognized for its ability to provide fiber with prebiotic activity, polyunsaturated fatty acids (PUFAs) with anti-inflammatory properties, bioactive compounds with antioxidative properties, vitamins, and minerals, the plant-based Mediterranean diet, which primarily consists of cereals, legumes, vegetables, fruits, olive oil, and nuts, also helps to modulate the microbiota and activate the host immunity, ultimately promoting health and preventing disease. Conversely, the Western diet is known for being low in fiber and high in fat, digestible carbohydrates, and animal protein. Observational studies have connected the Western diet to an increase in autoimmune diseases as well as the risk and development of immune system disorders like asthma and allergies, as well as inflammatory bowel diseases (IBDs) like Crohn's disease (CD) and ulcerative colitis (UC). Moreover, the obesity pandemic around the world, which results in low-grade immune system activation, is largely attributable to the Western diet. Thus, the variations in the impacts of the Western and Mediterranean diets show that the immune system and overall immune fitness are impacted differently by different dietary patterns. People with compromised immune systems are referred to as immunocompromised. These individuals are less able to fight off infections and other illnesses. Immunocompromised individuals can be caused by a variety of conditions, including cancer, diabetes, AIDS, certain genetic disorders, and even simple malnourishment. Furthermore, immunosuppression is frequently a side effect of radiation therapy and stem cell therapy, two common treatments for diseases like cancer. Since the immunocompromised have a weakened immune system and are therefore more susceptible to immunological challenges, it is necessary to increase their immune fitness.

We investigate how functional foods affect the immune fitness of three key vulnerable groups in our society: the immunocompromised, the obese, and the elderly. Fighting viruses like COVID-19 presents difficulties for the elderly, the obese, and people with long-term underlying medical conditions because these groups already have compromised immune systems. Older people and those with underlying medical conditions are more vulnerable to severe infections because their immune systems are already compromised. The immune systems of the elderly and obese are prone to low-grade inflammation, which leaves them open to infection. While aging is an inevitable part of life, there are other things one can do to maintain optimal physical fitness, manage weight, and strengthen the immune system to fend off viral invasion. Immune fitness is the sum total of an individual's immune system's morbidity from chronic immune disorders, resilience, and fragility. Numerous factors, including biological ones like the microbiome and epigenome, lifestyle ones like food, exercise, and sleep patterns, and other psychosocial ones like stress management and social surroundings, can all have an impact on immune fitness. Thus, we must search for natural means of enhancing immune fitness in the elderly, obese, and immunocompromised to guarantee that, as a result of upholding a healthy lifestyle, the immune system is in optimal condition for combating viral infection.

7.1.1.1. Immune Response in the Elderly

It has been demonstrated that, even in pandemic epicenters, individuals under 65 years of age have a lower risk of dying from COVID-19 infection, whereas the elderly are considered to be among the most vulnerable groups of people when it comes to viral infections, including COVID-19 infection. The elderly are known to have lower vaccination efficacy in addition to being more susceptible to infectious diseases. Due to the accumulation of damage over time at a molecular, cellular, and organ-based level, there is an increased dysregulation of the immune system as we age. This leads to a higher risk of disease and higher rates of morbidity and mortality. Immunosenescence is the term used to describe how much aging affects the immune system. A person's immune fitness can be maximized to slow down the rate of immunosenescence.

The innate immune system is greatly impacted by aging. The ability of the GIT's mucosal immune system to mount a quick, protective immune response against dangerous pathogens or to distinguish between harmless antigens and develop tolerance to them—which serves as the first line of defense against ingested pathogens—becomes severely impaired in the elderly. Because of the gut-associated lymphoid tissue (GALT), which is made up of mesenteric lymph nodes, lymphoid follicles, and Peyer's patches, the intestinal mucosa provides innate immunity. Reduced secretion of mucus and defensins in aged

GALT has been reported; these findings suggest compromised gut function. As people age, their innate cells in the gut become more numerous and have higher levels of pro-inflammatory mediators, such as C-reactive protein and pro-inflammatory cytokines (IL-6, TNF-α, IFN-γ, and IL-1β), which are linked to both lower cognitive function and low-grade inflammation that is a common aspect of aging. The ability of aged dendritic cells (DCs) to present antigens that activate cells of the adaptive immune system is compromised by marked reductions in IL-12p70 and IL-15 production as well as decreased expression of co-stimulatory factors CD80/CD86. Moreover, there is a connection between older people's increased intestinal barrier permeability and lower expression of the cytokinesIL-22 and IL-17. Furthermore, elevated levels of TNF-α and IL-6 are linked to metabolic endotoxemia and increased microbiome paracellular permeability, both of which exacerbate the low-grade chronic inflammatory state that is associated with aging. An increase in pro-inflammatory commensals at the expense of beneficial microbes and a decrease in microbiota diversity are two common changes in the gut microbiome that are linked to aging. An increase in leaky gut, compromised gastrointestinal tract (GIT) integrity and function, and changes in microbiota-associated metabolite levels are all consequences of this type of dysbiosis. Additionally, as we age, changes in the gut microbiota can be caused directly or indirectly by reductions in chewing ability, dentition, taste, digestion, and intestinal transit time. Consequently, the gut's integrity and functionality deteriorate with age, increasing the gut's permeability as a result of its diminished integrity. As a result, the elderly are more vulnerable to viral infection because their first line of defense is weakened.

Aging has a significant impact on the adaptive arm of the immune system. Some notable alterations include reduced T cell function, the central defect of immunosenescence, and a decrease in T cell population production. Aging is linked to a decrease in T cell proliferation, a reduction in IL-2 production, a decrease in T cell receptor expression, and changes to signaling pathways that activate T cells, such as the NF-κB pathway. An older person's T cell function is already compromised, so getting COVID-19 and the resulting T cell exhaustion (Tex) increases the likelihood of a bad outcome for these people. In a similar vein, aging impacts B cell function, rendering the elderly less capable of producing optimal antibody responses. Reduced binding, reduced class switch recombination—which produces a secondary response of class-switched antibodies—reduced somatic hypermutation of the antibody variable region, decreased neutralization ability and binding specificity of secreted antibodies, increased frequencies of inflammatory B cell subsets and shorter telomers, and increased epigenetic modifications linked to decreased antibody responses are some of these defects. Furthermore, an aging immune system has a clear impact on B cell function. While the percentage of terminally differentiated and senescent memory CD27-B cells rises with age, the number of competent B cells dramatically declines. Because of this, older people have lower antibody levels, which increases their risk of catching viral infections.

Aging causes the gut's adaptive immune response to alter in a number of ways. Cellularly, antigen-specific T cell responses in the gut are decreased, but DCs' capacity to start T cell responses is hampered by poor priming by mucosal tissue DCs. The elderly have significantly lower levels of expression of the co-inhibitory molecules CTLA-4 and PD-1 in lamina propria $CD4^+$ T cells, which regulate homeostasis and antigen-specific responses. Moreover, compared to their younger T subset counterparts, older Th1 and Th17 cells proliferate less, and older $CD4^+$ T cells exhibit higher rates of spontaneous death.

The predominant class of antibodies secreted by the intestinal mucosa is called immunoglobulin A (IgA), and it is essential for the control of the gut microbiota. The gut mucosa secretes a major class of antibodies called IgA, which is essential for gut immunity, intestinal homeostasis, controlling the mucosal immune response, preserving the microbiome, and stimulating the gut microbiome to support inflammation defense. Via IgA-committed B cell responses, T cells control the quantity and kind of microbiota-specific IgA production. But as we age, these turn senescent, and for that reason, it is suggested that they play a role in the declining antigen-specific IgA responses linked to aging. It is therefore believed that as we age, this decrease in IgA response plays a role in the deterioration of intestinal and gut immunity. More specifically, this drop in IgA is connected to the decline in gut immunity and advanced aging-related small intestinal CCL25 and colonic CCL28 increases. In general, older people have a distinct pro-inflammatory inclination. They suffer from persistent low-grade inflammation (LGI), which develops into chronic systemic inflammation. This chronic inflammation is closely linked to the aging population and is the root cause of numerous age-related illnesses, including stroke and cardiovascular disease (CVD), as well as autoimmune disorders like rheumatoid arthritis. An older person will therefore already have an increased inflammatory response in the event of a viral infection, such as COVID-19, and is considerably more likely to experience a more severe outcome from viral infection (Finnegan et al., 2023).

7.1.1.2. Immune Response in Obese Individuals

Clinical definitions of obesity include an excessive build-up of adipose tissue and a body mass index (BMI) of greater than 30 kg/m^2. Because of impaired ventilation in the base of the lungs, obesity is associated with a lower blood oxygen saturation level. It is distinguished by persistent low-grade systemic inflammation, a pro-inflammatory cytokine profile that is elevated in adipose tissue, and leukocyte infiltration into the adipose tissue, including macrophages. Chronic inflammation of this kind impairs insulin signaling in adipocytes, leading to insulin resistance, and accelerating the onset of metabolic diseases like type 2 diabetes, hypertension, and CVD.

It has long been established that obesity increases a person's vulnerability to viral infections, such as the swine flu and influenza A virus. Obesity-related comorbidities can worsen the prognosis for COVID-19 patients and possibly necessitate mechanical breathing. Due to their increased weight, poorer diet, and decreased exercise, obese people are more likely than normal-weight-bearing people to experience heart disease, lung disease, and diabetes, among other independent risk factors for severe COVID-19. These factors compromise their metabolic health. Adults who are obese are more likely than those of healthy weight to get viral infections, even after receiving vaccinations. For instance, following an influenza vaccination, adults who are obese have a twofold increased risk of developing influenza or influenza-like illness in comparison to those who are healthy weight. As a result, obese people are more likely to get COVID-19 and other viral infections because they already have reduced lung function, breathing difficulties, and persistent low-grade inflammation. They run a higher chance of becoming infected with a virus, which increases their susceptibility to illness, their prognosis, the severity of their illness, and the death rate.

Obese people's gut innate immunity is significantly compromised, which increases their susceptibility to viral infections. An important component of innate immunity is macrophages. They are the most prevalent immune cells that cause obesity through infiltrating adipose tissue and subsequently secreting inflammatory cytokines in reaction to obesity. They are important mediators of inflammation within adipose tissues. The M2 macrophage phenotype, which is "alternatively activated," is linked to a lean profile, whereas the M1 macrophage phenotype, which is "classically activated," is associated with an obese profile. The intestinal cytokine profile linked to diet-induced obesity and frequently insulin resistance is caused by the breach of the intestinal barrier, which triggers microbe-associated molecular patterns that stimulate intestinal epithelial cells (IECs), macrophages, and DCs to produce proinflammatory cytokines, such as IL-1, IL-6, IL-12, IL-18, and IL-23. Due to a high-fat diet that causes an imbalance in the diversity of the gut microbiota and changes the microbial composition, obesity increases intestinal permeability. An innate immune response is set off by this imbalance, which occurs when pathogens can more easily pass through the intestinal barrier. The decreased expression of epithelial tight junction proteins, such as claudins and zonula occludens 1 (ZO-1), is the cause of this increased intestinal permeability. Immune cells that secrete IFN-γ, i.e., M1 macrophages have a role in barrier permeability because IFN-γ decreases IECs expression of ZO-1. In a similar vein, there exists a correlation between IL-1β and increased intestinal epithelial tight junction permeability. Chemokines and pattern-recognition receptors (PRRs), like toll-like receptors (TLRs), which recognize distinct danger signals for the differentiation of pathogens for the neutralization of pathogens or clearance of stressed cells induced by obesity, are responsible for the recruitment of macrophages (McLaughlin et al., 2017).

Increased accumulation of macrophages and elevated levels of TNF-α and IL-6 cytokines have been observed in obese adipose tissues. Therefore, obesity is linked to an accumulation of immune cells that ultimately leads to a pro-inflammatory state that promotes dysregulated metabolism, insulin resistance, and a state of low globulin inflammation. Additionally, obesity affects the development of immune cells in the bone marrow by causing ectopic fat to accumulate there, which interferes with immune cell development. It has been demonstrated that circulating PBMCs in obese people secrete more TNF-α and less IL-10, an anti-inflammatory cytokine, which supports this persistent chronic inflammatory state. Furthermore, obese individuals have impaired TLR activation of PBMCs, which results in a decreased production of antiviral type I IFNs, IFN-α, and IFN-β. Adipose tissue and macrophages secrete the chemoattractant monocyte chemoattractant protein-1 (MCP-1), which is more prevalent in obese people than in lean people. IL-1β, TNF-α, IL-8, IL-4, and IL-6 all stimulate MCP-1, which helps with the recruitment of macrophages to adipose tissue in obesity.

It has been demonstrated that pro-inflammatory T and B cell phenotypes in the gut are closely associated with the course and severity of obesity-related diseases. Th1 cells, Th2 cells, Th17 cells, Tfh cells, and Treg cells are among the $CD8^+$ and $CD4^+$ T cells that are gut resident T cells. In mucosal tissue, $CD4^+$ Th17 and Treg cells are the most prevalent types. IgA+ antibody-secreting cells that produce secretory IgA to interface within the lamina propria, as well as tolerogenic and mucosal-barrier-maintaining cells like IL-10-producing Treg cells, IL-22-producing lymphoid cells, and IL-17-protective Th17 cells, are the predominant immune cells within the intestinal immune environment under normal lean homeostatic conditions. The intestinal immune environment is altered when mice are fed an obesity-causing diet. This results in a decrease in Treg and Th17 tolerogenic cell types, an increase in Th1 and $CD8^+$ T cells, a decrease in intestinal homing $CCR2^+$ macrophage, a decrease in intestinal intra-epithelial $CD8\alpha\beta^+$ T cells, and an increase in IL-2 through small intestinal group 2 innate lymphoid cells. Through intestinal dysfunction that permits dysregulated glucose homeostasis, all of these contribute to diet-induced obesity. The microbiome's inability to properly metabolize glucose is caused by obesity, which is linked to a decrease in intestinal IgA^+ immune cells as well as a decrease in secretory IgA and IgA-promoting immune mediators. IgA is an essential B-cell-induced antibody that regulates microbial invasion, intestinal permeability, inflammation of the intestines and adipose tissue, and the makeup of the intestinal microbiome. Restricting IgA secretion has a significant effect on the immune system of the gut, highlighting the harmful consequences of obesity on the gut. IgA is necessary for gut homeostasis, and decreased levels have been linked to obesity, which alters the gut microbiota and further points to intestinal immunity's critical role as a major modulator of the microbiome involved in systemic glucose metabolism.

The body suffers when it is attacked by a virus because of the elevated proinflammatory state caused by the increased Th1 and Th17 subpopulations in obese people. Therefore, obese people have an increased risk of a worse prognosis if they contract viral diseases, such as COVID-19, because they are already in a state of LGI, cytokine dysregulation, and T cell pro-inflammatory activation.

7.1.1.3. Immune Response in People with Underlying Chronic Conditions

Due to their increased risk of death from the severity of the illness, people with underlying medical conditions are particularly vulnerable to viral infections that could require hospitalization, intensive care, ventilators, or mechanical machinery for them to survive. These people are highly susceptible to viral infections due to their compromised immune systems, which are caused by medical conditions and/or treatments that compromise the immune system. Individuals with CD or UC, HIV patients, cancer patients, and organ transplant recipients are examples of immune-compromised cohorts whose underlying conditions only exacerbate the effects of COVID-19.

HIV is a virus that targets the immune system, impairing its capacity to fight off infection and disease. As a result, people who are HIV-positive are more likely to get other viral infections and experience worse clinical outcomes. HIV weakens the immune system by infecting CD4 cells, which aid in the body's defense against illness and infection. HIV affects about 37.7 million people worldwide, and in 2020, 680,000 people died from HIV-related illnesses for which there is currently no treatment.

HIV infection modifies the gut microbiome's constituents and the host immune system's reaction to gut microbes, indicating that the gut is essential to HIV-positive people's immune systems. Even after antiretroviral therapy, HIV is linked to a chronic inflammatory state that is manifested by elevated soluble IL-6 and high-sensitivity levels of D-dimer, CRP, and cystatin C. Increased levels are strong independent predictors of mortality. HIV infection has a significant impact on the integrity of the gut epithelial barrier, including intestinal fatty acid binding protein and zonulin-1 levels, as well as innate immune activation and inflammation through markers like soluble CD14 levels, kynurenine/tryptophan ratio, and TNF receptor 1 levels. HIV causes a progressive loss of CD4 T cells, which compromises the integrity of the gut mucosal barrier and impairs the innate immunity supplied by the gut due to poor CD4 cell recovery within the lamina propria. As a result, the initial line of defense against invasive pathogens is compromised, and as a result, the amount of cytokines secreted—which are essential for maintaining proper barrier function—decreases.

Because of the increased bloodstream circulation of microbiological components and the increased exposure of the resident gut mucosal T cell population to new antigens, a leaky gut barrier causes systemic inflammation. Therefore, the gut barrier dysfunction associated with HIV infection may start in the gut lamina propria and its resident CD4 T cells. Because lamina propria T cells express and activate HIV receptors such as CCR5, it is believed that they are more vulnerable to HIV infection. HIV infection is linked to a progressive loss of peripheral $CD4^+$ T cells, primarily as a result of T cells proliferating, expanding, and dying more quickly. This high turnover of T cells causes the hyperactive immune system's regenerative capacity to be exhausted, which in turn causes opportunistic infections, cancer, and, eventually, death. The immune system's weariness causes acquired immunodeficiency syndrome (AIDS), which is brought on by the immune system's inability to continue producing T cells at the high rate necessary for healthy functioning.

Whether antigen-specific, cytokine—or viral gene-induced, this immune system hyperactivation on naïve T cells may increase the consumption of both $CD4^+$ and $CD8^+$ naïve T cells through the death of activated T cells or differentiation into memory phenotypes. HIV infection is also associated with decreases in sigmoid IL-22-producing $CD4^+$ T cells, which are crucial for maintaining the integrity of the sigmoid mucosa. This worsens epithelial barrier dysfunction and increases microbial translocation.

Due to immunological abnormalities in the intestinal mucosa linked to cells of the adaptive immune system, IBD is linked to discomfort in the gut. IBDs develop when patients with conditions like UC and CD experience chronic inflammation brought on by the immune system's reaction to self-antigens. IBDs are chronic digestive diseases that affect over 10 million people worldwide; their cause and cure are unknown. Their prevalence is rising.

The studies have connected the increased consumption of meat and animal products in the modern diet to the development of IBDs, including CD and UC. The intestinal lumen bacteria's epithelial cells, which are required for immune system communication, are disrupted in IB). IECs play a critical role in the activation of cellular innate and adaptive immune responses, the production of cytokines and chemokines, and the maintenance of the epithelial barrier. They also prevent the influx of antigens, the invasion of pathogens and commensal microorganisms, and maintain tolerance to alimentary antigens and commensal microbiota. Chronic intestinal inflammation results from disruptions in the function of macrophages, which are vital in maintaining gut homeostasis within the intestinal mucosa. The production of IL-10 by M2 macrophages plays a major role in the regulation of intestinal gut inflammation. This anti-inflammatory cytokine is secreted less frequently when IL-10 function is dysregulated, and this is why it is associated with a decreased risk of IBDs, especially UC. The protein known as nucleotide-binding oligomerization domain 2 (NOD2) is encoded by the intracellular microbial sensor known as caspase recruitment domain-containing protein (CARD). CARD is a powerful activator and regulator of inflammation. IBDs, including CD, are frequently linked to NOD2 mutation or deficient expression due to elevated expression of inflammatory factors. TLR signaling contributes to tolerance to commensal bacteria and protects the epithelial barrier; however, a malfunction in TLR signaling linked to inflammatory bowel disease (IBD) causes an intestinal inflammatory response by activating the NF-κB transcription factor, which controls the expression of important inflammatory cytokines, including TNF-α, IL-1, IL-2, IL-6, and IL-12.

The majority of inflammatory disorders, including inflammatory bowel disease (IBD), are linked to problems with the adaptive immune system and include modifications to cell autophagy, antigen processing, cell signaling regulation, and T cell homeostasis. The development and endurance of inflammatory responses that result in chronic inflammatory disease are determined by the intestinal mucosa's unbalanced release of Th1 and Th2 cytokines. Important Th1 cytokines associated with the formation of IDBs include IL-6, IL-12, IL-18, TNF-α, TGF-β, IFN-γ, and the reaction to self-antigens. Similar to this, Th17-related cytokines like IL-17 and IL-22 contribute to the growth and manifestation of IBDs. As a result, chemokines and cytokines that cause or sustain mucosal inflammation may be overexpressed in T cells and their subsets. Treg cells have a role in the etiology of IBD. In mice models of IBD, $CD4^+$ and $CD25^+$ Treg cells are involved in immune regulation and treatment. These cells are recruited to the intestinal lymphatic tissues and lamina propria, where they are essential for preserving intestinal homeostasis. IBD may arise as a result of a decrease in Treg cells. There is a dysregulation in the quantity of antibodies secreted and produced by B cells in patients with IBDs. The hallmark of UC is the infiltration of IgG-producing plasma cells through the mucosa's CD184 receptor, which intensifies inflammation by stimulating intestinal $CD14^+$ macrophages. High serum and intestinal mucosa concentrations of IgG1, IgG2, and IgG3 are characteristics of CD. In a similar vein, increased IgM levels are linked to the pathophysiology of IBD. People who have uncontrolled HIV and immune-compromised individuals with IBD experience persistent low-grade inflammation and T cell exhaustion. These disorders highlight the extent to which the gut is impacted when the immune system is compromised and highlight the host's susceptibility to additional viral infection. Individuals with these illnesses may benefit from dietary interventions that improve immune fitness.

7.2. FUNCTIONAL FOODS AS IMMUNE FITNESS BOOSTERS IN THE CONTEXT OF VI-RAL INFECTION

It is believed that the idea of functional foods originated in Japan less than 40 years ago. The Japanese developed the field of functional food science by drawing inspiration from the ancient Chinese statement that "food and medicine are isogenic." Though there has been a significant increase in interest and popularity in this field of food science research, the FDA still does not recognize the term "functional food" as a distinct regulatory product category, and it is not legally defined. Although there is currently no legal definition for the rapidly developing field of functional foods in either EU or Irish food legislation, it is regulated by currently enacted food laws (Martirosyan and Singh, 2015).

Zhang et al. highlighted the role of iron, zinc, selenium, omega-3 PUFAs, vitamins A, B2, B3, B6, C, D, and E, and selenium in the fight against viral infections. These are important traditional functional food ingredients that may be able to aid in the defense against viral infections. Here, we highlight functional foods that may help fight viral infections, such as milk proteins, fermentation products, other plant-derived products, zinc, selenium, and vitamin D. In this segment, we emphasize their relationships with the immune system as well as the processes that underlie their immune-stimulating effects.

7.2.1. Whole Milk Proteins and Hydrolysates

Casein and whey are the two protein groups found in milk. Whey protein makes up the remaining 20% of the total protein in cow's milk, with casein making up the remaining 80%. Whey is the principal by-product produced by the cheese manufacturing sector and is made up of immunoglobulins, serum albumin, β-lactoglobulin, α-lactalbumin, lactoferrin, and transferrin. Conversely, casein is made up of different protein fractions, such as αs1, αs2, β-, and κ caseins. There are several ways in which proteins derived from milk can function as antiviral agents. Among these conventional antiviral strategies are binding to structural viral proteins to block host-cell interactions, obstructing viral entry by interfering with viral and/or cell surface interactions, and blocking specific viral enzymes that are necessary for viral replication. Although casein has also been demonstrated to have some antiviral activity against viruses, the majority of the antiviral qualities of milk are linked to whey proteins, particularly lactoferrin.

7.2.1.1. Whey

Numerous studies have demonstrated the antiviral properties of whey proteins. It has been demonstrated that whey protein from human breastmilk efficiently inhibits the growth of infectious viruses and SARS-CoV-2 and its related pangolin coronavirus by preventing viral attachment and replication at cytoplasmic entry and post-entry sites. More specifically, the RNA-dependent RNA polymerase (rdRp) activity of SARS-CoV-2 was markedly and dose-dependently inhibited by the whey protein of human breastmilk. This is assumed to be caused by the high concentration of lactoferrin, which has been shown to have antimicrobial properties, as well as other elements present in breastmilk. A naturally occurring, nontoxic glycoprotein called lactoferrin has been shown to help fend off viral infections such as SARS-CoV, which is closely linked to SARS-CoV-2, the virus that causes COVID-19. Numerous viruses, such as the hepatitis B and C viruses (HBV and HCV), herpes simplex viruses 1 and 2, HIV, human cytomegalovirus, human papillomavirus (HPV), enteroviruses, adenoviruses, influenza viruses, parainfluenza viruses, and rotaviruses, have been shown to be inhibited by lactoferrin. Lactoferrin can prevent HIV from replicating because, among other things, it binds to the enzymes that allow for viral replication, such as reverse transcriptase, protease, integrase, and HIV-1.

Because of its iron-binding ability, direct interaction with cell surfaces, ability to promote immune cell activation, differentiation, and proliferation, and ability to downregulate immune responses through anti-inflammatory cytokine activity, lactoferrin has immunomodulatory and anti-inflammatory properties that can be used to confer protection in host systems. Lactoferrin, for instance, inhibits viral replication and causes the expression of type I interferon IFN-α/β, which is known to be a powerful antiviral cytokine and immunomodulator. It has also been demonstrated to reduce TNF-α and IL-6, two important cytokine storm players.

Inhibition of ACE2 and S glycoprotein is another possible mechanism via the spike protein, ACE2 serves as the receptor and primary landing site for SARS-CoV-2 on host cells. The S glycoprotein, which is this spike protein, is crucial for the attachment, fusion, and penetration of viruses into host cells. Therefore, it is possible to inhibit ACE2 receptor binding and thereby prevent viral attachment and subsequent infection by inhibiting the surface S glycoprotein. A research project by Fan and colleagues demonstrated that whey can marginally inhibit the affinity between ACE2 and S glycoprotein.

In the study, Serrano et al. were able to determine a potential dose of liposomal lactoferrin, Lactyferrin™, for the prevention and treatment of COVID-19 infection. This dose was determined to be 64–96 mg (20–30 mL) every 6 hours daily (256–384 mg/d), which can be increased to 128 mg every 6 hours (512 mg) if necessary to cure COVID-19, or 64 mg two to three times a day (128–192 mg/d). Smaller doses shielded those in close contact with the patient from infection, allowing all 75 patients to recover fully and quickly within the first 4-5 days of treatment. In a different study, the lactoferrin found in breastmilk and the popular infant formulas containing lactoferrin in this population were linked to lower incidence rates of COVID-19 and less severe cases in children and infants.

7.2.1.2. Casein

It has been demonstrated that bovine κ-casein directly inhibits the ability of viral particles to bind to human rotavirus (HRV) via glycan residues. Viral activity against HRV is reduced by 50–70% as a result of direct binding of viral particles; the remaining 30–50% of uninhabitable activity is thought to be caused by the possibility of multiple important molecules being involved in the process of viral attachment and replication during cell entry. Separate investigations, however, have demonstrated that casein—in its unaltered state—had no inhibitory effect on HIV-1. On the other hand, chemically altered casein blocked HIV-1 by directly binding to the CD4 cell receptor and the HIV-1 gp 120 envelope glycoprotein.

7.2.2. Fermented Dairy Products

Numerous studies have shown that fermented foods, including kefir, can enhance and bolster immune responses in humans. As an instance, kefir, a fermented dairy product, has been recognized for its antiviral and anti-inflammatory properties. It can hinder ACE levels and cholesterol metabolism, promote healing of wounds, impede tumor growth, change the immune system to alleviate allergy symptoms, diminish viral activity by adjusting immune responses, and disrupt viral adhesion, among other functions. Additionally, it acts as an anti-inflammatory by inhibiting proinflammatory cytokines such as IL-1β, TNF-α, and IL-6. These properties are significant in the low-grade inflammation observed in elderly, obese, and immunocompromised individuals, as well as being primary factors in the cytokine storm associated with COVID-19 infections. Consequently, kefir's potential antiviral activity against COVID-19 could be explored, primarily due to its ACE inhibitory properties and capacity to decrease proinflammatory cytokines. Kefir is believed to exert its antiviral effects through direct probiotic–virus interaction and entrapment, the production of antiviral inhibitory metabolites, and/or stimulation of the immune system to generate bacteriocins, lactic acid, and hydrogen peroxide as antiviral agents. Kefir is a potentially potent functional food for the elderly, people with impaired immune systems from IBD, obese people whose gut immunity is weakened, and anyone else whose gut microbiota composition is altered, low-grade inflammation is regulated, intestinal permeability is controlled, and gut homeostasis is regulated. Kefir raises blood levels of zonulin, which is important for controlling intestinal permeability and tight junction modulation. Additionally, kefir may prevent obesity by blocking enzymes that aid in the breakdown of fats and carbohydrates, which reduces the release of energy.

Fermented milk products like yogurt are made from milk and contain cultures of *Lactobacillus bulgaricus* and *Streptococcus thermophilus*. Since yogurt-derived peptides are known to have ACE inhibitory properties, they might be useful in fighting viral infection. Numerous studies conducted both *in vitro* and *in vivo* have demonstrated the direct antiviral effects of the bioactive peptides found in yogurt. Yogurt has also been connected to enhanced intestinal barrier function, decreased chronic inflammation through the enhancement of innate and adaptive immune responses, and improved gut health in addition to these antiviral effects. Through its interactions with the intestinal epithelium, yogurt regulates the expression of genes related to anti-microbial peptides, tight junction proteins, and autophagy—all of which are essential for the proper operation of the gut barrier. Yogurt helps to restore gut homeostasis and has inhibitory effects on colon cancer, which helps to prevent and control inflammatory bowel disease. The ingestion of LAB is linked to reductions in TNF-α. Yogurt is therefore thought to be helpful in reducing low-grade inflammation, which is common in the elderly, obese, and immunocompromised populations, such as those with type 2 diabetes. Yogurt also plays an important anti-inflammatory role in the elderly, obese, and immunocompromised, especially in those with IBDs. It does this by increasing the anti-inflammatory cytokine IL-10 and concurrently reducing the proinflammatory cytokines IL-17 and IL-12.

Originating in Mongolia, koumiss is a traditional fermented dairy product made from fermented mare's milk. Koumiss's capacity to lower TNF-α, a major contributor to the cytokine storm associated with COVID-19 infection and a major cause of low-grade inflammation in the elderly, obese, and immunocompromised, has demonstrated its immunomodulatory potential. It has been demonstrated that koumiss increases IFN-γ, and these cells that secrete IFN-γ are essential for preserving the integrity of the gut barrier. Moreover, Koumiss influences both the innate and adaptive immune responses because it can stimulate the production of sIgA, which in turn induces gut mucosal responses. By preventing bacteria and viruses from attaching to the GIT, SIgA guards against infection. All things considered, fermented dairy products may be regarded as functional foods that offer protection against viral infection. Through the modulation of gut microbiota composition and their overall antiviral abilities by virtue of their ACE inhibitory role, their direct viral inhibitory mechanisms, their maintenance of the gastrointestinal system, and their contribution to enhanced epithelial gut barrier function, these fermented foods can be extremely beneficial for the elderly, obese, and immune-compromised individuals.

Furthermore, fermentates have recently become a noteworthy food component. In general, a fermentate is defined as "a powdered preparation that is derived from a food product that has undergone fermentation and may contain a variety of metabolites and bioactive components as well as the fermenting microorganisms and their components, culture supernatants, and fermented substrates." As an illustration, the oral fermentation product EpiCor, which is made from *Saccharomyces cerevisiae* (*S. cerevisiae*), demonstrated the ability to strengthen immunity and support defense against symptoms similar to the common cold or flu. It was demonstrated that this oral over-the-counter fermentate can lower the incidence of cold and flu-like symptoms in people with and without a history of influenza vaccination in these two 12-week randomized, double-blind, placebo-controlled trials. Further research into fermentates' potential applications against COVID-19 and other viral infections is warranted given the promising results of these studies regarding their ability to protect against and prevent viral infections (Finnegan et al., 2023).

Table 7.1. Summary of Immune Mechanisms Enhanced by Fermented Dairy Products

Immune-Active Components	**Immune-Boosting Functions**	**Mechanism**	**Reference**
Kefir	Antiviral	· Inhibits ACE levels and suppresses viral activity · Directs probiotic–virus interaction and trapping, production of antiviral inhibitory metabolites, and development of lactic acid and hydrogen peroxide as antiviral agents · Antioxidant	Tochikura et al. (1987)

	Immuno-modulator	· Enhances cholesterol metabolism, aids in wound healing, suppresses tumor growth, and improves allergy · Modulates gut microbiota composition, controls intestinal permeability, and regulates gut homeostasis · Improves zonulin levels and regulates intestinal permeability and modulation of tight junctions	Hamida et al. (2021)
	Anti-inflammatory	· Inhibits IL-1β, TNF-α, and IL-6	Hamida et al. (2021)
Yogurt	Antiviral	· ACE inhibitor	Nielsen, (2009)
	Immuno-modulator	· Antithrombotic · Improves gut health and intestinal barrier function · Upregulates expression of autophagy, tight junction proteins, and anti-microbial peptide-related genes for gut barrier health	Gouda, (2021)
	Anti-inflammatory	· Decrease TNF-α · Decreases IL-17 and IL-12 and increases IL-10	Wilkins and Sequoia (2017)
Koumiss	Antiviral	· Enhanced SigA production, inhibiting the attachment of viruses in the gastrointestinal tract	Ya et al. (2008)
	Immuno-modulator	· Maintains healthy gastric intestinal systems, regulates cholesterol and sugar levels, controls blood pressure, and produces important vitamins · Increases IFN-γ secreting cells to maintain gut barrier function	Kırdar (2021)
	Anti-inflammatory	· Decreases TNF-α	Ya et al. (2008)

Source: Finnegan D, Tocmo R, Loscher C. (2023). Targeted application of functional foods as immune fitness boosters in the defense against viral infection. Nutrients, 15(15):3371. https://doi.org/10.3390/nu15153371.

7.2.3. Plant-Derived Functional Food

The popularity of plant-based functional foods is rising along with the interest in vegetarian and vegan diets. Plant-based functional foods can come from raw, natural plant foods or from plant foods that have undergone biotechnological modification. Due to their immune-modulating qualities, plants have long been recognized to have therapeutic benefits that lower the risk of contracting a variety of diseases, such as diabetes, cancer, CVD, hyperlipidemia, and hyperuricemia.

The coconuts of the Cocos nucifera plant yield virgin coconut oil (VCO), which is abundant in nutrients, vitamins, and minerals such as monolaurin, vitamin E, palmitic acid, lauric acid, and plant sterols. It is also rich in bioactive compounds, such as polyphenols, sterols, and tocopherols. Anti-inflammatory, analgesic, gut microbiota modulator, anti-stress, antioxidant, and antimicrobial properties of VCO have been reported. As a result, VCO is a powerful functional food with a host of advantageous characteristics that may help improve immune fitness in the elderly, obese, and immune-compromised as well as protect against viral infection and support intestinal homeostasis.

VCO has recently come to light as a possible antiviral functional food that can help patients with suspected or confirmed COVID-19 infection recover from their viral infection more quickly by lowering CRP levels. The innate immune macrophage's phagocytotic activity can be enhanced by VCO, which has also been demonstrated to inhibit and suppress important inflammatory cytokines like TNF-α, IFN-γ, IL-6, IL-8, and IL-5. As a result, it might be helpful in reducing low-grade inflammation in the elderly, obese, and immune-compromised people as well as in managing the cytokine storm that results from COVID-19 infection. When VCO is supplemented for 3×15 mL/day for 6 weeks, HIV-positive individuals have been observed to have an increased $CD4^+$ T cell concentration, which has been observed to have a positive effect on the adaptive immune response. This highlights the importance of VCO as a functional food for immune-compromised individuals, including HIV-positive individuals. In doxorubicin-induced immunosuppressed rats, VCO has also been demonstrated to increase $CD4^+$ and $CD8^+$ T cells. These findings suggest that VCO may be useful for older, immunocompromised, and obese people whose T cell levels are frequently lowered. More research on animals has also demonstrated the connection between higher adaptive immunity and VCO consumption, with higher VCO consumption resulting in higher levels of IgA in the spleen and small intestine Peyer's patch cells.

Extra virgin olive oil (EVOO), which is made from the olives of the olive tree (*Olea europaea*), is the least processed type of olive oil. Vitamin E, vitamin K, PUFAs, oleic acid, and phenolic compounds such as hydroxytyrosol and oleuropein are among the many vitamins and minerals that are abundant in

EVOO. A naturally occurring secoiridoid glucoside oleuropein compound from Oleaceae plants has been the subject of a US patent application for the treatment of viral diseases, including hepatitis, mononucleosis, shingles, herpes, influenza, the common cold, and viral types that cause leukemia. Similar results have been observed when 50 g of EVOO is consumed daily by elderly HIV-positive individuals without the need for antiretroviral therapy. This improves the lipid profiles and alpha diversity of the intestinal microbiota, primarily due to an increase in *Bifidobacteriaceae* and *Gardnerella* species, and a decrease in proinflammatory genera like Dethiosulfovibrionaceae. In a different study, daily ingestion of 50 mL EVOO reduced the concentrations of high-sensitivity C reactive protein (CRP) in HIV-positive patients receiving antiretroviral therapy. When EVOO is consumed, the gut microbiota benefits from the decrease of pathogenic bacteria, the increase of beneficial bacteria, and the production of microbially produced short-chain fatty acids (SCFAs), which have a variety of anti-inflammatory properties.

By promoting the production of intestinal IgA, EVOO affects the mucosa of the gut and maintains gut homeostasis. Polyphenolic compounds found in EVOO have been associated with decreased T cell activation and proliferation as well as decreased secretion of proinflammatory cytokines. There is a significantly higher concentration of bioactive polyphenols in the gut because other molecules in EVOO, like oleuropein, enter the large intestine as unmodified compounds that are subsequently catabolized to hydroxytyrosol by the human colonic microbiota. Thus, EVOO may be essential for managing viral infection in immunocompromised individuals, such as HIV patients, as well as the elderly and obese, whose immune systems are already compromised by viral infection.

7.2.4. Polyunsaturated Fatty Acids (PUFA)-Rich Foods

PUFAs function as substrates for specialized pro-resolving lipid mediators (SPMs), which are essential chemical mediators required for the stimulation of the resolution of inflammatory responses. Proinflammatory and anti-inflammatory mediators such as prostaglandins, leukotrienes, thromboxanes, protectins, and resolvins are also mediated by PUFAs. Omega-6 PUFA arachidonic acid (AA) is the substrate for eicosanoids, including leukotrienes and prostaglandins, produced through the lipoxygenase and cyclooxygenase pathways, whereas omega-3 PUFA EPA and DHA act as the substrate for SPM. Oily fish, like salmon, mackerel, and trout, are important providers of these omega-3 fatty acids; meat, poultry, and eggs are good sources of omega-6 fatty acids. One serving of lean fish, like cod, may provide approximately 0.2–0.3 g of these omega-3 fatty acids, whereas one serving of oily fish, like salmon or mackerel, may provide 1.5–3.0. Nevertheless, despite their widespread availability, omega-3 PUFAs are frequently lacking in Western diets. A dosage of 60–90 mg of omega-3 polyunsaturated fat has been proposed to support immune system function and the gut microbiota's healing.

The innate and adaptive immune systems are impacted by omega-3 PUFAs, which helps combat invasive viral particles. Omega-3 PUFAs enhance immune cell activation and function. For instance, omega-3 PUFAs can encourage macrophage phagocytosis and stimulate the release of chemokines and cytokines. The production of antibodies in B cells, the stimulation of antigen-presenting cells (APCs), which in turn activate T cells, the enhancement of migration, phagocytic capacity, and the generation of reactive free radicals to kill microbes are all additional effects of omega-3 PUFAs. Additionally, the activation of DCs, natural killer (NK) cells, mast cells, basophils, and eosinophils enhance the first-line defense. It has been demonstrated that long chain AA, EPA, and DHA improve the integrity of the epithelial barrier and decrease IL-4-mediated permeability in the gut. For 12 weeks, a diet containing 18 g of fish oil daily raised the levels of EPA and DHA in the colon while lowering the amount of AA in the mucosa of people with IBD.

The gut microbiota can be modulated by omega-3 polyunsaturated fats (PUFAs), which have also been demonstrated to enhance the abundance of various genera of gut microbes, such as Roseburia and *Bifidobacterium* (a decrease in Lactobacillus and *Bifidobacterium* is linked to numerous metabolic disorders) and maintain a lean phenotype. As a result, omega-3 polyunsaturated fats (PUFAs) can help control and treat obesity. It has also been demonstrated that Lactobacillus and *Bifidobacterium* reduce clinical symptoms in IBDs. The intestinal resident macrophages' constant stimulation to release IL-10, which promotes Treg cells and inhibits excessive Th17 cell activity, depends on these gut microbiotas. It has been demonstrated that omega-3 PUFAs raise triglyceride levels in HIV patients, thereby preventing lipid disorders that could increase the already at-risk person's susceptibility to other diseases, such as CVD. This rise in triglycerides brought about by omega-3 supplementation may therefore also apply to the elderly.

Further research is necessary to determine whether omega-3-derived pro-resolving mediator protectin D1 has an additive antiviral effect and can inhibit influenza viral replication in experimental models. This makes it a promising treatment option for COVID-19 and other deadly infections. The antiviral properties of omega-3 polyunsaturated fats (PUFAs) include protectins derived from DHA and RvE1 derived from EPA. The protectin D1 isomer (PDX) inhibits the nuclear export of viral mRNA, thereby suppressing influenza virus replication. A connection has been discovered between acute lung injury and ARDS, as well as the supportive function of specialized pro-resolving mediators (SPM). It has been demonstrated that ARDS patients who take omega-3 supplements have much better outcomes, such as shorter ICU stays, shorter periods of mechanical ventilation, and a significant reduction in ARDS mortality, while infectious complications do not change. Due to their direct inhibition of viral replication, these studies demonstrate the potential of omega-3 PUFAs as natural therapeutics for the treatment and prevention of viral infections, such as influenza and COVID-19. As a result, they are crucial for elderly, obese, and immune-compromised individuals who are already at risk (Finnegan et al., 2023).

It is hypothesized that one can skew the immune response in favor of the resolution of inflammation by favoring higher concentrations of resolvin/protectin rather than leukotriene/prostaglandin by increasing omega-3 PUFAs and decreasing omega-6 PUFAs. Since omega-3 FAs are known to produce fewer pro-inflammatory cytokines, increasing diet-related intake of them may reduce viral entry, enhance immune function, and potentially lessen the severity of disease in COVID-19 patients by modifying the immune response's overdrive, which is manifested as a cytokine storm. NF-kB is a transcription factor that is expressed in nearly all cell types and regulates the activation of genes that code for proinflammatory mediators.

Omega-3 PUFAs activate the anti-inflammatory transcription factor peroxisome proliferator-activated receptor (PPAR)-γ, which in turn inhibits NF-kB activation and prevents the transcription of proinflammatory mediators. DHA, EPA, and omega-3 FAs all lower NF-κB transcriptional activity and upstream cytoplasmic signaling events. In THP-1-derived macrophages, the production of proinflammatory cytokines IL-1β, IL-6, and TNF-α is downregulated by omega-3 FAs, EPA, and DHA. These cytokines are linked to the etiology of metabolic syndrome. Specifically, DHA has been associated with a more effective anti-inflammatory profile than EPA. It has been demonstrated that omega-3 PUFAs inhibit peripheral blood monocytes' production of TNF-α, IL-2, IL-1α, and IL-1β as well as the proliferation of mononuclear cells. Thus, omega-3 PUFAs can reduce some of the major pro-inflammatory cytokines that are present in the gut of the elderly, obese, and immunocompromised people. These cytokines manifest as the chronic low-grade inflammation that makes these people more vulnerable to viral infection. The ability of omega-3 PUFAs to raise the IFN-γ/IL-4 ratio is especially strong. DHA and EPA can counteract stress-induced abnormalities in the gut by increasing the expression of ZO-1, Z0-3 occluding, and E-cadherin and decreasing proinflammatory markers such as IFN-γ, TNF-α, IL-1β, and IL-6. Thus, by raising DHA in particular, one can target the transcription factors of these proinflammatory genes and inhibit their transcription, helping to modulate the inflammatory process. This can block the pathway and reduce the cytokine storm associated with COVID-19 infection or the chronic low-grade inflammation associated with aging, obesity, and immune compromise. At 60–90 mg doses of omega-3 PUFA, mucous SIgA and serum IL-10 are elevated, further demonstrating their powerful anti-inflammatory effect. Additional research on the impact of omega-3 PUFAs on DC function has revealed these functions in upregulating IL-10, reducing IL-12, and improving the expression of MHC II, CD40, CD80, and CD86. In the gut of the elderly, obese, and immunocompromised, this suggests that omega-3 PUFAs may help reduce proinflammatory cytokines and increase anti-inflammatory IL-10. They may also help manage the chronic low-grade inflammation that is seen in these populations by inhibiting signaling pathways that control the hyperactivation of the inflammatory response.

In addition to the cytokine storm caused by COVID-19, it is believed that the so-called eicosanoid storm—which is defined by elevated levels of proinflammatory lipid mediators—also plays a significant role in the development of severe infection by exacerbating the host's hyperactive immune response. The recruitment of inflammatory cells, vasodilation, broncho—and vasoconstriction, as well as increased vascular permeability, are just a few of the ways that eicosanoids contribute to inflammation. Research indicates that the proinflammatory lipid mediators known as the eicosanoid storm, in addition to the cytokine storm, play a role in the widespread and harmful hyperinflammation associated with COVID-19 infection. Thus, blocking the proinflammatory cyclooxygenase and eicosanoid lipoxygenase signaling pathways may offer a protective strategy against COVID-19 infection.

7.2.5. Vitamin-D-Enriched Foods

Vitamin D is an essential vitamin that plays a key role in controlling the body's calcium and phosphate levels, protecting bones and teeth from the damaging effects of excess vitamin A, and preventing diseases like osteoporosis and rickets. In addition, vitamin D is required for nerve and muscle function as well as for the immune system to help fend off bacteria and viruses that invade the body. The primary source of vitamin D is exposure to sunlight through the skin; nevertheless, it can also be obtained naturally from foods like eggs and oily fish like salmon and sardines. Immune cells, such as B cells, T cells, and APCs, express the vitamin D receptor. Since a vitamin D deficiency is linked to an increased risk of infection, these cells can synthesize the active vitamin D metabolite and, consequently, regulate both the innate and adaptive immune responses. Low levels of vitamin D, specifically 1,25-dihydroxyvitamin D (1,25OHD), have been linked to proinflammatory molecules that can cause ARDS in patients with COVID-19-associated pneumonia. Similarly, low levels of vitamin D in the active form of 1,25-dihydroxyvitamin D (1,25OHD) have been linked to severe disease and the progression of ARDS in some COVID-

19-infected patients. To reach the crucial 50 nanomoles per liter blood of vitamin D, where immunity against COVID-19 can be enhanced, McCartney recommends that Irish adults take 20–25 micrograms (800–1,000 iu) of vitamin D per day, taken with food, during the COVID-19 pandemic. According to these studies, vitamin D is especially important for the elderly, obese, and immunocompromised because it strengthens their innate and adaptive immune systems.

Since the skin contains the majority of vitamin D, it works in its active form, 1,25-dihydroxyvitamin D, in conjunction with vitamin D receptor (1,25OHD or VDR), to support the innate immune system by preserving adherens junctions, tight junctions, and gap junctions. Through increased expression of VDR-associated intracellular junction proteins—occludin, claudin, vinculin, ZO-1, and ZO-2—which form tight junctions between epithelial cells, vitamin D supports the integrity of the epithelial barrier. Since VDR is widely expressed throughout the body, including the gut, it may be a therapeutic target in situations where gut immunity is compromised. It is expressed in a variety of tissues, including the skin, parathyroid gland, adipocyte, small intestines, and colon. Inflammatory responses are linked to elevated claudin-2 junction protein, which is why vitamin D and VDR deficiency are linked to the pathogenesis of IBDs and play a crucial role in intestinal barrier function. Individual bacterial taxa are influenced by VDR, such as Parabacteroides, which is found in far lower abundance in UC and CD patients. Reduced Lactobacillus and increased Proteobacteria in the gut are linked to the downregulation of VDR, or the inability to produce the active form of vitamin D, indicating the impact of vitamin D on gut microbiota. When combined, lower levels of VDR and vitamin D are linked to dysfunctional intestinal integrity, intestinal barrier function, and gut microbiota composition. As a result, higher vitamin D intake as a functional food ingredient may help at-risk groups such as the elderly, obese, and immune-compromised people maintain gut health and immunity against viruses.

Th1-mediated immune responses are suppressed by active vitamin D, which also promotes a Th2 response by producing anti-inflammatory cytokines like IL-4 and IL-10 that indirectly inhibit Th1 cells. Inhibitory cytokines include IL-2 and IFN-γ.

Moreover, it activates Treg cells to suppress inflammation and thwart a viral assault in its entirety. Vitamin D deficiency inhibits Treg function and negatively affects Treg differentiation, which can set off autoimmune diseases like IBDs. Correcting vitamin D deficiency has been linked to the ability to attenuate IFN-γ and IL-6 inflammatory responses, which are both highly correlated with critically ill, ventilated COVID-19 patients as well as the elderly, obese, and immune-compromised. It has also been linked to suppressed CD26 adhesion molecules, which are used for COVID-19 cell adhesion and invasion.

When combined, these mechanisms of antiviral activity through suppression of proinflammatory markers may be used to treat the cytokine storm that results from COVID-19 infection or the chronic low-grade inflammation that is observed in the elderly, obese, and immune-compromised. Through the targeting of cell surface adhesion molecules, these mechanisms suppress and/or inhibit the otherwise dangerously proinflammatory state that can lead to viral infection or the persistence of chronic disease (Finnegan et al., 2023).

Table 7.2. Summary of Immune Mechanisms Enhanced by Vitamin-D-Enriched Foods

Immune-Active Components	**Immune-Boosting Functions**	**Mechanism**
1,25-dihydroxyvitamin D and vitamin D receptor (1,25OHD or VDR)	Antiviral	· Suppresses CD26 adhesion molecules, inhibits COVID-19 cell adhesion and invasion
	Immunomodulator	· Maintains tight, gap, and adherens junctions · Supports integrity of epithelial barrier and increases expression of VDR-associated intracellular junction proteins, including occludin, claudin, vinculin, ZO-1, and ZO-2 · Improves gut barrier function · Influences gut microbiota · Induces B cell proliferation and the secretion of IgE and IgM, enables formation of memory B cells and B cell apoptosis promotion
	Anti-inflammatory	· Suppresses Th1-mediated immune responses (inhibits IL-2 and IFN-γ), promotes Th2 response (produces IL-4 and IL-10) · Induces Treg cells · Attenuates IL-6

Source: Finnegan D, Tocmo R, Loscher C. (2023). Targeted application of functional foods as immune fitness boosters in the defense against viral infection. Nutrients, 15(15):3371. https://doi.org/10.3390/nu15153371

7.2.6. Zinc-Enriched Foods

A vital micronutrient that influences both the innate and adaptive immune responses, zinc plays a direct role in immune system maintenance. Foods high in zinc include oysters, red meat, and poultry. Whole grains, beans, and nuts also contain trace amounts of zinc. Individuals with chronic illnesses like inflammatory bowel disease (IBD) and the elderly and obese are more likely to be zinc deficient. It has been demonstrated that taking zinc supplements can protect against viruses that cause common colds and reduce the risk of infections, including pneumonia in the elderly. Around 60% of all deep respiratory infections globally are caused by zinc deficiency, indicating a possible correlation between zinc deficiency and the severe prognosis and infection risk of COVID-19. This shows that supplements may have a place in antiviral therapy or as a prophylactic measure.

Zinc is necessary for maintaining the integrity of the tissue barrier and is crucial in preventing viral entry into a host. It improves mucociliary clearance of viruses, such as coronaviruses, eliminating the viral particle and lowering the risk of secondary infections. Supplementing with zinc has been shown to increase first responder cellular chemotaxis and phagocytosis, which have been linked to zinc deficiency. NK cells can attack cells with abnormal or unusual proteins in their plasma by infecting the cells, releasing the microorganisms inside, and then phagocytosing them by neutrophils and macrophages. Zinc has the ability to enhance the cytotoxic activity of NK cells. Moreover, low zinc levels are associated with changed NK cell recognition of MHCI, which affects NK lytic potential. It takes MHCI recognition for NK cells to perform at their peak efficiency to eradicate the invasive virus. A person with low zinc levels and impaired oxidative burst has decreased macrophage function; neutrophil granulocytes also exhibit decreased chemotactic activity and quantity. The pathophysiology of CD is linked to zinc deficiency because of the small intestine's gastrointestinal lumen's inadequate absorption of zinc. A zinc deficiency is linked to reductions in transepithelial resistance and changes in the tight and adherens junctions, including ZO-1, occludin, β-catenin, and E-cadherin. These changes cause the integrity of the membrane barrier to be disrupted, which allows neutrophils to infiltrate. Zinc binding to the ACE2 active center becomes essential for enzymatic activity. Zinc homeostasis may affect ACE2 expression, which is regulated by Sirt-1 and which zinc decreases. This may decrease ACE2 expression and subsequent viral entry into cells. Furthermore, zinc-dependent changes in gene expression by pneumocytes also affect viral entry.

Furthermore, zinc has been shown to directly block the replication of numerous viruses, such as influenza, HIV, papillomaviridae, picornaviridae, herpesviridae, metapneumovirus, and coronavirus (SARS-CoV); given their similarities, it is anticipated that this will also likely be the case for SARS-CoV-2. It is believed to accomplish this by inhibiting the fusion of the virus with the host membrane, lowering the activity of the viral polymerase, deteriorating the translation and processing of proteins, obstructing the release of particles, and destabilizing the viral envelope. In HIV-positive patients, long-term zinc supplementation at nutritional levels reduces the risk of opportunistic infection, delays immunological failure, and lessens diarrhea over time. It is believed that zinc inhibits HIV reverse transcriptase by competitively dislodging one or more Mg^{2+} ions that are bound to the enzyme, thereby facilitating the formation of a very stable, slowly advancing reverse transcriptase complex. By directly inhibiting RNA-dependent RNA polymerase (rdRp), low-dose zinc supplementation in conjunction with zinc ionophores, such as pyrithione and hionkitol, can reduce RNA synthesis in influenza, poliovirus, picornavirus, equine arteritis virus, and SARS-CoV.

T cell lymphopenia is a result of the adaptive immune system being impacted by zinc deficiency. The inhibition of nicotinamide adenine dinucleotide phosphate (NADPH) oxidases, which facilitate the eradication of invasive pathogens, has been associated with either excessively high or low levels of zinc. To prevent immunosuppression caused by supplying zinc in either excess or insufficient amounts, it is crucial to achieve an optimal zinc homeostasis by maintaining a balance in the body's zinc levels. An increase in proinflammatory cytokines such as IL-1β, IL-6, and TNF-α—all of which are elevated during COVID-19 infection—and chronic low-grade inflammation in the gut of the elderly, obese, and immunocompromised are characteristics of zinc deficiency. In a similar vein, a zinc deficiency leads to an increase in IL-8, which in turn is crucial for gut inflammation. By promoting the development of Treg cells and inhibiting the growth of proinflammatory Th17 and Th9 cells, zinc functions as an anti-inflammatory to preserve immune tolerance. This reduces the inflammatory response and manages low-grade inflammation, which is common in the elderly, obese, and immunocompromised. Compared to antiretroviral therapy alone, zinc supplementation has been demonstrated to elevate $CD4^+$ T cell counts in HIV patients.

Controlling proinflammatory cytokines is essential for avoiding immunocompromised, aged, and obese people's cytokine storm as well as chronic low-grade inflammation. Because zinc can inhibit NF-κB signaling and modulate regulatory T-cell functions, it has antiviral and anti-inflammatory properties. This means that zinc can reduce the cytokine storm associated with COVID-19 and chronic low-grade inflammation.

A direct correlation has been observed between viral infections, such as COVID-19 and HIV, and zinc deficiency. Lower zinc levels have been strongly associated with patients in the at-risk group for acquiring COVID-19 and who are at risk of a worse prognosis of COVID-19. Individuals with comorbidities such as obesity, diabetes, cancer, atherosclerosis, liver cirrhosis, immunosuppression, and known liver damage are among these groups, as are those with chronic obstructive pulmonary disease (COPD), bronchial asthma, CVD, autoimmune diseases like UC and CD, kidney diseases, and dialysis patients. Since zinc deficiency is already known to be strongly correlated with worse clinical outcomes, it is imperative to take into account the potential role that zinc homeostasis may play in preventing and protecting against COVID-19 and other viral infections. This is especially important for populations that are already at risk, such as the elderly, obese, and immune-compromised.

Table 7.3. Summary of Immune Mechanisms Enhanced by Zinc-Enriched foods

Immune-Active Components	**Immune-Boosting Functions**	**Mechanism**
Zn^{2+}	Antiviral	· Enhances mucociliary clearance of viruses, removes the viral particle, reduces risk of secondary infections, preserves tissue barrier integrity to prevent viral entry · Inhibits ACE2 · Inhibits viral fusion with host membrane, decreases viral polymerase function, impairs protein translation and processing, blocks particle release, and destabilizes the viral envelope · Inhibits HIV reverse transcriptase · Decreases RNA synthesis of viruses by direct inhibition of rdRp
	Immunomodulator	· Increases first responder cellular chemotaxis and phagocytosis · Increases cytotoxic activity of NK cells · Influences NK lytic abilities via MHCI recognition by NK cells · Regulates transepithelial resistance and tight and adherens junctions, including ZO-1, occluding, β-catenin, and E-cadherin, thus influencing membrane barrier integrity · Modulates NADPH oxidases · Stimulates production of IgG · Increases premature and immature B cells and affects antibody production

	Anti-inflammatory	· Reduces IL-1β, IL-6, and TNF-α · Decreases IL-8 · Induces Treg cell development, mitigates Th17 and Th9 · Inhibits NF-κB signaling and modulates Treg cell function · Increases $CD4^+$ T cell counts in HIV patients

Source: Finnegan D, Tocmo R, Loscher C. (2023). Targeted application of functional foods as immune fitness boosters in the defense against viral infection. Nutrients, 15(15):3371. https://doi.org/10.3390/nu15153371.

7.2.7. Selenium-Enriched Foods

One of the most common elements in sulfur, selenium is obtained naturally through food. The 25 selenoproteins that makeup selenium are essential for reproduction, thyroid hormone metabolism, protein folding, DNA synthesis, mitochondrial health, and—above all—oxidative damage and viral infection resistance. Selenium deficiency is linked to obesity and is a risk factor for several chronic illnesses, including IBDs, that are brought on by oxidative stress and inflammation. Because of its selenocysteine-active center, selenium functions. Selenium is a vital component of all living things and can be found in abundance in eggs, fish, corn products (such as wheat, maize, and rice), chicken liver, garlic, onions, broccoli, yeast bran, coconut fruits, Brazil nuts, and seafood. It is estimated that 500 million to 1 billion people worldwide suffer from selenium deficiency, primarily as a result of eating too little.

Increased dietary selenium has been linked to an increased diversity of the gut microbiota, which in turn influences the gut microflora and selenium bioavailability and selenoprotein expression. Selenium regulates the intestinal microflora. IBD is linked to increases in proinflammatory taxa, such as Dorea and Turicibacter. Moderate consumption of selenium can improve intestinal damage and regulate microbiota, including Dorea and Turicibacter. Lack of selenium impairs NK cells' capacity to kill.

Selenium deficiency increases viral pathogenesis through oxidative stress and redox signaling, which in turn impacts cytokine expression, cell proliferation, and apoptosis. Reactive oxygen species (ROS) and their scavenging systems are out of balance due to viral infections, which leads to an imbalance between ROS and the antioxidant defense system of the cell. This imbalance is what causes oxidative stress. Oxidative stress brought on by viral infections promotes the virus's ability to replicate and accumulate mutations in its RNA genome, which through this amplification loop increases the virus's virulence and causes harm to the host. Selenium deficiency has been connected to elevated susceptibility and pathogenicity of viral infections, as well as to mutations in the viral genome that produce extremely virulent forms of the viral particle. Due in large part to the vital selenium proteins glutathione peroxidase (GPXs) and thioredoxin reductases, which offer antiviral defense through redox signaling and homeostatic functions, selenium proteins are necessary for a successful immune response to infections.

The antiviral activity of selenium is predominantly governed by antioxidant factors, such as the regulation of GPXs by selenocysteine. Additionally, it has been demonstrated that selenium inhibits HIV through the antioxidative properties of GPX and other selenoproteins; low selenium levels are associated with HIV infection and subsequent disease progression. Selenium deficiency is also observed in hepatitis B and C virus patients, and elevated levels of selenium may improve treatment response. Selenium functions as an oxidant, which has significant effects on selenium's antiviral properties. Selenium easily combines with sulfhydryl groups in the active site of viral protein disulfide isomerase (PDI), turning them into inactive disulfide. As a result, the hydrophobic spike of the virus is unable to exchange with the disulfide groups of cell membrane proteins, which stops the virus from entering the cytoplasm of healthy cells and inhibits viral entry.

Although overall selenium levels are lower in COVID-19 patients than in healthy control counterparts, selenium status has been found to positively correlate with patients' survival when compared to non-survivors. This may indicate that selenium is relevant as an antiviral for other viral infections and that maintaining sufficient levels of the mineral is crucial for preventing COVID-19. There is a clear correlation between selenium status and cure rates for COVID-19, as demonstrated by the analysis of population selenium statuses across several Chinese cities based on city, and these viral mechanisms also contribute to the oxidative stress linked to many RNA viral infections, the increased viral replication and consequently increased mutation rate, and the higher pathogenicity or even higher mortality observed in selenium-deficient patients with COVID-19. These results have also been clinically validated in Germany, where serum selenium levels were found to be strongly correlated with COVID-19 outcomes in hospitalized patients; in that country, 44% of patients had very low selenium levels, and 65% of patients who died had low selenium, compared to 39% of survivors. Crucially, the lowest selenium levels were highly correlated with death, emphasizing the role of selenium in COVID-19 patient defense and protection against adverse clinical outcomes.

Additionally, because selenium is a well-known NF-kB inhibitor, it plays a crucial role in reducing virally-induced apoptosis. Because it interrupts the signaling pathway that causes the chronic proinflammatory state, it may also have an impact on mitigating the cytokine storm in COVID-19 infection and chronic low-grade inflammation seen in the gut of the elderly, obese, and immunocompromised. The primary immune and inflammatory mediator, NF-kB, is largely in charge of the proinflammatory cytokine production that contributes to the potentially fatal cytokine storm and persistent low-grade inflammation in the gut. The management of the cytokine storm may benefit from supplementing at-risk groups, such as the elderly, obese, and immunocompromised, with 200 mcg of selenium per day for three weeks, followed by a maintenance dose of less than or equal to 200 mcg μg for the duration of COVID-19's active circulation. Documenting serum selenium levels in COVID-19 hospitalized patients would also help, as it would allow for the systemic addition of selenium as soon as possible upon hospitalization.

In many tissues, including the GIT, deficiencies in selenium have been associated with reductions in anti-inflammatory cytokines such as IL-2, IL-10, IL-17, IL-1β, IFN-α, and IFN-β. Proinflammatory cytokines such as IL-6, IL-8, IFN-γ, and TNF-α have also been linked to deficiencies in selenium. Supplementing with selenium enhances macrophage polarization from the M1 to M2 phenotype, which promotes inflammatory resolution and is crucial in inflammatory bowel disease (IBD). This points to selenium's function as an anti-inflammatory that can control the proinflammatory immune response by regulating cytokine production and related signaling pathways. Chronic low-grade inflammation is commonly observed in the gut of the elderly, obese, and immune-compromised.

The development and proliferation of $CD4^+$ Th cells depend heavily on selenium. Through TCR stimulation, elevated selenium levels promote the differentiation of Treg cells from naïve $CD4^+$ T cells. Because selenium regulates T cell differentiation, it may therefore help manage the low-grade inflammation found in the guts of the elderly, obese, and immunocompromised. Additionally, increased selenium may help manage the chronic inflammation associated with IBDs (Finnegan et al., 2023).

REFERENCES

1. Atherton, J. G., Kratzing, C. C., & Fisher, A. (1978). The effect of ascorbic acid on infection of chick-embryo ciliated tracheal organ cultures by coronavirus. *Archives of Virology, 56*, 195–199.
2. Bach, J. F. (2002). The effect of infections on susceptibility to autoimmune and allergic diseases. *New England Journal of Medicine, 347*, 911–920.
3. Chihara, G. O. R. O. (1993). Medical aspects of lentinan isolated from *Lentinus edodes* (Berk.) Sing. In *Mushroom Biology and Mushroom Products* (pp. 261–266). Chinese University Press.
4. Christ, A., Lauterbach, M., & Latz, E. (2019). Western diet and the immune system: An inflammatory connection. *Immunity, 51*, 794–811.
5. Cordain, L., Eaton, S. B., Sebastian, A., Mann, N., Lindeberg, S., Watkins, B. A., O'Keefe, J. H., & Brand-Miller, J. (2005). Origins and evolution of the Western diet: Health implications for the 21st century. *American Journal of Clinical Nutrition, 81*, 341–354.
6. Dillon, S. M., Liu, J., Purba, C. M., Christians, A. J., Kibbie, J. J., Castleman, M. J., McCarter, M. D., & Wilson, C. C. (2020). Age-related alterations in human gut CD4 T cell phenotype, T helper cell frequencies, and functional responses to enteric bacteria. *Journal of Leukocyte Biology, 107*, 119–132.
7. Eder, W., Ege, M. J., & von Mutius, E. (2006). The asthma epidemic. *New England Journal of Medicine, 355*, 2226–2235.
8. Faria, A. M. C., Mucida, D., McCafferty, D. M., Tsuji, N. M., & Verhasselt, V. (2012). Tolerance and inflammation at the gut mucosa. *Clinical and Developmental Immunology, 2012*, 738475.
9. Finnegan, D., Tocmo, R., & Loscher, C. (2023). Targeted application of functional foods as immune fitness boosters in the defense against viral infection. *Nutrients, 15*(15), 3371. https://doi.org/10.3390/nu15153371.
10. Franceschi, C., Garagnani, P., Parini, P., Giuliani, C., & Santoro, A. (2018). Inflammaging: A new immune–metabolic viewpoint for age-related diseases. *Nature Reviews Endocrinology, 14*, 576–590.
11. Gouda, A. S., Adbelruhman, F. G., Sabbah Alenezi, H., & Mégarbane, B. (2021). Theoretical benefits of yogurt-derived bioactive peptides and probiotics in COVID-19 patients—A narrative review and hypotheses. *Saudi Journal of Biological Sciences, 28*, 5897–5905.
12. Hamida, R. S., Shami, A., Ali, M. A., Almohawes, Z. N., Mohammed, A. E., & Bin-Meferij, M. M. (2021). Kefir: A protective dietary supplementation against viral infection. *Biomedicine & Pharmacotherapy, 133*, 110974.

13. Hayashi, T., Hayashi, K., Maeda, M., & Kojima, I. (1996). Calcium spirulan, an inhibitor of enveloped virus replication, from a blue-green alga *Spirulina platensis*. *Journal of Natural Products, 59*, 83–87.
14. Jeffery, I. B., Lynch, D. B., & O'Toole, P. W. (2016). Composition and temporal stability of the gut microbiota in older persons. *ISME Journal, 10*, 170–182.
15. Khan, Z., Bhadouria, P., & Bisen, P. S. (2005). Nutritional and therapeutic potential of spirulina. *Current Pharmaceutical Biotechnology, 6*, 373–379.
16. Koithan, M., & Devika, J. (2010). New approaches to nutritional therapy. *Journal of Nurse Practitioners, 6*, 805–806.
17. Lewis, J. D. (2016). The role of diet in inflammatory bowel disease. *Gastroenterology & Hepatology, 12*, 51–53.
18. Mabbott, N. A., Kobayashi, A., Sehgal, A., Bradford, B. M., Pattison, M., & Donaldson, D. S. (2015). Aging and the mucosal immune system in the intestine. *Biogerontology, 16*, 133–145.
19. Magrone, T., & Jirillo, E. (2013). The interaction between gut microbiota and age-related changes in immune function and inflammation. *Immunity & Ageing, 10*, 31.
20. Martirosyan, D. M., & Singh, J. (2015). A new definition of functional food by FFC: What makes a new definition unique? *Functional Foods in Health and Disease, 5*, 209.
21. McLaughlin, T., Ackerman, S. E., Shen, L., & Engleman, E. (2017). Role of innate and adaptive immunity in obesity-associated metabolic disease. *Journal of Clinical Investigation, 127*, 5–13.
22. Mohammad, S., Aziz, R., Al Mahri, S., Malik, S. S., Haji, E., Khan, A. H., Khatlani, T. S., & Bouchama, A. (2021). Obesity and COVID-19: What makes obese hosts so vulnerable? *Immunity & Ageing, 18*, 1.
23. Müller, L., & Pawelec, G. (2014). Aging and immunity—Impact of behavioral intervention. *Brain, Behavior, and Immunity, 39*, 8–22.
24. Nielsen, M. S., Martinussen, T., Flambard, B., Sørensen, K. I., & Otte, J. (2009). Peptide profiles and angiotensin-I-converting enzyme inhibitory activity of fermented milk products: Effect of bacterial strain, fermentation pH, and storage time. *International Dairy Journal, 19*, 155–165.
25. Pae, M., & Wu, D. (2017). Nutritional modulation of age-related changes in the immune system and risk of infection. *Nutrition Research, 41*, 14–35.
26. Pae, M., Meydani, S. N., & Wu, D. (2011). The role of nutrition in enhancing immunity in aging. *Aging and Disease, 3*, 91–129.
27. Palmer, D. B. (2013). The effect of age on thymic function. *Frontiers in Immunology, 4*, 316.

28. Ragonnaud, E., & Biragyn, A. (2021). Gut microbiota as the key controllers of "healthy" aging of elderly people. *Immunity & Ageing, 18*, 2.

29. Schandalik, R., Gatti, G., & Perucca, E. (1992). Pharmacokinetics of silybin in bile following administration of silipide and silymarin in cholecystectomy patients. *Arzneimittelforschung, 42*, 964–968.

30. Sørensen, H. M., Rochfort, K. D., Maye, S., MacLeod, G., Brabazon, D., Loscher, C., & Freeland, B. (2022). Exopolysaccharides of lactic acid bacteria: Production, purification and health benefits towards functional food. *Nutrients, 14*, 2938.

31. Suardi, C., Cazzaniga, E., Graci, S., Dongo, D., & Palestini, P. (2021). Link between viral infections, immune system, inflammation and diet. *International Journal of Environmental Research and Public Health, 18*, 2455.

32. Sugahara, H., Okai, S., Odamaki, T., Wong, C. B., Kato, K., Mitsuyama, E., Xiao, J. Z., & Shinkura, R. (2017). Decreased taxon-specific IgA response in relation to the changes of gut microbiota composition in the elderly. *Frontiers in Microbiology, 8*, 1757.

33. Takehara, M., Kuida, K., & Mori, K. (1979). Antiviral activity of virus-like particles from *Lentinus edodes* (shiitake). *Archives of Virology, 59*, 269–274.

34. Thevaranjan, N., Puchta, A., Schulz, C., Naidoo, A., Szamosi, J. C., Verschoor, C. P., Loukov, D., Schenck, L. P., Jury, J., Foley, K. P., et al. (2017). Age-associated microbial dysbiosis promotes intestinal permeability, systemic inflammation, and macrophage dysfunction. *Cell Host & Microbe, 21*, 455–466.e4.

35. Thomas, S. J., Moreira, E. D., Kitchin, N., Absalon, J., Gurtman, A., Lockhart, S., Pérez, J. L., Pérez Marc, G., Polack, F. P., Zerbini, C., et al. (2021). Safety and efficacy of the BNT162b2 mRNA COVID-19 vaccine through 6 months. *New England Journal of Medicine, 385*, 1761–1773.

36. Tochikura, T. S., Nakashima, H., Hirose, K., & Yamamoto, N. (1987). A biological response modifier, PSK, inhibits human immunodeficiency virus infection *in vitro*. *Biochemical and Biophysical Research Communications, 148*, 726–733.

37. van de Loo, A. J. A. E., Kerssemakers, N., Scholey, A., Garssen, J., Kraneveld, A. D., & Verster, J. C. (2020). Perceived immune fitness, individual strength and hangover severity. *International Journal of Environmental Research and Public Health, 17*, 4039.

38. Villar-Álvarez, F., de la Rosa-Carrillo, D., Fariñas-Guerrero, F., & Jiménez-Ruiz, C. A. (2022). Immunosenescence, immune fitness, and

vaccination schedule in the adult respiratory patient. *Open Respiratory Archives, 4*, 100181.

39. Walker, E. M., Slisarenko, N., Gerrets, G. L., Kissinger, P. J., Didier, E. S., Kuroda, M. J., Veazey, R. S., Jazwinski, S. M., & Rout, N. (2019). Inflammaging phenotype in rhesus macaques is associated with a decline in epithelial barrier-protective functions and increased pro-inflammatory function in CD161-expressing cells. *Geroscience, 41*, 739–757.
40. Wilkins, T., & Sequoia, J. (2017). Probiotics for gastrointestinal conditions: A summary of the evidence. *American Family Physician, 96*, 170–178.
41. Wu, D., Lewis, E. D., Pae, M., & Meydani, S. N. (2018). Nutritional modulation of immune function: Analysis of evidence, mechanisms, and clinical relevance. *Frontiers in Immunology, 9*, 3160.
42. Wu, Y., Xu, J., Rong, X., Wang, F., Wang, H., & Zhao, C. (2021). Gut microbiota alterations and health status in aging adults: From correlation to causation. *Aging Medicine, 4*, 206–213.
43. Ya, T., Zhang, Q., Chu, F., Merritt, J., Bilige, M., Sun, T., Du, R., & Zhang, H. (2008). Immunological evaluation of *Lactobacillus casei Zhang*: A newly isolated strain from koumiss in Inner Mongolia, China. *BMC Immunology, 9*, 68.
44. Zimecki, M., Właszczyk, A., Zagulski, T., & Kübler, A. (1998). Lactoferrin lowers serum interleukin 6 and tumor necrosis factor alpha levels in mice subjected to surgery. *Archives of Immunology and Therapy Experimental, 46*, 97–104.

CHAPTER 8

BENEFICIAL EFFECTS OF PLANT-BASED DIETS ON SKIN HEALTH AND INFLAMMATORY SKIN DISEASES

Contents

INTRODUCTION

Since the skin is an essential organ that is exposed to the outside world, it serves a variety of purposes, including controlling body temperature, warding off infections, and maintaining internal organs by serving as a barrier of defense. Both extrinsic and intrinsic factors can result in skin damage. Intrinsic factors, resulting from genetic factors, cause wrinkles, dermal atrophy, angiogenesis, dry skin, and physiological aging of the skin. They also adversely affect sweat production and immune function. Conversely, extrinsic factors consist of smoking, humidity, pollution, UV radiation, and poor diet. Moreover, inflammatory skin diseases are the most common type of chronic conditions that can affect the health of the skin. Immunosuppressant medications are a well-established pharmacological treatment for these skin conditions; however, most patients do not fully recover or may experience side effects. As a result, patients are regularly looking for both topical and non-topical natural treatments. Dietary modifications, like switching to a plant-based diet, are becoming more and more popular as an adjunctive treatment for skin conditions (Kent et al., 2022).

A plant-based diet consists of a wide range of fruits, vegetables, legumes, nuts, seeds, fungi, whole grains, and legumes with little to no intake of processed foods, animal products, or sweets. "Appropriately planned vegetarian diets, including total vegetarian or vegan diets, are healthful, nutritionally adequate, and may provide health benefits in the prevention and treatment of certain diseases," according to the American Dietetic Association. Vegetarian diets that are carefully thought out are suitable for people at every stage of life, including lactation, infancy, childhood, adolescence, and pregnancy, as well as for athletes. Diets based on plants can help to modulate oxidative and inflammatory processes, which are the primary causes of inflammatory skin diseases. They also include a range of fruits and vegetables that are categorized as plant-based functional foods, which contain extra ingredients that may have a positive impact on general health.

This chapter's main goal is to investigate how a plant-based diet and functional foods can support good skin health and mitigate the symptoms of skin conditions (Figure 8.1). This section also looks at the gastrointestinal connection between skin health and plant-based diets and the gut microbiome. This link, known as the gut-skin axis, has been linked to several inflammatory and autoimmune skin disorders, such as psoriasis and acne (Flores-Balderas et al., 2023).

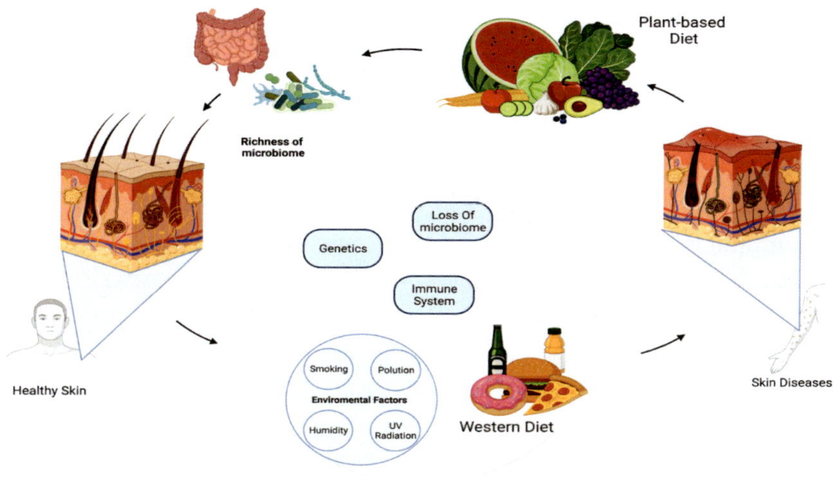

***Figure 8.1**. Association between a plant-based diet and health/diseases of the skin.*

Source: Flores-Balderas X, Peña-Peña M, Rada KM, Alvarez-Alvarez YQ, Guzmán-Martín CA, Sánchez-Gloria JL, Huang F, Ruiz-Ojeda D, Morán-Ramos S, Springall R, et al. Beneficial Effects of plant-based diets on skin health and inflammatory skin diseases. Nutrients, 2.

8.1. ASSOCIATION BETWEEN PLANT-BASED DIET AND PSORIASIS

Psoriasis is a long-term immune-mediated skin condition characterized by erythematous, scaly plaques with distinct margin lesions. With a 3% prevalence, over 7.5 million adult Americans were registered as having psoriasis in 2021. Psoriasis was long thought to be limited to a skin condition, but it is now understood to be a systemic pathology affecting multiple organs and causing a persistent inflammatory state that significantly lowers quality of life.

The pathophysiological mechanisms responsible for psoriasis typically entail immunological response dysregulation and epidermic cell differentiation. Furthermore, the most distinctive circulating inflammatory markers in psoriasis, interleukins, and tumor necrosis factor-alpha (TNF- α), are elevated in psoriatic patients. Psoriasis and its comorbidities are caused by a multitude of environmental factors, including diet, stress, and smoking. In particular, obesity, metabolic syndrome, and CVD are closely linked to dietary-related pathologies and psoriasis. To enhance their quality of life, psoriasis patients are now looking for complementary or alternative therapies, such as natural ways to reduce symptoms and promote overall health. Nowadays, it is widely known that some diets are beneficial for psoriasis in many ways. In this context, a 61-question survey was administered to 1206 patients in the US as part of a cohort study that examined food habits in psoriasis patients. As per the findings, 86% of the participants altered their diet. It was observed that the vegan, Pagano, and Paleolithic diets yielded the most favorable outcomes in terms of skin improvement when compared to the gluten-free, Mediterranean, vegetarian, and diets high in carbohydrates and protein diets. Meat and eggs were listed as one of the less frequent common triggers for psoriasis, but dairy and sugar consumption was listed as one of the most common triggers.

Furthermore, they saw that in 42.5% of patients, adding only vegetables to the regular diet also produced a favorable skin response; in 40.6% of patients, probiotics; in 38.4% of patients, organic foods; and in 34.6% of patients, fruits. However, it was also noted that avoiding particular foods produced comparable outcomes, such as junk food (50%), dairy products (49.9%), white flour products (49.9%), pork (35.6%), and red meat (30.4%). Furthermore, significant weight loss was reported by over half of the patients (69%) who followed any of these types of diets, indicating a connection between plant-based diets and a lower BMI (body mass index). Consequently, a large body of research has demonstrated that certain diets, like the vegan diet, reduce some psoriasis symptoms. However, the study does not specify the precise dietary modifications made in psoriasis patients.

Beneficial effects on the skin are typically reported; nonetheless, additional research is required to fully comprehend how plant-based diets contribute to the management of disease.

Obesity and overweight have been linked to severe stages of the disease and are undeniable risk factors for the development of psoriasis. The effects of a low-energy diet coupled with physical exercise on the severity of psoriasis in patients with an elevated BMI who had just started treatment and whose response to conventional systemic therapy with etanercept, infliximab, adalimumab, ciclosporin, and PUVA (8-MOP) psoralen combined with ultraviolet A therapy (8-methoxypsoralen) was not favorable were studied in a randomized controlled trial (N = 303). The Psoriasis Area and Severity Index (PASI) score decreased by 48%, according to the results. Men, patients over 50 with a BMI between 25 and 29.9 kg/m^2, and patients with PASI scores greater than or equal to 10 showed a greater increase in this beneficial effect. During the 20-week intervention follow-up period in this study, 29.8% of participants achieved a significant weight loss of nearly 5%. A minority of patients reported feeling hungry and constipated during the low-energy diet period; however, no serious side effects related to vitamin deficiencies were observed during the study. It is commonly known that obesity-related increases in adipose tissue tend to release excessive amounts of proinflammatory cytokines, such as TNF, IL-6, and IL-8, which are involved in the pathophysiology of psoriasis. It would be reasonable to suggest that plant-based diets, like the vegan diet, could be the cornerstones of psoriasis management if they can significantly lower BMI and weight loss is strongly associated with clinical improvement in psoriasis patients. More proof is needed, though, to support the theory that diet may be more beneficial than traditional treatments.

In reference to the aforementioned research, a case study involving a patient with psoriatic arthritis has been published. The patient's whole-food plant-based diet (WFPB) was sufficient to achieve a favorable disease response, including reduced scalps, articular pain, stiffness, capsulitis, articular discomfort episodes, and uveitis. This report describes a 40-year-old woman with psoriatic arthritis who was diagnosed in 2003. She underwent 15 years of methotrexate treatment before changing her diet to a WFPB. Because there was no longer any disease activity, she was able to stop her immunosuppressor drug-based pharmacology treatment. As side effects, the patient experienced mouth ulcers and a decrease in white blood cell count while receiving a weekly dose of 20 mg of methotrexate as the base treatment. In contrast to the previously used gluten-free diet, which showed no clinical improvement, the woman reported that her symptoms initially improved when she avoided foods derived from animals, such as dairy and red meat. As a result of this positive change, she adopted a strictly vegan diet and decreased her weekly dose of methotrexate to 12.5 mg. But it took the establishment of a WFPB to bring the illness under complete control. Except for soy milk and soy sauce, it included a lot of fruits and vegetables, whole grains, and no added sugar, oil, or salt. Furthermore, in March 2018, following roughly six months of a plant-based diet, the results of a blood test revealed an ESR of 10 mm/h and a CRP of 2 mg/L. Further research is required to support the recommendation that medication be avoided while treating psoriasis or psoriatic arthritis, but evidence in this regard suggests that plant-based diets may modulate inflammation response. When used as a treatment, this could reduce medication use and prevent side effects related to anti-inflammatory drugs.

In a different case study, a 47-year-old man with severe plaque psoriasis who had been diagnosed with the condition 28 years prior experienced remission of the illness after fasting and switching to a plant-based diet. Long-term topical corticosteroids were the mainstay of his treatment, but the lesions did not completely disappear. Due to concerns about long-term drug use, the patient stopped taking drugs. As an alternative treatment, he adopted a plant-based diet on his own and saw a slight improvement in his clinical condition. As a result, the patient decided to go through a 13-day water-only fast under medical supervision, followed by six days of refeeding on a vegan and vegetarian diet plan, until a WFPB was implemented. The benefits of the dietary intervention included a notable improvement in the severity of the psoriasis plaques, a reduction in the BMI, which remained within the normal range, and a significant improvement in the pain and arthritis associated with the nailbed psoriasis. Following treatment, the patient continued to improve the condition of the existing plaques, reported no new lesions, and followed a whole-plant diet free of salt, sugar, and oil. Concerning doubts regarding the safety of this kind of diet, the only side effect experienced was a slight discomfort in the epigastrium on day eight of a water-only fast that was broken up with vegetable broth and juice; on day nine, he resumed fasting to finish the diet adjustment. As a result, this case strengthens the case for more investigation into the medicinal advantages of plant-based diets and their ability to reduce inflammation in the treatment of skin conditions like psoriasis.

Based on the information provided, it appears that skin health can be improved by switching to a plant-based diet that includes fewer or no animal-derived products and a higher intake of fruits, vegetables, legumes, nuts, and cereal products. Therefore, using plant-based diets as a potential therapeutic option for the management of some psoriatic patients is advised. This diet is low in saturated fat, trans fat, and arachidonic acid (ARA) and high in antioxidants and omega-3 fatty acids. Together with its direct therapeutic effects of reducing inflammation and cutaneous symptoms and decreasing the risk of severe stages of the disease, as well as its indirect effects of promoting weight loss, these patients' general quality of life has improved (Flores-Balderas et al., 2023).

8.2. ASSOCIATION BETWEEN PLANT-BASED DIET AND ATOPIC DERMATITIS

The most prevalent inflammatory skin condition, atopic dermatitis (AD), is typified by itchy, scaly plaques and erythema papules that are mostly found in skin folds on the hands, neck, and head. Its prevalence in the adult population is 2.3% worldwide. Its etiology is complex, arising from the interplay of immune function, and environmental, and genetic factors. AD is caused by a variety of factors, including immune system dysregulation, apoptosis, changes in the skin microbiota, eosinophil and T-lymphocyte infiltration, and IgE (immunoglobulin-E) sensitization. These factors ultimately result in the degradation of the stratum corneum and the epidermal barrier that it forms. Systemic or topical corticosteroids, emollients, anti-inflammatory medications, antihistamines, and immunosuppressive medications like cyclosporine-A are among the conventional treatments for AD. Given that diet and AD are first linked to food allergies, avoiding particular foods would seem to be a good place to start for treatment. Plant-based diets, like the vegetarian diet, are now acknowledged as an additional therapeutic option for AD.

A recent examination of correlations examined the link between adult AD sufferers and lifestyle aspects, including alcohol intake, stress, obesity, physical exercise, sleep length, and diet. In this research (consisting of 56,896 participants), no connection was discovered between diet, specifically vegetarian and vegan diets, and the prevalence or severity of AD. However, Class I obesity, a disease connected to diet, was revealed as a positive factor associated with moderate to severe AD. Information on lifestyle factors was gathered between 2006 and 2013. Previous studies have suggested the advantages of plant-based diets in managing AD, but this study did not yield the same findings. Consequently, further studies appear to be necessary for a comprehensive understanding of the relationship between plant-based diets and AD, enabling the establishment of this type of diet as part of the treatment for the condition, by exploring the effects of a vegetarian diet on AD and the underlying immunological mechanisms of this therapy.

A vegetarian diet dramatically reduced the SCORAD (scoring atopic dermatitis, comprising erythema, edema, crusts, and excoriation) index from 49.9 to 27.4 at the conclusion of the intervention in this open-trial study (N = 20), with ciclosporin-A treatment having a comparable effect. The reduction of skin inflammation was accompanied by a noteworthy decline in serological parameters, including PGE2 synthesis, LDH5 activity, and the quantity of neutrophils and eosinophils in peripheral blood. Neutrophils also showed a similar effect. In the case of eosinophils, the primary inflammatory cells in AD, their count dropped to half of the basal value, reaching its lowest level two weeks after the diet. This decrease in cells occurred before any improvement in the skin's appearance, designating it as the initial mechanism in the pathway to reduce skin inflammation. Furthermore, since AD patients typically produce excessive amounts of PGE2, the synthesis of PGE2 was observed to be inhibited by peripheral blood mononuclear cells (PBMCs). Serum IgE levels, in contrast, did not alter during the intervention; however, they did drop following the study period, during which time the quantity of peripheral eosinophils also recovered to its typical level. A vegetarian low-energy diet comprising fresh vegetable juice, brown rice with kelp powder, tofu, and sesame paste was administered for two months as part of the study's intervention. No negative effects associated with the vegetarian diet were noted. The evidence that has been observed thus far indicates that a vegetarian diet plays a role in the mechanism that lowers AD immunological parameters like LDH5 activity, peripheral eosinophil counts, and PGE2 synthesis. These parameters appear to be accountable for skin inflammation, and their suppression leads to a reduction in skin symptoms. However, additional research and larger sample sizes are needed to elucidate the role of plant-based diets in reducing inflammation in AD. Furthermore, regardless of the severity of the disease, the results demonstrated a significant decline in the SCORAD index of all 19 subjects (N = 19) enrolled, with the index reaching its lowest score at week 2. The BMI, systolic blood pressure, LDH-5 activity, and the quantity of peripheral eosinophils were among the other parameters that were assessed; all of them showed a noteworthy decline. Furthermore, a significant decrease in the urinary levels of 8-OHdG, a biomarker of cellular oxidative stress that is released in the urine following DNA repair, was noted in patients with severe AD, indicating that a vegetarian diet can lessen oxidative DNA damage. The dietary intervention in this study included fresh vegetable juice, brown rice with seaweed powder, tofu, and sesame. This represented 55% of the daily recommended energy intake, but it also included high intakes of iron (approximately 130% of the daily requirement), vitamin A (150%), vitamin C (250%), and vitamin E. The primary worry about plant-based diets in general, including vegetarian diets, is that they may result in some degree of malnutrition. However, no laboratory or clinically significant adverse events were reported during the study period. This suggests that vegetarian and low-energy diets may help with inflammatory skin conditions like AD. Research has shown that these diets can improve skin quality, allowing patients to lower their SCORAD and lessen the severity of their condition. They can also significantly reduce oxidative damage to DNA by providing significant amounts of nutrients and antioxidants, like vitamins C and E, which are present in well-planned vegetarian and low-energy diets. Thus, lowering BMI and consuming antioxidants may help to lessen skin inflammation (Kouda et al., 2000).

Table 8.1. Summary Table of the Main Research in Which Plant-Based Diets are Associated with Inflammatory Skin Diseases

Study Design	**Participant Characteristics**	**Intervention Details**	**Outcomes**	**References**
Exploratory survey study	1206 psoriasis patients. The mean age of the sample population was 50.4, and 73% of respondents were female.	The survey contained 61 questions; the first 30 questions were from National Health and Nutrition Examination Survey; the last 31 questions focused on patient-report skin responses to dietary changes.	72% of Pagano, 70% of vegan, and 68.9% of Paleolithic diets reported favorable skin response. The addition of fruits to the diet was protective against psoriasis, OR 0.22, $p < 0.0001$.	Afifi et al. (2017)

Assessor-blind, randomized controlled trial	303 patients between 18 and 80 years, overweight or obese, with moderate to severe chronic plaque psoriasis.	Each participant underwent an introductory session with a dietitian. Subjects were instructed to perform sessions of continuous aerobic physical exercise for at least 40 min three times a week. Compared how psoriasis severity was affected by either a 20-week quantitative or qualitative dietary plan associated with physical exercise for weight loss.	Reduction of PASI score by 48% in the dietary intervention and 25.5% in the information-only.	Naldi et al. (2014)
Cross-sectional study	6,896 participants. The mean age was 55.8, and 22,577 were female.	First, the researchers sent out a digital questionnaire focused on atopic dermatitis (AD). After collecting the responses, they administered another questionnaire to extract data on lifestyle factors, such as smoking, alcohol consumption, stress, obesity, physical activity, and diet (participants were dichotomized into vegetarian/vegan and nonvegetarian/nonvegan), and finally analyzed the association between lifestyle factors and presence of AD.	No associations were observed with abdominal obesity, physical activity, diet quality, or a vegetarian/vegan diet.	Zhang et al. (2022)

Controlled clinical trial	19 patients with atopic dermatitis. The mean age was 25.5.	The patients were exposed to a low-energy diet, no pharmacotherapy. The SCORAD indices were noted by 2 dermatologists. Blood and urine samples were collected at weeks 0.4 and 8. The 24 h void urine was collected and kept frozen at −80 °C.	SCORAD significant reduction from 6.1 ± 2.8 at week 0 to 2.7 ± 2.8 at week 8.	Kouda et al. (2000)
Clinical trials	Round 20 patients (6 males and 14 females) aged 15 to 36 years (average: 25 years) with AD varying in severity from mild to serious. All patients fulfilled the criteria for atopic dermatitis (AD) established by Hanifin and Rajka.	All patients followed the same meal plan throughout the study. The diet included a glass of fresh vegetable juice for breakfast, followed by brown rice porridge, kelp powder, tofu, and sesame paste for both lunch and dinner. In addition, a daily intake of 2.5 g of non-refined salt was included in the diet. Instead of regular water, the patients were provided with persimmon leaf tea. This treatment regimen was followed for two months.	Reduction in SCORAD index from 49.9 ± 18.6 to 27.4 ± 16.8 (p = 0.001), diminution in peripheral eosinophil count 423 ± 367 to 213 ± 267 (p = 0.01), reduction in monocyte PGE2 synthesis 2886 ± 1443 to 1390 ± 773 (p = 0.001).	Tanaka et al. (2001)
Case-control cross-sectional questionnaire study	57 acne vulgaris patients as cases and 57 participants as controls. Aged 14 and above and were seeking medical consultations at a private clinic.	The Comprehensive Acne Severity Scale (CASS) grading system was used to grade acne severity. Cases were defined as patients with CASS grade of two to five and controls with CASS grade 0 or 1. Thereafter, controls were recruited and matched by age, gender, and ethnicity. A self-administered questionnaire was used to collect information regarding the participants' dietary intake and cigarette smoking habits.	No significant associations were found between dietary patterns and acne; on the other hand, they only found an association between milk and chocolate consumption and acne.	Suppiah et al. (2018)

Case-control study	279 acne patients and 279 controls aged 10–24 years.	Acne severity was determined by a dermatologist using the Global Acne Severity Scale. Epidemiological data were collected with a pre-structured questionnaire. The food consumption habits were recorded using a food frequency questionnaire. Investigated food included whole milk, low-fat milk, cream of milk, ice cream, cheese, chocolate, cake, potatoes, fresh fruit, fresh vegetable, meat, chicken, and egg. Consumption habit data were collected for 8 months.	The study found that consuming whole milk at least three days a week was significantly associated with moderate to severe acne (OR = 2.36, 95% CI, 1.39–4.01). The association was slightly weaker for low-fat milk (OR 1.95, 95% CI, 1.10–3.45).	Aalemi et al. (2019)
A prospective case-control study	460 subjects of both sexes, 230 patients with acne vulgaris, and 230 healthy volunteers were included as controls.	Acne severity in patients was classified as mild, moderate, and severe according to the classification of the American Academy of Dermatology. Two formulated tools were used in the study. First tool: interview questionnaires that included two parts (sociodemographic data, anthropometric measurements, and questions related to food habits, drug intake, and diet history). Second tool: a recommended diet regimen.	Significantly decreased consumption of vegetables was noticed among severe and moderate acne patients compared with mild ones.	Youssef et al. (2014)

Clinical-trial study	Healthy post-menopausal women aged 50 to 70 ($N = 28$). Inclusion criteria were Fitzpatrick skin type I, II, or III and a body mass index (BMI) between 18.5 and 35 kg/m^2.	Participants were randomized by block design into an open-label, two-arm parallel clinical trial consuming either 85 g or 250 g of Ataulfo mango, four times per week for 16 weeks.	0.5 cup frozen Ataulfo mango consumption decreases facial wrinkle depth and severity and photodamage induced by UVB. Serum triglycerides decreased. A higher amount of mango (1.5 cups) showed the contrary effect.	Fam et al. (2020)
Random-ized clinical trial	31 participants (all postmeno-pausal females) with a Fitzpat-rick skin type 1 or 2.	Consisted of a total of five study visits following a 4-week dietary washout period: baseline, 4, 8, and 16 weeks. The patients were randomized into two intervention groups: almond group and control group. Al-mond group received an average of 340 kcal/day of almonds.	There were no sig-nificant differences in sebum production and transepidermal water loss between the almond and con-trol groups. Wrinkle severity and wrinkle width were signifi-cantly decreased in the almond group compared with the control intervention.	Foolad. et al. (2019)
Random-ized-blinded	56 postmeno-pausal women with a Fitzpat-rick skin type I or II.	A total of 56 female par-ticipants were randomly assigned to two interven-tion groups: the almond group, which received almonds, and the control group, which received a calorie-matched snack. In the almond group, the dose of almonds provided accounted for 20% of the total energy in the participants' diet. The almond snack group re-ceived approximately 2 g of sugars per day, whereas the control snack group received approximately 8 g of sugars per day that was additional to their regular food intake. 24 weeks.	Reduction in the average sever-ity of wrinkles compared to the baseline. Addi-tionally, intensity of facial pigmen-tation decreased in the almond group.	Rybak et al. (2021)

Randomized parallel group	39 women with a Fitzpatrick skin type II–IV.	The patients were randomized into two intervention groups: avocado and control. The participants either consumed one avocado daily or no avocado based on randomization performed. 8 weeks.	Daily consumption of one avocado per day for 8 weeks improved firmness and elasticity and reduced the tiring of repeat stretching of the forehead skin.	Henning et al. (2022)
Case-control	30 patients with AD (according to the criteria for the diagnosis of AD established by the Japanese Dermatological Association) less than 20 years of age and age—and sex-matched control subjects (68 healthy individuals).	A questionnaire survey was conducted for 1 week, and then fecal specimens and 24 h skin secretion specimens were collected from all subjects. Fecal microflora, fecal IgA levels, and IgA concentrations in the skin were analyzed. The data for the 2 groups (i.e., patients with AD and healthy control subjects) were compared.	The counts of *Bifidobacterium* were significantly lower in patients with AD than in healthy control subjects (9.75 ± 0.68 vs. 10.10 ± 0.50, $p < 0.05$). Percentages of *Bifidobacterium* were significantly lower in patients with severe skin symptoms than in those with mild skin symptoms (40 ± 6% vs. 19 ± 6%, $p < 0.05$). In addition, the frequency of occurrence of *Staphylococcus* was significantly higher in patients with AD than in healthy control subjects (83% vs. 59%, $p < 0.05$).	Watanabe et al. (2003)

Case-control	21 psoriasis patients from a Brazilian referral dermatology service and 24 controls.	A stool sample was collected from each participant at the time of inclusion in the study, and the samples were analyzed by sequencing the 16S rRNA gene.	Patients with psoriasis showed higher levels of the *Dialister* genus and the species *Prevotella copri* compared to the control group. Conversely, there was a decrease in the *Ruminococcus, Lachnospira*, and *Blautia genera*, as well as the *Akkermansia muciniphila* species, in the psoriasis group compared to the control group. Additionally, individuals with psoriasis had lower diversity in their gut microbiota compared to the control group.	Schade et al. (2022)
Case report	63-year-old woman with ulcerations on both lower legs and diagnosed livedoid vasculopathy by biopsy.	The patient adopted a whole-food plant-based diet (WFPB) as a potential treatment with no other form of treatment.	The symptoms were initially completely resolved, but they reappeared later due to a lack of adherence to a proper diet.	Smith et al. (2021)
Cross-sectional	They measured skin autofluorescence (SAF) in 332 adult hemodialysis patients who were dialyzing in north central London.	SAF was measured in 332 adult dialysis patients.	SAF was lower in the 27 vegetarians than it was in the non-vegetarians.	Nongnuch et al. (2015)

Source: Flores-Balderas X, Peña-Peña M, Rada KM, Alvarez-Alvarez YQ, Guzmán-Martín CA, Sánchez-Gloria JL, Huang F, Ruiz-Ojeda D, Morán-Ramos S, Springall R, et al. (2023). Beneficial effects of plant-based diets on skin health and inflammatory skin diseases. Nutrients; 15(13):2842. https://doi.org/10.3390/nu15132842.

8.3. ASSOCIATION BETWEEN PLANT-BASED DIET AND ACNE

The most prevalent type of acne, known as acne vulgaris, is characterized as a persistent inflammatory skin condition that affects the pilosebaceous unit. Comedones, papules, pustules, nodules, and cysts are typical skin lesions; the face, neck, and chest are the most common places for acne. With a prevalence of roughly 85%, adolescents are the most affected demographic, accounting for 94% of cases worldwide. Excessive sebum production, hyperproliferation of the bacteria *Cutibacterium acnes,* hyperkeratinization of the pilosebaceous follicles, and inflammatory mechanisms are some of the recognized mechanisms for the development of acne. Insulin-like growth factor 1 (IGF-1) appears to be a key factor in potentiating pathogenesis mechanisms that lead to sebum overproduction and acne severity by sebocyte and keratinocyte proliferation. Moreover, IGF-1 is closely linked to hyperinsulinemia and has a positive feedback loop with androgen levels. The pathophysiology of acne is influenced by both genetic and environmental factors, including diet and hormones. For example, a particular population's adoption of a Western diet can lead to the subsequent development of acne. That being said, the connection between diet and acne is changing.

One study looked into the possible link between eating dairy products and acne. Regular consumption of low-fat milk was linked to an increased risk of developing moderate to severe acne, while whole milk and egg intake had similar effects in this case-control study (N = 279 both groups). Unexpectedly, there was a negative correlation between acne and chicken consumption. Although the study did not specifically look into the relationship between plant-based diets and acne, it did find that the control group had a higher correlation than the case group between vegetable consumption, a low BMI, and following a diet, though it was not made clear what kind of diet was followed. Data on consumption patterns were gathered for eight months. The possible explanation for the correlation between milk consumption and acne is that some of its constituents, like specific amino acids, have the ability to promote the synthesis of IGF-1. In terms of the connection to the egg, this could be explained by the leucine in the food, which stimulates sebaceous gland activity by promoting the synthesis of lipids and proteins. As a result, there is unmistakable evidence linking the consumption of milk and eggs to the onset of acne, even though this study did not find any connection between the two. Adhering to these guidelines, which prohibit foods derived from animals like milk and eggs, may be a helpful adjunctive treatment for acne.

A different study found a connection between dietary factors and young adult acne. The findings revealed that acne patients consumed more meat, beef, fish, butter, honey, corn, rice, chicken, and pizza than did controls; on the other hand, the control group consumed more vegetables and potatoes. Furthermore, it was discovered that there was a variation in the quantity of vegetables consumed by the acne patients, with mild acne patients consuming more vegetables than severe acne patients. There have been documented notable variations in plant-based diets between acne sufferers and controls. In patients with acne, the percentage of patients who adhered to a vegetarian diet was hardly higher than 10%. In contrast, over half of the controls followed this diet. Furthermore, there was a statistically significant rise in vitamin A supplementation among the vegetarian patients in the control group. For two years, this case-control study (N = 460) was conducted. Thus, since the control group in this study showed a higher propensity to adhere to this dietary pattern, the evidence suggests that the vegetarian diet plays a role in the treatment and prevention of acne. To be clear, more research is still required to determine whether plant-based diets completely control acne, prevent acne, or just support traditional pharmaceutical treatment for acne.

The available data on eating patterns and the consumption of particular foods in connection to the onset of acne is debatable and occasionally contradictory. For example, both positive and negative correlations between the consumption of chicken and milk are reported in the studies examined in this section. However, the majority of research indicates that low-energy diets, high-vegetable diets, and diets with restrictive patterns—particularly those involving foods derived from animals—help patients with acne lesions and improve the quality of their skin Table 8.1 (Flores-Balderas et al., 2023).

8.4. PLANT-BASED DIET EFFECTS ON OTHER SKIN CONDITIONS

Morgen Smith and associates reported a case of a 63-year-old woman who had ulcerations on both lower legs, which were later confirmed by biopsy to be livedoid vasculopathy. A rare and chronic skin condition called livedoid vasculopathy is characterized by the development of painful ulcers or sores on the lower extremities, especially the ankles and feet. With no other course of treatment, the patient adopted a WFPB as a potential course of action. Remarkably, the symptoms were totally resolved; however, inadequate dietary compliance caused the symptoms to return. Lastly, the authors hypothesized that dietary changes may have a direct effect on the health of vascular endothelial cells, which may in turn affect the risk of a prothrombotic state. Conversely, Vini Kamlesh Solanki gave a case study of a 58-year-old female patient who had symptoms of pemphigus vulgaris (PV), a persistent autoimmune blistering illness. She experienced burning, itching, and blisters all over her body that were filled with fluid. She also mentioned that she had recently received a type 2 diabetes diagnosis and that her skin condition was causing her anxiety, indigestion, and malaise. The patient participated in a yoga and naturopathy-based lifestyle modification program to treat her condition. This program ran for 53 days in 2019, 10 days in 2020, and 15 days in 2021, with follow-up appointments in between. A variety of techniques were included in the program, such as massage, hydrotherapy, yoga, vegetarian food, and herbal preparations. By the end of 2020, the patient's PV had completely resolved and he had been weaned off all medications. Given the intricate nature of PV, the all-encompassing and patient-centered lifestyle approach presented in this case study seems to be helpful in PV management.

Arkom Nongnuch and associates' other research demonstrates the advantageous effects of plant-based diets on skin. They examined 332 adult hemodialysis patients and measured their skin autofluorescence (SAF). SAF is a crucial instrument used in hemodialysis to evaluate the skin's advanced glycation end product (AGE) content. They discovered that SAF levels were lower in vegetarians than in non-vegetarians, and both univariate and multiple linear regression analyses revealed a negative correlation between SAF and a vegetarian diet. This implies that tissue AGE deposition, as measured by SAF levels, may be reduced by a vegetarian diet. It is hypothesized that tissue AGE and SAF levels may be lowered by diet modification, which would lower the risk of CVD.

On the other hand, the study conducted by Marta Fusano and associates, 21 omnivore patients and 21 vegan patients who had undergone surgical excision for nonmelanoma skin cancer were enrolled in a prospective observational study to compare surgical scars between patients who followed an omnivore diet and those who followed a vegan diet. The modified Scar Cosmesis Assessment and Rating (SCAR) scale was used by the researchers to rate scar quality and post-surgical complications. They discovered that vegans were more likely than omnivores to have wound diastasis and that after six months, vegan patients had higher modified SCAR scores, which indicated more marked scar spread, a higher incidence of atrophic scars, and a worse overall impression. Ultimately, they came to the conclusion that a vegan diet might negatively affect how surgical scars turn out (Table 8.1) (Flores-Balderas et al., 2023).

8.5. THE ROLE OF PLANT-BASED FUNCTIONAL FOODS ON SKIN HEALTH

Since the face receives the most exposure to sunlight of any part of the body, research into methods of protecting and repairing it is growing in importance. Continuous exposure to UV light induces photoaging, which results in the breakdown of collagen and a deficiency of elastin fibers in the dermis. This leads to wrinkle formation, a reduction in skin elasticity, and dyspigmentation. People are generally becoming more interested in skincare and rejuvenation, and some groups are specifically searching for treatment options based on diet. It is now recognized that diet can affect the condition of the skin. "Functional foods" are defined by the International Life Sciences Institute as foods that, because of the potency of their active ingredients, offer health advantages above and beyond those of basic nutrition (Figure 8.2). As a result, there are many different plant-based functional foods in a plant-based diet, and if they are eaten consistently and in the right amounts, their positive effects on the skin can be seen (Hasler et al., 2000).

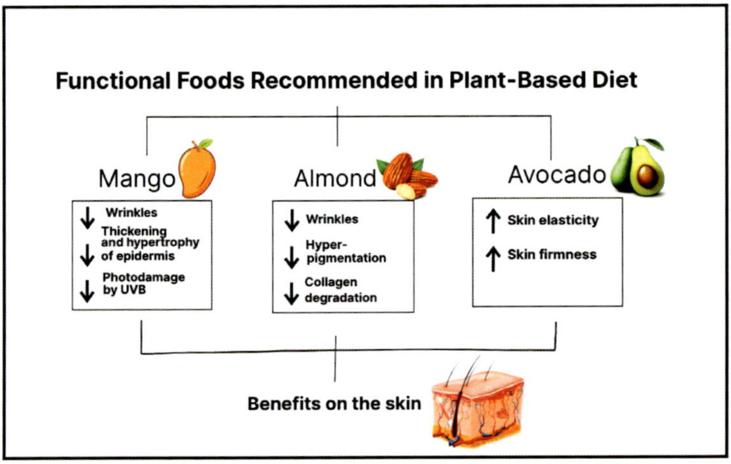

Figure 8.2. *Specific plant-based functional foods recommended to add to a plant-based diet for the greatest benefits.*

Source: Flores-Balderas X, Peña-Peña M, Rada KM, Alvarez-Alvarez YQ, Guzmán-Martín CA, Sánchez-Gloria JL, Huang F, Ruiz-Ojeda D, Morán-Ramos S, Springall R, et al. (2023). Beneficial effects of plant-based diets on skin health and inflammatory skin diseases. Nutrients; 15(13):2842. https://doi.org/10.3390/nu15132842.

8.5.1. Mango

Vitamin C, B-carotene, xanthones like mangiferin, and phenolic acids (primarily gallic, chlorogenic, protocatechuic, and vanillic acids) are abundant in mangos. Furthermore, skin cells in the human body contain high levels of vitamin C, a well-known antioxidant. Ascorbic acid, also known as vitamin C, has several benefits, including the capacity to prevent photoaging by reducing UV-induced reactive oxygen species (ROS) damage, protecting against oxidative stress, and promoting the synthesis of collagen and elastin. Because of their increased epidermal proliferation, vitamin A and B-carotene can both prevent photodamage and offer photoprotection. Furthermore, some varieties of mango, particularly the Ataulfo mango, have higher vitamin C content than what is needed daily in the diet.

In reference to the protective qualities of mangoes for the skin, a study found that mango extract made from dried mangos improves wrinkles, inhibits the loss of collagen fibers, and lessens the thickness and hypertrophy of the epidermis caused by UVB rays. Thus, the findings support mango's evident antiaging and photoprotective properties, of which mangiferin appears to be a major component. The mango intervention was administered orally to the mice used in this investigation. This data lays the groundwork for upcoming human interventions involving mango consumption.

The impact of eating Ataulfo mangoes on the onset of facial photodamage was examined in a different study. The results of this 8 to 16-week clinical trial (N = 28) showed that consuming 0.5 cups of frozen Ataulfo mangos can reduce the depth and severity of facial wrinkles. Serum triglycerides also decreased in comparison to basal values. Additionally, this study showed that eating 1.5 cups of mango increased the amount of wrinkles on the face. While the unexpected effects of consuming large amounts of mango may be related to the total sugar intake resulting in the glycation process of collagen fibers and consequent disruption of its structure, the beneficial effects that have been reported may be attributed to compounds found in mangos, such as mangiferin, carotenoids, and flavonoids. The primary issue with this intervention was the rise in blood glucose levels, but this parameter did not change. The findings thus demonstrated that the effects of a whole active compound complex in mangos reduce wrinkles and photodamage caused by UVB rays, underscoring the significance of eating a sufficient amount of food and avoiding overindulging in it, which can exacerbate skin conditions. Although there is currently little data supporting the benefits of mango consumption for skin health, the antioxidant qualities of the fruit's constituents, along with some encouraging new findings, should be enough to encourage consuming mangos as part of a healthy, well-balanced diet.

Additionally, it has been demonstrated in a different study that the antioxidant and bioavailability of mango phenolic acids remain intact regardless of whether the pulp is consumed or turned into juice, with the potential for improved properties occurring in the juice presentation. According to this data, eating mango in a variety of ways does not seem to impair its nutritional value.

8.5.2. Almond

Almonds are a food with antioxidant qualities because they are a rich source of fatty acids, polyphenols, and alpha-tocopherol, also known as vitamin E. Negar et al. examined how eating almonds affected the amount, depth, and duration of wrinkles on the face; the findings indicated a notable reduction in wrinkles. For 16 weeks, women (n = 31) in this control-trial study ate 2.1 oz (340 kcal) of packaged almonds. It was reported that the almond group did not experience any increase in sebum production or any other negative effects from their intake during the intervention period.

In another study, 56 postmenopausal women who consumed almonds every day saw a significant decrease in the average severity of wrinkles and the intensity of facial pigmentation without an increase in sebum production on the face. The antioxidant and photoprotective qualities of vitamin E, which is incorporated into the cell membrane and has a protective effect against free radicals and the damage they cause, maybe the cause of alpha tocopherol's beneficial effects against skin aging. Vitamin E may also be able to stop collagen deterioration because it reduces matrix metalloproteinase activity. Thus, consuming enough almonds each day may offer an alternative to stop or slow the development of wrinkles and facial pigmentation, which in turn slows down the aging process of the skin.

8.5.3. Avocado

Recent research has demonstrated that the components of avocados—carotenoids, monounsaturated fatty acids, and phenols—have health benefits. Avocado is well known for its topical application as a cream, which increases skin elasticity. However, recent research has shown that oral consumption of avocado also has skincare benefits. One example is Henning et al. revealed notable alterations in the facial skin of overweight women who included avocados in their daily diet in a pilot study (N = 39). The forehead and under-eye areas of the skin showed improvements in firmness and elasticity. Avocado consumption had no negative effects, such as weight gain or increased sebum production on the face. This better state of skin health could be the consequence of a process whereby dietary changes trigger the expression of specific genes, like those for collagen and elastin. Accordingly, the data indicates that including avocados in one's regular diet may help to preserve and enhance the skin's firmness and elasticity while also slowing down the aging process (Flores-Balderas et al., 2023).

8.6. ASSOCIATION BETWEEN PLANT-BASED DIET ALTERNATIVES AND GUT MICROBIOME

The last few decades have seen a significant increase in interest in the effects of various diets on the skin. However, because diets also significantly affect the gut microbiome, which can also have an impact on skin health, it is challenging to assess the direct effects of diets on skin health. In fact, a wealth of data has shown that the skin and intestines interact in both directions to either change or preserve their biological balance. For example, different gut microbial profiles are seen in animal models of patients with common skin disorders like psoriasis and AD, as well as less common conditions like rosacea.

The largest endocrine organ, the gut microbiome is situated between the food we eat and different body organs like the skin. It generates a variety of metabolites, such as bile acids, neurotransmitters, tryptophan derivatives, and short-chained fatty acids (SCFA). These metabolites have pleiotropic effects on distant organs and affect gut physiology, including gut barrier integrity. It is interesting to note that intestinal barrier integrity has been linked to skin health. Furthermore, a growing body of research is currently being conducted on the potential of certain bacterial metabolites to modulate the immune system at distant organ sites, such as the skin. Certain dietary metabolites can enter the skin directly, while others enter the skin through the metabolism of gut microbes. These metabolites may have an impact on the health of the skin as well as several diseases and disorders related to it.

Diet, among many other factors, has a significant impact on the gut microbiome's composition, which in turn affects the organism's immune and metabolic processes. Variations in the macronutrient composition and source (plant or animal) may have a special impact on the variety of microbiomes found in the gut. Furthermore, the composition and diversity of the human gut microbiota can be significantly influenced by the quantity and source of dietary fiber. For example, eating whole grains or cereal fibers can increase the amount of *Bifidobacterium*, which is reduced in certain diseases, such as AD.

Studies that show a higher bacterial diversity and abundance of potentially helpful bacteria with a high-quality diet or a diet rich in "healthy foods" further support the significance of the host dietary patterns on the gut microbiota. According to other research, diets that are low in animal protein and saturated fat and high in plants (a significant source of natural fiber) and omega-3 fatty acids, such as the Mediterranean diet, encourage the growth of bacteria that break down dietary fiber and produce SCFA. On the other hand, a high-fat diet causes changes in the gut microbiota's composition known as dysbiosis, which raises intestinal permeability by causing an increase in the release of pro-inflammatory cytokines (Cani et al., 2008).

Studies conducted decades ago found a link between eating a Western diet and several skin conditions, including psoriasis, acne, and AD. The physiopathology of these diseases has been linked to the infiltration of molecules resulting from increased intestinal permeability, which in turn introduces proinflammatory mediators into the circulation and can disturb skin homeostasis. Furthermore, the consumption of Western-style diets can induce bacterial metabolism, which can lead to the production of toxic metabolites like p-cresol, which are also linked to compromised epidermal barrier function.

Surprisingly, phytochemicals with anti-inflammatory and antioxidant properties are released and transformed by microbial metabolism during the consumption of plant-based diets. Furthermore, the integrity of the gut barrier may benefit from the generation of SCFA that is linked to a Mediterranean diet, which is likewise high in plants. Additionally capable of enhancing the skin barrier and reducing skin inflammation are SCFA, especially butyrate. In fact, eczema and AD patients have been shown to have low levels of fecal SCFA, and probiotics that produce SCFA can be used to treat other skin conditions by reducing inflammation.

Overall, the data from the latter study emphasizes how crucial the diet-gut-skin axis is. In fact, it has been suggested that certain diets may also alter the intestinal microbiome, leading to skin inflammation. For instance, following a gluten-free diet can help patients with dermatitis herpetiformis, a cutaneous sign of celiac disease, alleviate their skin rashes.

Studies have indicated that plant-based functional foods may be advantageous in the treatment of inflammatory diseases, particularly by enhancing skin health. The relationship between the intestinal microbiota and a variety of bacteria unique to these skin conditions, such as psoriasis, AD, and acne, has been connected to this improvement (Figure 8.3).

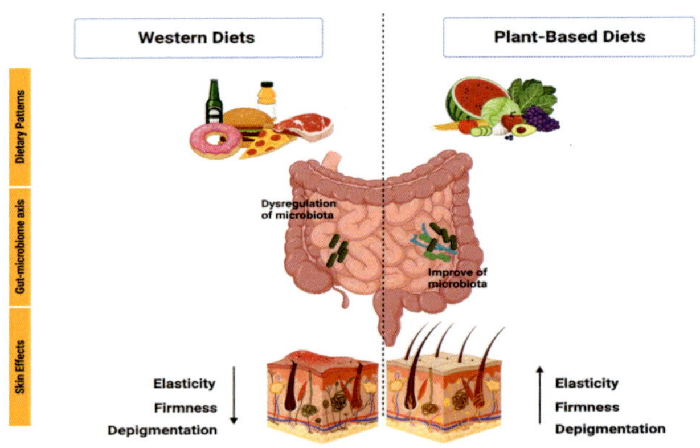

Figure 8.3. *Following a plant-based diet and maintaining it over time have demonstrated positive effects on the skin.*

Source: Flores-Balderas X, Peña-Peña M, Rada KM, Alvarez-Alvarez YQ, Guzmán-Martín CA, Sánchez-Gloria JL, Huang F, Ruiz-Ojeda D, Morán-Ramos S, Springall R, et al. (2023). Beneficial effects of plant-based diets on skin health and inflammatory skin diseases. Nutrients; 15(13):2842. https://doi.org/10.3390/nu15132842.

In a 12-week crossover design study, for instance, 27 participants had a 4-week washout period and consumed 100 kcal per day of either low-fat cookies or mangos. It is interesting to note that eating mangos raised the numbers *of Prevotella maculosa*, *Mogibacterium timidum*, and *Corynebacterium pyruviciproducens* while lowering *Prevotella copri*. It will be interesting to see if eating mangos can help treat psoriasis and other skin conditions because it has been demonstrated that subjects with psoriasis have higher levels of *Prevotella copri*.

In a study where 48 subjects received a daily dose of toasted almonds (56 g) or fructooligosaccharides (as a control) for six weeks to observe the effect of almond intake, it was found that almond or almond skin supplementation increased the populations of *Bifidobacterium* spp. Consuming almonds may help treat AD because it has been demonstrated that patients with the condition have lower levels of *Bifidobacterium*.

Lastly, it is critical to remember that the gut microbiome's components significantly regulate the skin microbiota and are involved in the physiopathology of several dermatoses. Prebiotics and probiotics can help maintain, prevent, or restore skin health by improving the gut microbiome and reducing intestinal inflammation brought on by dietary changes to the intestinal microflora. Considering everything mentioned above, restoring the gut microflora through supplementation and dietary modifications may aid in the treatment of certain skin conditions by restoring the microbiome (Flores-Balderas et al., 2023).

REFERENCES

1. Aalemi, A. K., Anwar, I., & Chen, H. (2019). Dairy consumption and acne: A case control study in Kabul, Afghanistan. *Clinical, Cosmetic and Investigational Dermatology, 12*, 481–487.
2. Afifi, L., Danesh, M. J., Lee, K. M., Beroukhim, K., Farahnik, B., Ahn, R. S., Yan, D., Singh, R. K., Nakamura, M., Koo, J., et al. (2017). Dietary behaviors in psoriasis: Patient-reported outcomes from a U.S. national survey. *Dermatology and Therapy, 7*, 227–242.
3. Armstrong, A. W., & Read, C. (2020). Pathophysiology, clinical presentation, and treatment of psoriasis: A review. *JAMA, 323*, 1945–1960.
4. Asuncion, P., Liu, C., Castro, R., Yon, V., Rosas, M., Hooshmand, S., Kern, M., & Hong, M. Y. (2023). The effects of fresh mango consumption on gut health and microbiome—Randomized controlled trial. *Food Science & Nutrition, 11*, 2069–2078.
5. Baldwin, H., & Tan, J. (2021). Effects of diet on acne and its response to treatment. *American Journal of Clinical Dermatology, 22*, 55–65.
6. Beylot, C., Auffret, N., Poli, F., Claudel, J. P., Leccia, M. T., Del Giudice, P., & Dreno, B. (2014). Propionibacterium acnes: An update on its role in the pathogenesis of acne. *Journal of the European Academy of Dermatology and Venereology, 28*, 271–278.
7. Bonjour, M., Gabriel, S., Valencia, A., Goldhamer, A. C., & Myers, T. R. (2022). Challenging case in clinical practice: Prolonged water-only fasting followed by an exclusively whole-plant-food diet in the management of severe plaque psoriasis. *Integrative Complementary Therapies, 28*, 85–87.
8. Cani, P. D., Bibiloni, R., Knauf, C., Waget, A., Neyrinck, A. M., Delzenne, N. M., & Burcelin, R. (2008). Changes in gut microbiota control metabolic endotoxemia-induced inflammation in high-fat diet-induced obesity and diabetes in mice. *Diabetes, 57*, 1470–1481.
9. Chambers, E. S., & Vukmanovic-Stejic, M. (2020). Skin barrier immunity and ageing. *Immunology, 160*, 116–125.
10. Coates, M., Lee, M. J., Norton, D., & MacLeod, A. S. (2019). The skin and intestinal microbiota and their specific innate immune systems. *Frontiers in Immunology, 10*, 496479.
11. Craig, W. J., & Mangels, A. R. (2009). Position of the American Dietetic Association: Vegetarian diets. *Journal of the American Dietetic Association, 109*, 1266–1282.
12. Dąbrowska, A. K., Spano, F., Derler, S., Adlhart, C., Spencer, N. D., & Rossi, R. M. (2018). The relationship between skin function, barrier

properties, and body-dependent factors. *Skin Research and Technology, 24*, 165–174.

13. Dattola, A., Silvestri, M., Bennardo, L., Passante, M., Scali, E., Patruno, C., & Nisticò, S. P. (2020). Role of vitamins in skin health: A systematic review. *Current Nutrition Reports, 9*, 226–235.
14. David, L. A., Maurice, C. F., Carmody, R. N., Gootenberg, D. B., Button, J. E., Wolfe, B. E., Ling, A. V., Devlin, A. S., Varma, Y., Fischbach, M. A., et al. (2014). Diet rapidly and reproducibly alters the human gut microbiome. *Nature, 505*, 559–563.
15. Dowlatshahi, E. A., Van Der Voort, E. A. M., Arends, L. R., & Nijsten, T. (2013). Markers of systemic inflammation in psoriasis: A systematic review and meta-analysis. *British Journal of Dermatology, 169*, 266–282.
16. Dreher, M. L. (2021). A comprehensive review of almond clinical trials on weight measures, metabolic health biomarkers and outcomes, and the gut microbiota. *Nutrients, 13*, 1968.
17. Dreher, M. L., & Davenport, A. J. (2013). Hass avocado composition and potential health effects. *Critical Reviews in Food Science and Nutrition, 53*, 738.
18. Dréno, B. (2017). What is new in the pathophysiology of acne, an overview. *Journal of the European Academy of Dermatology and Venereology, 31*(Suppl. S5), 8–12.
19. Fam, V. W., Charoenwoodhipong, P., Sivamani, R. K., Holt, R. R., Keen, C. L., & Hackman, R. M. (2022). Plant-based foods for skin health: A narrative review. *Journal of the Academy of Nutrition and Dietetics, 122*, 614–629.
20. Fam, V. W., Holt, R. R., Keen, C. L., Sivamani, R. K., & Hackman, R. M. (2020). Prospective evaluation of mango fruit intake on facial wrinkles and erythema in postmenopausal women: A randomized clinical pilot study. *Nutrients, 12*, 3381.
21. Fang, Z., Pan, T., Li, L., Wang, H., Zhu, J., Zhang, H., Zhao, J., Chen, W., & Lu, W. (2022). *Bifidobacterium longum* mediated tryptophan metabolism to improve atopic dermatitis via the gut-skin axis. *Gut Microbes, 14*, 2044723.
22. Flores-Balderas, X., Peña-Peña, M., Rada, K. M., Alvarez-Alvarez, Y. Q., Guzmán-Martín, C. A., Sánchez-Gloria, J. L., Huang, F., Ruiz-Ojeda, D., Morán-Ramos, S., Springall, R., et al. (2023). Beneficial effects of plant-based diets on skin health and inflammatory skin diseases. *Nutrients, 15*(13), 2842. https://doi.org/10.3390/nu15132842.
23. Foolad, N., Vaughn, A. R., Rybak, I., Burney, W. A., Chodur, G. M.,

Newman, J. W., Steinberg, F. M., & Sivamani, R. K. (2019). Prospective randomized controlled pilot study on the effects of almond consumption on skin lipids and wrinkles. *Phytotherapy Research, 33*, 3212–3217.

24. Frazier, W., & Bhardwaj, N. (2020). Atopic dermatitis: Diagnosis and treatment. *American Family Physician, 101*, 590–598.
25. Fusano, M. (2022). Veganism in acne, atopic dermatitis, and psoriasis: Benefits of a plant-based diet. *Clinical Dermatology, 41*(1), 122–126.
26. Fusano, M., Fusano, I., Galimberti, M. G., Bencini, M., & Bencini, P. L. (2020). Comparison of postsurgical scars between vegan and omnivore patients. *Dermatologic Surgery, 46*, 1572–1576.
27. Ghosh, T. S., Rampelli, S., Jeffery, I. B., Santoro, A., Neto, M., Capri, M., Giampieri, E., Jennings, A., Candela, M., Turroni, S., et al. (2020). Mediterranean diet intervention alters the gut microbiome in older people reducing frailty and improving health status: The NU-AGE 1-year dietary intervention across five European countries. *Gut, 69*, 1218–1228.
28. Giampieri, F., Alvarez-Suarez, J. M., Mazzoni, L., Forbes-Hernandez, T. Y., Gasparrini, M., Gonzàlez-Paramàs, A. M., Santos-Buelga, C., Quiles, J. L., Bompadre, S., Mezzetti, B., et al. (2014). Polyphenol-rich strawberry extract protects human dermal fibroblasts against hydrogen peroxide oxidative damage and improves mitochondrial functionality. *Molecules, 19*, 7798–7816.
29. Hao, Y., Zhu, Y. J., Zou, S., Zhou, P., Hu, Y. W., Zhao, Q. X., Gu, L. N., Zhang, H. Z., Wang, Z., & Li, J. (2021). Metabolic syndrome and psoriasis: Mechanisms and future directions. *Frontiers in Immunology, 12*, 711060.
30. Hasler, C. M., Kundrat, S., & Wool, D. (2000). Functional foods and cardiovascular disease. *Current Atherosclerosis Reports, 2*, 467–475.
31. Henning, S. M., Guzman, J. B., Thames, G., Yang, J., Tseng, C. H., Heber, D., Kim, J., & Li, Z. (2022). Avocado consumption increased skin elasticity and firmness in women—A pilot study. *Journal of Cosmetic Dermatology, 21*, 4028–4034.
32. Jefferson, A., & Adolphus, K. (2019). The effects of intact cereal grain fibers, including wheat bran, on the gut microbiota composition of healthy adults: A systematic review. *Frontiers in Nutrition, 6*, 33.
33. Jones, V. A., Patel, P. M., Wilson, C., Wang, H., & Ashack, K. A. (2020). Complementary and alternative medicine treatments for common skin diseases: A systematic review and meta-analysis. *JAAD International, 2*, 76–93.
34. Kamiya, K., Kishimoto, M., Sugai, J., Komine, M., & Ohtsuki, M. (2019).

Risk factors for the development of psoriasis. *International Journal of Molecular Sciences, 20*, 4347.

35. Kawai, T., Autieri, M. V., & Scalia, R. (2021). Adipose tissue inflammation and metabolic dysfunction in obesity. *American Journal of Physiology-Cell Physiology, 320*, C375–C391.
36. Kent, G., Kehoe, L., Flynn, A., & Walton, J. (2022). Plant-based diets: A review of the definitions and nutritional role in the adult diet. *Proceedings of the Nutrition Society, 81*, 62–74.
37. Kim, J., Kim, B. E., & Leung, D. Y. M. (2019). Pathophysiology of atopic dermatitis: Clinical implications. *Allergy and Asthma Proceedings, 40*, 84–92.
38. Koponen, K. K., Salosensaari, A., Ruuskanen, M. O., Havulinna, A. S., Männistö, S., Jousilahti, P., Palmu, J., Salido, R., Sanders, K., Brennan, C., et al. (2021). Associations of healthy food choices with gut microbiota profiles. *American Journal of Clinical Nutrition, 114*, 605–616.
39. Kouda, K., Tanaka, T., Kouda, M., Takeuchi, H., Takeuchi, A., Nakamura, H., & Takigawa, M. (2000). Low-energy diet in atopic dermatitis patients: Clinical findings and DNA damage. *Journal of Physiology and Anthropology, 19*, 225–228.
40. Lazar, V., Ditu, L. M., Pircalabioru, G. G., Gheorghe, I., Curutiu, C., Holban, A. M., Picu, A., Petcu, L., & Chifiriuc, M. C. (2018). Aspects of gut microbiota and immune system interactions in infectious diseases, immunopathology, and cancer. *Frontiers in Immunology, 9*, 1830.
41. Lebaka, V. R., Wee, Y. J., Ye, W., & Korivi, M. (2021). Nutritional composition and bioactive compounds in three different parts of mango fruit. *International Journal of Environmental Research and Public Health, 18*, 741.
42. Lee, H. J., & Kim, M. (2022). Skin barrier function and the microbiome. *International Journal of Molecular Sciences, 23*, 13071.
43. Levy, M., Blacher, E., & Elinav, E. (2017). Microbiome, metabolites and host immunity. *Current Opinion in Microbiology, 35*, 8–15.
44. Lewandowska, M., Dunbar, K., & Kassam, S. (2021). Managing psoriatic arthritis with a whole food plant-based diet: A case study. *American Journal of Lifestyle Medicine, 15*, 402–406.
45. Li, Y., Wang, D. D., Satija, A., Ivey, K. L., Li, J., Wilkinson, J. E., Li, R., Baden, M., Chan, A. T., Huttenhower, C., et al. (2021). Plant-based diet index and metabolic risk in men: Exploring the role of the gut microbiome. *The Journal of Nutrition, 151*, 2780.
46. Mahmud, M. R., Akter, S., Tamanna, S. K., Mazumder, L., Esti, I. Z.,

Banerjee, S., Akter, S., Hasan, R., Acharjee, M., Hossain, S., et al. (2022). Impact of gut microbiome on skin health: Gut-skin axis observed through the lenses of therapeutics and skin diseases. *Gut Microbes, 14*, 2096995.

47. Mann, E. A., Bae, E., Kostyuchek, D., Chung, H. J., & McGee, J. S. (2020). The gut microbiome: Human health and inflammatory skin diseases. *Annals of Dermatology, 32*, 265.

48. Melnik, B. C. (2012). Dietary intervention in acne: Attenuation of increased mTORC1 signaling promoted by Western diet. *Dermatoendocrinology, 4*, 20–32.

49. Michalek, I. M., Loring, B., & John, S. M. (2017). A systematic review of worldwide epidemiology of psoriasis. *Journal of the European Academy of Dermatology and Venereology, 31*, 205–212.

50. Mohania, D., Chandel, S., Kumar, P., Verma, V., Digvijay, K., Tripathi, D., Choudhury, K., Mitten, S. K., & Shah, D. (2017). Ultraviolet radiations: Skin defense-damage mechanism. *Advances in Experimental Medicine and Biology, 996*, 71–87.

51. Naldi, L., Conti, A., Cazzaniga, S., Patrizi, A., Pazzaglia, M., Lanzoni, A., Veneziano, L., Pellacani, G., & the Psoriasis Emilia Romagna Study Group. (2014). Diet and physical exercise in psoriasis: A randomized controlled trial. *British Journal of Dermatology, 170*, 634–642.

52. Nayak, R. R. (2021). Western diet and psoriatic-like skin and joint diseases: A potential role for the gut microbiota. *The Journal of Investigative Dermatology, 141*, 1630–1632.

53. Nayak, S. N., Aravind, B., Malavalli, S. S., Sukanth, B. S., Poornima, R., Bharati, P., Hefferon, K., Kole, C., & Puppala, N. (2021). Omics technologies to enhance plant-based functional foods: An overview. *Frontiers in Genetics, 12*, 742095.

54. Nongnuch, A., & Davenport, A. (2015). The effect of vegetarian diet on skin autofluorescence measurements in haemodialysis patients. *British Journal of Nutrition, 113*, 1040–1043.

55. Nutten, S. (2015). Atopic dermatitis: Global epidemiology and risk factors. *Annals of Nutrition and Metabolism, 66*(Suppl. S1), 8–16.

56. Pasparakis, M., Haase, I., & Nestle, F. O. (2014). Mechanisms regulating skin immunity and inflammation. *Nature Reviews Immunology, 14*, 289–301.

57. Pessemier, B., De Grine, L., Debaere, M., Maes, A., Paetzold, B., & Callewaert, C. (2021). Gut-skin axis: Current knowledge of the interrelationship between microbial dysbiosis and skin conditions. *Microorganisms, 9*, 353.

58. Peters, B. A., Xing, J., Chen, G. C., Usyk, M., Wang, Z., McClain, A. C., Thyagarajan, B., Daviglus, M. L., Sotres-Alvarez, D., Hu, F. B., et al. (2022). Healthy dietary patterns are associated with the gut microbiome in the Hispanic Community Health Study/Study of Latinos. *American Journal of Clinical Nutrition, 117*, 540–552.

59. Pinela, J., Carocho, M., Dias, M. I., Caleja, C., Barros, L., & Ferreira, I. C. F. R. (2016). Wild plant-based functional foods, drugs, and nutraceuticals. In *Wild Plants, Mushrooms and Nuts: Functional Food Properties and Applications* (pp. 315–351). Wiley Online Library.

60. Quirós-Sauceda, A. E., Oliver Chen, C. Y., Blumberg, J. B., Astiazaran-Garcia, H., Wall-Medrano, A., & González-Aguilar, G. A. (2017). Processing "Ataulfo" mango into juice preserves the bioavailability and antioxidant capacity of its phenolic compounds. *Nutrients, 9*, 1082.

61. Rendon, A., & Schäkel, K. (2019). Psoriasis pathogenesis and treatment. *International Journal of Molecular Sciences, 20*, 1475.

62. Rhodes, L. E., Darby, G., Massey, K. A., Clarke, K. A., Dew, T. P., Farrar, M. D., Bennett, S., Watson, R. E. B., Williamson, G., & Nicolaou, A. (2013). Oral green tea catechin metabolites are incorporated into human skin and protect against UV radiation-induced cutaneous inflammation in association with reduced production of pro-inflammatory eicosanoid 12-hydroxyeicosatetraenoic acid. *British Journal of Nutrition, 110*, 891–900.

63. Richmond, J. M., & Harris, J. E. (2014). Immunology and skin in health and disease. *Cold Spring Harbor Perspectives in Medicine, 4*, a015339.

64. Rybak, I., Carrington, A. E., Dhaliwal, S., Hasan, A., Wu, H., Burney, W., Maloh, J., & Sivamani, R. K. (2021). Prospective randomized controlled trial on the effects of almonds on facial wrinkles and pigmentation. *Nutrients, 13*, 785.

65. Sánchez-Tapia, M., Tovar, A. R., & Torres, N. (2019). Diet as regulator of gut microbiota and its role in health and disease. *Archives of Medical Research, 50*, 259–268.

66. Schade, L., Mesa, D., Faria, A. R., Santamaria, J. R., Xavier, C. A., Ribeiro, D., Hajar, F. N., & Azevedo, V. F. (2022). The gut microbiota profile in psoriasis: A Brazilian case-control study. *Letters in Applied Microbiology, 74*, 498–504.

67. Seethaler, B., Nguyen, N. K., Basrai, M., Kiechle, M., Walter, J., Delzenne, N. M., & Bischoff, S. C. (2022). Short-chain fatty acids are key mediators of the favorable effects of the Mediterranean diet on intestinal barrier integrity: Data from the randomized controlled LIBRE trial. *American Journal of Clinical Nutrition, 116*, 928–942.

68. Shi, Z., Wu, X., Yu, S., Huynh, M., Jena, P. K., Nguyen, M., Wan, Y. J. Y., & Hwang, S. T. (2020). Short-term exposure to a Western diet induces psoriasiform dermatitis by promoting accumulation of IL-17A-producing γδ T cells. *The Journal of Investigative Dermatology, 140*, 1815–1823.
69. Sinha, S., Lin, G., & Ferenczi, K. (2021). The skin microbiome and the gut-skin axis. *Clinical Dermatology, 39*, 829–839.
70. Smith, M., Wright, N., McHugh, P., & Duncan, B. (2021). Remission of long-standing livedoid vasculopathy using a whole foods plant-based diet with symptoms recurrent on re-challenge with standard Western diet. *BMJ Case Reports, 14*, e237895.
71. Solanki, V. K., & Nair, P. M. K. (2023). Lifestyle medicine approach in managing pemphigus vulgaris: A case report. *Explore*, *19*(4), 617–620.
72. Solway, J., McBride, M., Haq, F., Abdul, W., & Miller, R. (2020). Diet and dermatology: The role of a whole-food, plant-based diet in preventing and reversing skin aging—A review. *Journal of Clinical and Aesthetic Dermatology, 13*, 38.
73. Song, J. H., Bae, E. Y., Choi, G., Hyun, J. W., Lee, M. Y., Lee, H. W., & Chae, S. (2013). Protective effect of mango (*Mangifera indica* L.) against UVB-induced skin aging in hairless mice. *Photodermatology, Photoimmunology & Photomedicine, 29*, 84–89.
74. Stewart, T. J., & Bazergy, C. (2018). Hormonal and dietary factors in acne vulgaris versus controls. *Dermato-Endocrinology, 10*, e1442160.
75. Suppiah, T. S. S., Sundram, T. K. M., Tan, E. S. S., Lee, C. K., Bustami, N. A., & Tan, C. K. (2018). Acne vulgaris and its association with dietary intake: A Malaysian perspective. *Asia Pacific Journal of Clinical Nutrition, 27*, 1141–1145.
76. Tanaka, T., Kouda, K., Kotani, M., Takeuchi, A., Tabei, T., Masamoto, Y., Nakamura, H., Takigawa, M., Suemura, M., Takeuchi, H., et al. (2001). Vegetarian diet ameliorates symptoms of atopic dermatitis through reduction of the number of peripheral eosinophils and of PGE2 synthesis by monocytes. *Journal of Physiology and Anthropology, 20*, 353–361.
77. Temple, N. J. (2022). A rational definition for functional foods: A perspective. *Frontiers in Nutrition, 9*, 957516.
78. Toyoda, M., & Morohashi, M. (2001). Pathogenesis of acne. *Medical Electron Microscopy, 34*, 29–40.
79. Trompette, A., Pernot, J., Perdijk, O., Alqahtani, R. A. A., Domingo, J. S., Camacho-Muñoz, D., Wong, N. C., Kendall, A. C., Wiederkehr, A., Nicod, L. P., et al. (2022). Gut-derived short-chain fatty acids modulate

skin barrier integrity by promoting keratinocyte metabolism and differentiation. *Mucosal Immunology, 15*, 908–926.

80. Wang, D. D., Nguyen, L. H., Li, Y., Yan, Y., Ma, W., Rinott, E., Ivey, K. L., Shai, I., Willett, W. C., Hu, F. B., et al. (2021). The gut microbiome modulates the protective association between a Mediterranean diet and cardiometabolic disease risk. *Nature Medicine, 27*, 333–343.
81. Watanabe, S., Narisawa, Y., Arase, S., Okamatsu, H., Ikenaga, T., Tajiri, Y., & Kumemura, M. (2003). Differences in fecal microflora between patients with atopic dermatitis and healthy control subjects. *The Journal of Allergy and Clinical Immunology, 111*, 587–591.
82. Xiao, X., Hu, X., Yao, J., Cao, W., Zou, Z., Wang, L., Qin, H., Zhong, D., Li, Y., Xue, P., et al. (2023). The role of short-chain fatty acids in inflammatory skin diseases. *Frontiers in Microbiology, 13*, 1083432.
83. Youssef, E. M. K., & Youssef, M. K. E. (2014). Diet and acne in Upper Egypt. *American Journal of Dermatology and Venereology, 3*, 13–22.
84. Zhang, J., Loman, L., Oldhoff, M., & Schuttelaar, M. L. A. (2022). Association between moderate to severe atopic dermatitis and lifestyle factors in the Dutch general population. *Clinical and Experimental Dermatology, 47*, 1523–1535.
85. Zhang, S., & Duan, E. (2018). Fighting against skin aging: The way from bench to bedside. *Cell Transplantation, 27*, 729–738.
86. Zheng, D., Liwinski, T., & Elinav, E. (2020). Interaction between microbiota and immunity in health and disease. *Cell Research, 30*, 492–506.

CHAPTER 9

FUNCTIONAL FOOD FOR ELDERLY HIGH IN ANTIOXIDANT AND CHICKEN EGGSHELL CALCIUM TO REDUCE THE RISK OF OSTEOPOROSIS

Contents

INTRODUCTION

The population of the elderly, defined as those 65 years of age or older, is growing annually. According to reports, there were 703 million elderly people in the world in 2019. Eastern and South-Eastern Asia had the largest population of elderly people, accounting for 260 million, followed by Europe and North America with over 200 million. The world's senior population is expected to double to 1.5 billion by 2050. The proportion of the elderly in the general population increased to 9% in 2019 from 6% in 1990 and is expected to reach 16% in 2050. Accordingly, one in six people will be 65 or older in 2050. More than one-fifth of the population in the European Union-27 (20.3%) was 65 years of age or older in 2019. This represents an increase of 20.9% when compared to the elderly population (17.4%) in 2009.

Because the number of elderly people is rising, efforts to improve their quality of life—such as addressing the public health issue of osteoporosis—must be taken seriously. It is currently estimated that over 200 million people worldwide suffer from osteoporosis. One in five men and one in three women over the age of 50 will suffer an osteoporotic fracture at some point in their lives. Osteoporosis affected about 10% of women in their 60s, 5% of women in their 70s, 2% of women in their 80s, and 2/3 of women in their 90s. About 30% of postmenopausal women in the US and Europe have osteoporosis, and 40%t of them have fragility fractures.

A review based on recent research found that functional foods containing dietary fiber, phytoestrogens, omega-3 polyunsaturated fatty acid, polyphenols, carotenoids, lutein, zeaxanthin, prebiotics, probiotics, synbiotics, plant sterols, and stanols are necessary for the elderly to reduce their risk of various health problems. It has been observed that antioxidants lessen oxidative stress, which may help lower the risk of osteoporosis.

Resolving environmental issues, such as food loss and waste, is just as crucial as health issues. Currently, the world's most difficult problems are food loss and waste, even to feed 9 billion people by the year 2050. According to a report by the Food and Agriculture Organization of the United Nations, minimizing food loss and waste has the potential to significantly reduce production costs, enhance food security and nutrition, and promote environmental sustainability by lessening the strain on natural resources and lowering greenhouse gas emissions. The edible portions of plants and animals that are produced or harvested for human consumption but are not ultimately consumed by humans are referred to as food loss and waste. Food waste is defined as good-quality, fit-for-human consumption food that is thrown away either before or after it spoils. Food loss is defined as the reduction in food quantity or quality that renders it unfit for human consumption. Edible material that is purposefully fed to animals or that is a byproduct of food processing that is taken away from food intended for human consumption is also considered waste. Food waste at the consumption stage ranged from 6–11 kg/person/year in sub-Saharan Africa and South/Southeast Asia to 95–115 kg/person/year in North America and Europe. Among the European nations, the Netherlands (579 kg), Belgium (399 kg), and Cyprus (334 kg) had the highest per capita amounts of food waste per year, while Greece (44 kg), Malta (62 kg), and the Czech Republic (71 kg) had the lowest amounts. Every year, about 235 kg of food is wasted per person in Poland. Food loss and waste in Arab nations amounted to 210 kg per person annually.

Food wastes with high calcium contents have been found to include fish bones, seashells, oyster shells, quail eggshells, chicken eggshells, and duck eggshells. Calcium is essential for preventing osteoporosis. Eggshells can be used as operating supplies or as a raw material for the production of new products. Eggshells can be used as a food additive, soil amendment, cosmetics, biomaterial composite, and purified calcium carbonate ($CaCO_3$) in their raw form. Eggshells are reusable as sorbents and catalysts while in operation.

Food is one of the key elements that determines a healthy lifestyle, so senior citizens need to pay attention to it. The body's food supply is the most effective way to supplement calcium deficiency. But often, it is insufficient, so researchers are always looking for new, easily absorbed forms of calcium from alternate sources. However, it is important to remember that chicken eggshells are a fantastic natural calcium source with a lot of culinary applications. To lower the risk of osteoporosis in the elderly, this section tries to investigate functional foods, especially those high in antioxidants and calcium from chicken eggshells (Arnold et al., 2021).

9.1. BONE AND OSTEOPOROSIS

The primary calcified tissue in vertebrates is called bone, and it has several uses including storage, protection, and mechanical support. Bone is composed of 30% organic matrix, 60% mineral crystals (crystalline hydroxyapatite), and 10% cells. The two primary cells that remodel bone are osteoblasts and osteoclasts. While the osteoclast is a type of multinucleated giant cell that resorbs bone, the osteoblast is a bone-forming cell that sends signals to the osteoclasts to form. Over the course of several weeks, osteoblasts create new bone while osteoclasts remove the old or damaged bone in the process of bone remodeling. Nutritional and cellular factors, such as the availability of oxygen, nutrients, endocrines, cytokines, growth factors, and free radicals, can affect the activities of bone cells. Osteoporosis will eventually result from an imbalance between the activities of osteoblasts and osteoclasts. Low bone mineral density, deteriorating bone microarchitecture, decreased bone mass, and increased bone fragility are the hallmarks of osteoporosis, a health issue that is thought to be age-related. Osteoporosis may raise the risk of fragility fracture and have a detrimental impact on the population's quality of life. Osteoporosis risk factors are classified as modifiable or unmodifiable. Modifiable factors include oxidative stress-related factors, such as low antioxidant status. Lack of calcium intake is a significant contributing factor to osteoporosis because it causes a suboptimal bone mass peak and low bone mineralization. Findings from Cao and Rana et al. show that severe co-morbidities, such as bone loss, are also linked to childhood obesity.

Previous research has reported the prevalence of osteoporosis. Even with available preventive treatments, osteoporosis still poses a greater risk to the elderly. Women experience osteoporosis at a higher rate than men do, and after menopause, the condition becomes much more common. Increased osteoclastic function causes bone loss in postmenopausal osteoporosis, which is linked to a deficiency of estrogen, a hormone that plays a significant role in determining bone mineral density. The immune system and bone diseases, including osteoporosis and rheumatoid arthritis, have been linked in recent years. It has been shown that ovariectomy triggers the T-cell costimulatory immunomodulatory cytokine LIGHT (lymphotoxin-like inducible protein), which in turn increases osteoblastogenesis and osteoclastogenesis by modifying the expression of osteoclastogenic cytokines. It was discovered that LIGHT mediated the bone loss caused by ovariectomy, indicating that LIGHT antagonistic therapy may benefit postmenopausal osteoporosis patients.

In the UK, 15% in France and Germany, 16% in the USA, and 38% in Japan, women 50 years of age and older had a higher prevalence of osteoporosis than men. It was 1% in the UK, 4% in Japan, 3% in Canada, and 8% in France, Germany, Italy, and Spain for men 50 years of age and older. Anemia and a history of fractures in older men are linked to an increased risk of osteoporosis. Approximately 80–90% of Polish patients with osteoporosis—including those with osteoporotic fractures—do not receive enough medication.

9.1.1. Role of Oxidative Stress in Osteoporosis

One of the risk factors for osteoporosis is oxidative stress. Oxidative stress can be brought on by an imbalance between the generation of free radicals and antioxidant capacity. This oxidative stress is linked to the pathophysiology of several chronic diseases. An atom or molecule that has one unpaired electron is called a free radical. Examples of these include lipoperoxide radical ($LOO^•$), hydroxyl radical($^•OH$), and superoxide anion ($O_2^{•-}$). The radicals generated *in vivo* are known as reactive oxygen species (ROS) or reactive nitrogen species (RNS), and they include peroxynitrite (ONOO−), nitric monoxide ($NO^•$), and nitrogen dioxide ($NO_2^{\bullet}$). Because oxidative stress increases the expression of cytokines in the bone, it can lead to osteoporosis. The process of oxidative stress-induced bone resorption is mediated by the activation of nuclear factor-κB protein, an essential mediator of osteoclastogenetic activity and tumor necrosis factor-α (TNF-α).

The activation of several enzymes, including the cytoplasmic enzyme superoxide dismutase, the membrane enzyme nicotinamide adenine dinucleotide phosphate oxidase, or NADPH oxidase, and several mitochondrial oxidases, results in the production of ROS during normal metabolism. Superoxide and hydrogen peroxide are two examples of ROS that regulate osteoclastic resorption bone activity, and superoxide produced by osteoclasts plays a role in bone deterioration. Certain conditions, such as vitamin D deficiency, fractures, and aging, can cause an increase in the production of ROS in osteoclasts, which in turn causes osteoporosis. Because ROS have a very short half-life, measuring them is challenging. The only way to quantify ROS is to look at the damage they do to DNA, lipids, and proteins—damage that manifests as chronic illnesses like osteoporosis (Arnold et al., 2021).

9.1.2. Effects of Antioxidants in Osteoporosis

Antioxidants have the ability to lower the risk of osteoporosis and provide protection against oxidative stress. Either directly or by opposing the effects of oxidants, antioxidants support the activation of osteoblast differentiation, mineralization, and the reduction of osteoclast activity. To reduce the risk of oxidative damage, research has focused on endogenous antioxidants found in fruits and vegetables (such as glutathione peroxidase (GPXs), catalase, superoxide dismutase, and metal-chelating proteins) and exogenous antioxidants found in diets (such as polyphenols, carotenoids, vitamin C, and vitamin E). Wilson et al. examined the benefits of functional foods and the effects of antioxidants on the health of the elderly. According to their research, antioxidants play a significant part in the treatment of chronic illnesses that affect the elderly, such as diabetes, cancer, heart disease, Alzheimer's disease, and digestive system issues. Nevertheless, this section did not review the impact of antioxidants on osteoporosis in the elderly. Studies have also been conducted on the role that lycopene and other antioxidants play in the prevention and management of osteoporosis, particularly in postmenopausal women.

9.1.2.1. Vitamin C

As the main antioxidant, vitamin C can eliminate RNS and ROS, which reduces oxidative stress and the risk of osteoporosis. Reduced ascorbate intake has been linked to increased fracture risk, decreased bone mass, and a faster rate of bone loss. Vitamin C supplementation may result in improvements to bone mineral density and oxidative stress. Higher vitamin C intake was found to be associated with a decreased risk of osteoporosis in Korean adults over 50 who did not engage in much physical activity, but no discernible effect was observed in those who did. This information was reported by Kim and Lee. Moreover, increased vitamin C intake through diet was linked to decreased hip fracture and osteoporosis risk as well as increased bone mineral density, especially at the lumbar spine and femoral neck.

Previous research has demonstrated the function of vitamin C in lowering the risk of osteoporosis by improving collagen synthesis, raising calcium absorption, potentially influencing the formation of new bone, and reducing the production of excessive free radicals. Higher quantities of collagen were obtained at 200 μg/mL of vitamin C as opposed to 25 and 100 μg/mL. The hydroxylation of proline and lysine residues within collagen fiber is facilitated by the activation of enzymes for which vitamin C is necessary. This hydroxylation reaction creates covalent bonds between the amino acid residues, which strengthens collagen overall.

In higher concentrations, vitamin C may not only have antioxidant qualities but also prooxidant qualities. A previous study indicated that supplementing with 500 mg of vitamin C daily for six weeks could increase oxidative DNA damage in men and women (aged 17 to 49), which may have implications for osteoporosis. Additionally, it was initially observed that vitamin C functioned as an oxidant during the osteoclastogenesis process, causing oxidative stress to speed up the formation of osteoclasts but ultimately leading to their death. The primary sources of vitamin C in the human diet are tomatoes, broccoli, kale, cabbage, and citrus fruits like pepper and kiwifruit. As a result, the food market also provides a wide range of vitamin C-fortified products.

9.1.2.2. Vitamin E

Tocopherols and tocotrienols are the two main subgroups of vitamin E, which is an antioxidant. There are four unique analogs for each of these (gamma, beta, delta, and alpha). Osteoprotective effects are demonstrated by both vitamin C and the antioxidant vitamin E. As a scavenger of free radicals, vitamin E deficiency lowers bone density and hinders the intestinal transit of calcium. Vitamins C and E prevent the body from producing too many free radicals, which in turn regulates the levels of malondialdehyde (MDA), a sign of lipid peroxidation. Lipid peroxidation is a result of increased free radical formation brought on by osteoclast activity.

One of the vitamin E homologs is α-tocopherol. Because α-tocopherol is a powerful lipoperoxyl radical scavenger, it is of great interest. Nevertheless, current research indicates that vitamin E homologs may have distinct effects on bone indices. While research has shown that vitamin E has anti-inflammatory benefits for bone, human studies have not examined this relationship.

Studies on older adults have documented both beneficial and detrimental effects of vitamin E. Shi and team investigated the connection between bone mineral density and vitamin E (in diet and serum) in middle-aged and elderly Chinese adults (40–75 years old) in Guangzhou. The findings showed that higher serum levels and higher vitamin E consumption are linked to higher bone mineral density in Chinese women, but not in Chinese men. Nonetheless, an additional investigation conducted on senior men residing in Uppsala, Sweden, revealed that vitamin E, or α-tocopherol, can preserve their bone mineral density. Additionally, it was reported that there was a negative correlation between the concentration of α-tocopherol in serum and the femoral neck bone mineral density in the elderly population of the United States (men and women 50 years of age and older), indicating a detrimental effect of α-tocopherol on bone health.

9.1.2.3. Polyphenols

Polyphenols and the carotenoid lycopene (beneficially inhibit the development of chronic diseases brought on by oxidative stress) such as diabetes, cancer, asthma, osteoporosis, and cardiovascular disease (CVD). You can increase your intake of these antioxidants by eating more fruits and vegetables.

Plants (fruits, vegetables, grains, spices, and herbs) naturally contain water-soluble polyphenols. Three primary subgroups of polyphenols are flavonoids, non-flavonoids, and phenolic acids. Benzoic and cinnamic acids are the two primary types of phenolic acids, which are hydroxyl derivatives of aromatic carboxylic acids with a single phenolic ring. About 30% of the bound and free forms of phenolic compounds found in food plants are derived from phenolic acids. Two phenolic rings are present in flavonoids, and they are connected by a three-carbon bridge, which is typically an oxygenated heterocycle. Due to their aromatic characteristics and highly conjugated system with multiple hydroxyl groups, polyphenols can neutralize free radicals and other ROS because they are good donors of electrons or hydrogen atoms. Consuming polyphenols has been shown to have positive effects on many chronic illnesses, such as cancer, osteoporosis, and cardiometabolic risk.

Research on polyphenols and their potential to lower osteoporosis risk was published. Consuming tea that contains polyphenols may lower the risk of osteoporosis in all subgroups studied, according to a meta-analysis based on 17 journal articles utilizing the keyword "tea and osteoporosis." In estrogen-deficient ovariectomized rats, dietary consumption of red wine extract polyphenols had a remarkable effect on improving bone strength and preventing osteopenia. Shen et al. examined the role that polyphenol-rich fruits, such as berries, grapes, citrus fruits, and dried plums, have in protecting bone health.

Hubert et al. reported on the antioxidative properties of polyphenols in berries to decrease age-related bone loss., suggesting that berries could be an inexpensive way to lower the risk of osteoporosis. The antioxidative properties of cranberries, blueberries, chokeberries, and goji berries have been shown to lower the risk of osteoporosis. Recent research on the beneficial effects of sweat cherries on bone impairment linked to childhood obesity has shown that this impairment is sustained in obese children and adolescents by spontaneous osteoclastogenesis, which can be inhibited *in vitro* by the presence of polyphenols. Studies using cultured peripheral blood mononuclear cells (PBMCs) from children with obesity revealed that sweet cherry extract inhibited the cells' natural tendency to form multinucleated osteoclasts. According to a report, using fruits like sweet cherries that have been properly extracted and applied as nutraceuticals has great potential for use as a preventive measure in healthy children as well as a therapeutic intervention in obese people.

Tea is a widely consumed beverage worldwide due to its appealing aroma, conventional taste, potential for health benefits, and potential medicinal properties. Tea is rich in catechins, such as (-)-epicatechin-3-gallate (EGCG), (-)-epicatechin (EGC), (-)-epicatechin-3-gallate (ECG), and (-)-epicatechin (EC), which have four main monomers. In middle-aged and older men who drink tea, there is a significant reduction in the risk of osteoporotic hip/femur fractures, according to a case-control study.

Polyphenols, which are abundant in green tea, have the ability to boost antioxidant activity. According to a cross-sectional study, senior Japanese women's consumption of green tea was linked to higher bone mineral density. One of the main flavonoids in green tea, EGCG, has been proposed to cause osteoclasts to undergo apoptosis, which in turn prevents bone resorption and may increase bone mineral density. Green tea contains polyphenols that have been shown to prevent bone loss in middle-aged female rats by reducing oxidative stress damage.

However, because tea contains a high amount of caffeine (1-4% of the dry weight of the tea), consuming too much tea during a tea-drinking habit weakens the protection against osteoporosis. In addition to lowering intestinal calcium absorption and raising urine calcium excretion, caffeine also affects bone by acting as an agonist of adenylate cyclase (route C) through phosphodiesterase activity inhibition. When older postmenopausal women consume more than 300 mg of caffeine per day, their spine bone loss accelerates. Nevertheless, there has been no evidence of a negative impact of caffeine-containing beverages on bone health as long as the participants consume the recommended daily amounts of calcium. Tannic acid also lessens the absorption of calcium. Because of this, food products' polyphenol content needs to be taken into consideration to have the best possible effect on osteoporosis.

9.1.2.4. Lycopene

One of the carotenoids, lycopene is an acyclic isomer of β-carotene that lacks the β-ionone ring structure, which prevents it from forming vitamin A. With 13 double bonds total—11 conjugated and 2 unconjugated—it is a highly unsaturated hydrocarbon. Their biological roles, which include acting as free radical scavengers, are dictated by this conjugated double bond chain. Lycopene's ability to quench singlet oxygen is twice as high as that of β-carotene and 10 times higher than that of α-tocopherol, making it one of the most potent carotenoids. Lycopene is found in watermelons, papayas, pink grapefruits, guavas, and tomatoes.

There have been several published studies on the effects of lycopene on osteoblasts and osteoclasts. Lycopene may prevent the development of both ROS-secreting and multinucleated osteoclast cells. Ishimi et al. investigated how carotenoids, such as lycopene, affected the development of osteoclasts and the resorption of bone in murine osteoclasts grown in coculture with calvarial osteoblasts. According to a rat study, ovariectomized rats treated with lycopene mainly suppressed bone turnover, which restored bone strength and microarchitecture. According to one study, ovariectomized rats that took 10 mg/kg of lycopene for 60 days experienced less bone loss in their femur epiphysis because their trabecular bone remained similar to that of the controls. Following a 17-year period of observation, it was found that 75-year-old men and women who consumed more lycopene had a decreased risk of hip fracture ($p = 0.01$) and non-vertebral fracture ($p = 0.02$). A cross-sectional study found that postmenopausal women (50–60 years old) with higher dietary records of lycopene intake had higher levels of lycopene in their serum ($p < 0.02$). Lycopene lowers markers of bone resorption and oxidative stress, and may therefore help lower the risk of osteoporosis (Arnold et al., 2021).

9.2. CHICKEN EGGSHELL AS SOURCE OF CALCIUM AND ITS APPLICATION

Figure 9.1 illustrates the global increase in egg production (above 30%) between 2009 and 2019. With over 6.5 million tons produced in 2008, the European Union ranked second among all countries producing eggs, behind only China in the top spot and the United States in third. In Poland, each person consumed over 200 units of chicken eggs annually on average. Eggshell waste amounted to a significant amount due to hatcheries, home, and food industries, in addition to the high egg consumption. Waheed and team discovered that 9–12% of the weight of an egg is made up of its shell. If these eggshells are thrown away and never used again, they could pose a risk to environmental pollution.

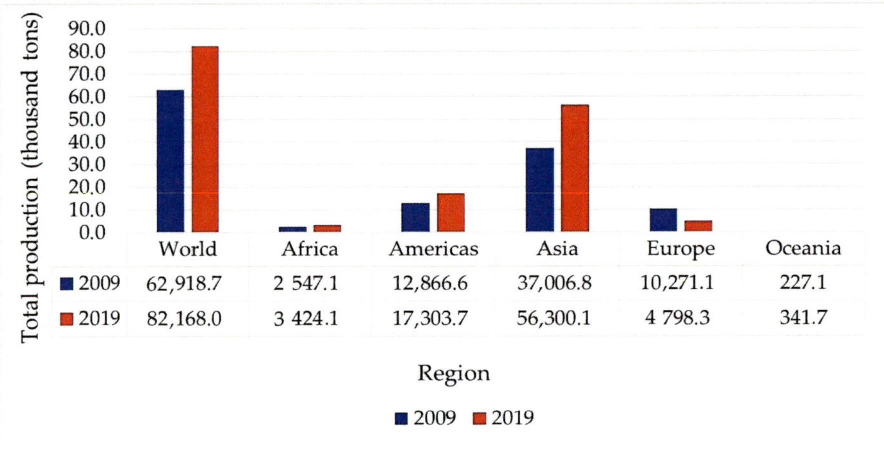

Figure 9.1. *Total chicken egg production in the world and some regions (2009 and 2019).*

Source: Arnold M, Rajagukguk YV, Gramza-Michałowska A. (2021). Functional food for elderly high in antioxidant and chicken eggshell calcium to reduce the risk of osteoporosis-a narrative review. Foods;10(3):656. doi: 10.3390/foods10030656. PMID: 33808726; PM.

Numerous studies have suggested that chicken eggshells have the potential to serve as a calcium source in an attempt to make use of the large amount of waste eggshells that are produced. Nutritional intake has long been recognized as a highly effective strategy for maintaining good bone status at all stages, beginning with early infancy. This includes adequate calcium intake, which should range from 200 to 1200 mg daily, especially after menopause. The eggshell has a 95% $CaCO_3$ content in addition to other important elements like boron and strontium, which are crucial in preventing osteoporosis. It is possible to separate about 37–39% of calcium from chicken eggshells. Previous research has reported on the composition of eggshell (Arnold et al., 2021).

Table 9.1. Composition of Chicken Eggshells

Specific Minerals Present in the Eggshell	**Unit of Measurement**	**According to the Prior Research**			
		Al-awwal et al. (2015)	**Hassan (2015)**	**Brun et al. (2013)**	**Schaafsma et al. (2000)**
Total ash content	g/100 g	89.9–91.1	90.2	N.D. [1]	N.D.
Calcium	mg/100 g	35,100–35,400	35,080	38,200	40,100
Magnesium	mg/100 g	370–400	262.0	N.D.	450
Iron	mg/100 g	N.D.	13.06	N.D.	2.24
Phosphorus	mg/100 g	120	150.2	N.D.	99
Zinc	mg/100 g	N.D.	145.1	N.D.	0.513
Sodium	mg/100 g	150–170	47.9	510	N.D.
Potassium	mg/100 g	100–130	50.00	140	N.D.
Copper	mg/100 g	N.D.	4.1	N.D.	0.77
Manganese	mg/100 g	N.D.	149.9	N.D.	N.D.
Strontium	µg/g	N.D.	N.D.	140	372
Fluorine	µg/g	N.D.	N.D.	N.D.	3.75
Selenium	ng/g	N.D.	N.D.	N.D.	23.5

[1] N.D. = not determined.

Source: Arnold M, Rajagukguk YV, Gramza-Michałowska A. (2021). Functional food for elderly high in antioxidant and chicken eggshell calcium to reduce the risk of osteoporosis-a narrative review. Foods, 10(3):656. doi: 10.3390/foods10030656. PMID: 33808726; PM

Numerous factors, such as the strain or age of the bird, induced molt, heat and general stress, diseases, production system, and addition of proprietary products to the diets, can affect the quality of eggshells. Therefore, the composition of the eggshell and the internal quality of the egg are strongly influenced by nutritional factors such as water quality and feed composition (calcium, phosphorus, vitamins D, C, E, and A, non-starch polysaccharides, enzymes, and contaminants, such as mycotoxins). Numerous studies examining the impact of plant-based ingredients on egg quality have confirmed their influence on the composition of the egg's inside, but the composition of the eggshell—aside from physical characteristics—has not been examined.

Given that more than 85% of rural households in sub-Saharan Africa keep chickens, powdered eggshells from chickens may be a useful way to boost calcium intake in this region. Calcium from chicken eggshells has a bioavailability that is comparable to that of $CaCO_3$. Calcium absorption from a diet containing chicken eggshell powder was 45.59 ± 14.43% in a study involving rats; there was no significant difference when supplementing with $CaCO_3$ (39.88 ± 16.07%) ($p > 0.05$). Table 9.2 lists the uses of chicken eggshell as a calcium source in different food products. Eggshells have a lot to offer nutritionally and, most importantly, in terms of increased calcium content. It should be mentioned, though, that the kind and pH of the food product determine whether or not adding eggshells will be successful. The appropriate amount of calcium, the acceptability of the product's organoleptic qualities, and—most importantly from a technological standpoint—the dough's increased rheological properties, volume, and final product texture all influence whether or not it is added to bread and biscuits. Eggshells can have a negative impact on the sensory qualities and acceptability of sweet confectionery products, particularly with regard to taste and aroma. It appears that adding eggshells to dairy products—like yogurt, fried cheese, or cranberry juice—is the most advantageous and aesthetically pleasing option. Due to high levels of microbiological contamination, such as that caused by Salmonella, and the requirement to properly grind the shells so that the consumer does not perceive their presence as "sand" in the product, the use of eggshells in the food industry is also restricted.

Table 9.2. Application of Chicken Eggshell Powder in Food Products

Food Products	**Main Findings**		**References**
	Chicken Eggshell Powder Concentration	**Notes for Recommendation**	
Biscuits	6% (*w/w*) of wheat flour	Calcium content, texture, sensory properties, calcium bioavailability	Hassan et al. (2015)
Bread	8% (*w/w*) of ingredients	Calcium content, specific volume of bread, sensory properties	Ali et al. (2019)
Bread	2% (*w/w*) of wheat flour	Increase of rheological characteristics of dough and nutritional properties, decrease of the general acceptability and odor score	Alsuhaibani, (2018)
Bread strips	10% (*w/w*) substitution of wheat flour	Minor changes in sensory properties	Ali et al. (2017)

Breaded fried meat, bread, pizza, spaghetti	500 mg Ca/person	Minor changes in texture, without flavor changes	Brun et al. (2013)
Chocolate cakes	6% (*w/w*) of wheat flour	Calcium content, texture, sensory properties	Ray et al. (2017)
Chokeberry juice, cranberry juice	1% of chokeberry and cranberry juice	Calcium content of chokeberry and cranberry juice, no significant change in color and sediment content	Lachowicz et al. (2020)
Muffin	8 g/500 g wheat flour	Mineral content, sensory properties	Afzal et al. (2020)
Nham (Thai-style fermented pork sausage)	150 mg Ca/100 g of *Nham* (eggshell powder was converted to eggshell calcium lactate)	No difference in sensory scores of sour taste, flavor, and overall acceptance	Daengprok et al. (2002)
Ser smażony (Polish bread spread)	265 mg/100 g of *ser smażony*	Increased calcium contents >2.5-fold, the calcium bioavailability was higher after the addition of lysine and vitamin K	Kobus-Cisowska et al. (2020)
White bread	1–1.5% (*w/w*) of ingredients	High total score in sensory evaluation	Platon et al. (2020)
White bread	2% (*w/w*) substitution of bread flour	Consumer acceptation in sensory evaluation	Chilek et al. (2018)
Yogurt	0.15–0.30% (*w/v*) of milk	No significant unfavorable effects on the physicochemical, microbial, and sensory properties	Al Mijan et al. (2014)
Yogurt (from cow milk and buffalo milk)	0.3% of yogurt (nanosized eggshell powder)	Acceptable composition, texture, and sensory attributes	El-Shibiny et al. (2018)

Source: Arnold M, Rajagukguk YV, Gramza-Michałowska A. (2021). Functional food for elderly high in antioxidant and chicken eggshell calcium to reduce the risk of osteoporosis-a narrative review. Foods, 10(3):656. doi: 10.3390/foods10030656. PMID: 33808726.

9.2.1. Correlation Between Tannic Acid and Calcium Absorption

Researchers looked into how tannin affected calcium's bioavailability. A study published recently claimed that tannin inhibited the bioavailability of calcium; however, Proulx et al. found that tannins had no effect on calcium bioavailability, refuting this claim. Amalraj and Pius claim that a higher tannic acid concentration causes a greater reduction in the bioavailability of calcium salts. With a 1:1 ratio between calcium salt and tannic acid, the largest percentage of decrease was observed in calcium formate (28.1%), while the lowest percentage was found in calcium gluconate (19.6%). Tannic acid by itself does not completely reduce the bioavailability of calcium from green leafy vegetables, according to research by Gupta et al. on the impact of the bioavailability of calcium and iron from specific green leafy vegetables. They discovered that the presence of various inhibitory factors (such as phytic acid, oxalic acid, polyphenols, and dietary fiber) affected the bioavailability of calcium. Still, Proulx et al. found that the tannin content of legumes has no effect on their bioavailability of calcium. It has been discovered that tannins may impede the bioavailability of iron more than they do calcium.

Lachowicz et al. previously reported on the use of chicken eggshell powder in tannic acid-rich fruits, such as chokeberry and cranberry juice. They were worried that the low pH in the juice of chokeberries and cranberries could increase the solubility of calcium from chicken eggshell powder, which could then increase the absorption of calcium. They did not address in their article the inhibition of calcium absorption by tannic acid from these berries. The same authors also came to the conclusion that natural $CaCO_3$ is far more soluble than artificial $CaCO_3$, and that the porous nature of $CaCO_3$ in chicken eggshell may also contribute to its higher calcium solubility (Arnold et al., 2021).

9.3. FACTORS SUPPORTING CALCIUM ABSORPTION

It is crucial to absorb calcium in addition to having a lot of it. A condition of acidity in the intestines (particularly for the absorption of $CaCO_3$), estrogen, vitamin D, soluble fiber/prebiotics (inulin, fructooligosaccharose, oligofructose), probiotics, and synbiotics are some of the factors that have been reported to positively influence the absorption of calcium. There are few studies on humans' bioavailability of calcium from eggshells, despite the possibility. Schaafsma and Pakan investigated the effects of consuming dairy-based supplements containing eggshell powder on the bone mineral density of the hip and lumbar spine in humans over a brief period of time. The lumbar spine, total proximal femur, and trochanter all showed significantly higher bone mineral densities as a result of the results, indicating that eggshell powder is a source of bioavailable calcium. Additional studies on the impact of eating eggshells revealed that within a year of starting a supplement, healthy late postmenopausal women with adequate baseline calcium intake may see an increase in hip bone mineral density.

9.3.1. Vitamin D

Hip fracture incidence is higher in Western nations where diets are high in calcium; in contrast, hip fracture incidence is lower in developing nations where diets are low in calcium. This phenomenon is known as the "calcium paradox." Taking into account the amount of vitamin D present in the local population's serum, this paradox can be partially explained.

Since vitamin D is the primary regulator of calcium homeostasis and absorption, it is essential for the metabolism of calcium and bone. A vitamin D deficiency can hasten fractures, osteoporosis, increased bone turnover, and bone loss. Due to fewer dietary sources of vitamin D, reduced sun exposure, a decreased body's ability to synthesize vitamin D in the skin, and a decreased kidney's ability to convert vitamin D into its active form, elderly people are more vulnerable to vitamin D deficiency. The two main forms of vitamin D are vitamin D2 (ergocalciferol), which is produced naturally in plants, fungi, and invertebrates from ergosterol, the precursor to vitamin D, exposed to UV-B light. The other is vitamin D3, or cholecalciferol, which is produced in the presence of sunlight from 7-dehydrocholesterol, a cholesterol precursor that can also function as provitamin D3. While 90% of vitamin D is produced endogenously—in the skin through UV irradiation—it is also present in certain food sources, such as eggs, meat, fish (salmon, tuna, sardines, herring), mushrooms, and dairy products. Because elderly people have a lower rate of cutaneous vitamin D synthesis, they are more susceptible to vitamin D deficiencies, which are linked to bone fractures. For this reason, it is critical to be concerned about the vitamin D sources found in food.

The human body requires vitamin D to be activated. Serum 25-hydroxyvitamin D (25OHD) is a marker that can be used to estimate an individual's vitamin D nutritional status. Serum 25OHD functions as intermediate compounds before forming the final biologically active vitamin D. While a level of 25OHD at least 50 nmol/L (20 ng/mL) is considered sufficient, even though the ideal level for intestinal calcium resorption is approximately 80 nmol/L, a level less than 27–30 nmol/L is linked to rickets. The most active metabolite of vitamin D, 1,25-dihydroxy vitamin D, is produced in the kidney after vitamin D is hydroxylated in the liver to 25OHD. This vitamin also affects bone cells by promoting the absorption of calcium and phosphate from the gut.

There have been studies on vitamin D, particularly concerning improving calcium absorption in the elderly. According to meta-analyses, vitamin D supplementation by itself is unlikely to lower fracture risk; calcium supplementation by itself can lower fracture risk somewhat, by about 10%, but its effectiveness is not very good over the long run. Still, the combo of calcium and vitamin D supplementation, particularly in those who are at risk of having a marginal or low vitamin D status (. <60–80 nmol/L serum 25OHD), reduces the total fractures, including hip fractures. A study reported that 1000 mg calcium citrate/day and vitamin D at a dose equivalent to 600 IU/day given over a year improved the bone mineral density or bone markers, not significantly different from the higher vitamin D dose of 3750 IU/day in overweight elderly with a mean 25OHD of 20 ng/mL. Taylor included vitamin D and calcium-enriched food as one of the functional food types for the elderly, such as fruit juice, soya milk, cheese, milk, and yogurt enriched with vitamin D and calcium. According to the European Food Safety Association, 800 IU or 20 µg of daily vitamin D intake and 1200 mg of daily calcium intake are recommended for women aged 50 years and older to reduce bone loss and maintain bone mineral density. Nowson reported that the elderly are recommended to consume > 1,100 mg of calcium per day in addition to keeping vitamin D levels sufficient (>60 nmol/L 25 OHD) to lower the risk of fracture.

9.3.2. Prebiotics, Probiotics, and Synbiotics

Nondigestible fibers known as prebiotics serve as a substrate for gut microbes to ferment food and energy sources. They can also increase the viability of gut microbes. Prebiotics have been linked to increased calcium absorption, which may lower the risk of osteoporosis, according to numerous studies. Prebiotics that increase the absorption of calcium include fructans (dosage 4 to 8 g/day), which include inulin, oligofructose, and fructooligosaccharide. The fermentation of these fructans by gut microbes lowers the pH in the colon, which can enhance the absorption of macrominerals like calcium. This is because calcium takes on an ionic form in an acidic environment, which makes it more soluble and bioavailable. Short-chain fatty acids are the fermentation products of prebiotic fiber that create an acidic environment. When combined with low pH, these factors cause the mucosal cells to hypertrophy, increasing the surface area of the intestine and enhancing calcium absorption.

Prebiotics have been shown to significantly reduce the amount of bifidobacteria in the gut, which is good for older people and may have an impact on their bone health. The use of prebiotics in foods high in calcium has the potential to lower the elderly population's risk of osteoporosis. *In vitro* experimentation was used by Krupa-Kozak et al. found that intestinal-like cellular calcium uptake by Caco-2 cells from calcium-enriched gluten-free bread digest was significantly enhanced by inulin-type fructans, particularly short-chain fructooligosaccharide. Additionally, intestinal bacteria (Lactobacillus, Enterococcus, and Enterobacteriaceae) applied in the cultures were stimulated to the intensive synthesis of organic acids (p. < 0.05). Jakeman et al. found that daily soluble corn fiber intake significantly increased bone calcium retention in postmenopausal women and improved the bone calcium balance by an estimated 50 mg/day. In rat studies, Arora and Patel compared the calcium absorption of skimmed milk enriched with fiber Blend-I (psyllium husk, oat fiber, MCC, inulin), fiber Blend-II (psyllium husk, oat bran, wheat fiber and inulin), and control (cellulose). The calcium absorption of fiber Blend-I, fiber Blend-II, and cellulose was $94.4 \pm 1.9\%$, $96.3 \pm 1.32\%$, and $88.8 \pm 4.9\%$, respectively. However, the difference in calcium absorption between groups was not significant ($p > 0.05$), suggesting that dairy products' nutritional value can be increased by adding these dietary fibers.

Studies have been done on how probiotics affect the absorption of calcium. Probiotics work through many mechanisms to increase the bioavailability of minerals, particularly calcium. These mechanisms include: (i) producing short-chain fatty acids that increase mineral solubility; (ii) producing phytase by bacteria to counteract the effect of phytate on minerals; (iii) reducing intestinal inflammation, which is followed by an increase in bone mass density; (iv) hydrolyzing glycoside bonds food in the intestines by *Lactobacillus* and *Bifidobacterium*; and (v) synthesizing vitamins that affect calcium absorption. After 24 hours ($p < 0.05$), *Lactobacillus salivarius* dramatically increased the calcium uptake into Caco-2 cell monolayers, according to Gilman and Cashman's findings. A three-strain probiotic (*Lactobacillus paracei* 8700:2, *Lactobacillus plantarum* Heal 9, and *Lactobacillus plantarum* Heal 19) group of postmenopausal women demonstrated a 0.71% average difference ($p < 0.05$) reduction in bone mineral density loss when compared to a placebo. Even though a lot of research produced important findings, Scholz-Ahrens et al. stated that the effects of probiotics on gut ecology and bone mineralization were not statistically significant.

Synbiotics—prebiotics and probiotics—have been shown to function exceptionally well together. As previously indicated, the probiotic did not have a major impact on gut ecology or bone mineralization on its own. Rats fed prebiotics had significantly higher amounts of cecal contents and lower pH in cecal and colonic contents, and their calcium balance tended to be increased ($p < 0.01$). However, synbiotics were shown to have a more synergistic effect on bone mineralization. As a result, the animals fed prebiotics had significantly higher calcium absorption values. But Klobukowski et al. a study conducted on rats found no discernible variation in the absorption of calcium between white cheese that contained probiotics (*Lactobacillus plantarum*), synbiotics

(*Lactobacillus plantarum* plus 2.5% inulin HPX), and a control group that did not contain any probiotics or prebiotics (Arnold et al., 2021).

9.4. EATING PREFERENCE AND DIETARY PATTERNS OF THE ELDERLY

Understanding the dietary habits and food preferences of the elderly is crucial when creating functional food products for them. Aging may have an impact on various facets of oral physiology and eating behavior in the elderly population. When it comes to food oral processing, older adults anticipate being able to form and swallow food boluses; when it comes to food sensory processing, they worry about texture, taste, and reduced pain perception. Around 7–10% of adults over 50 had dysphagia, or difficulty swallowing food. The amount of money, the makeup of the household, psychological factors (fears related to aging, such as losing a partner, shifting roles, moving, being rejected, meeting new friends, and getting sick), society, culture, and traditions, easy cooking techniques, and the sensory aspect of food all have an impact on the food preferences of the elderly. According to a cross-sectional survey study, one of the biggest factors influencing the food choices of the elderly in Europe is their attempt to eat healthily. The main advantages of eating healthily for the elderly are disease prevention, quality of life promotion, and maintaining health.

Many studies have reported on the dietary habits of the elderly in various parts of the world in relation to those factors. Briley stated that white bread, ground coffee, whole milk, sugar, potatoes, tea, orange juice, eggs, butter, and bacon were the most frequently consumed foods by the elderly in America. A cross-sectional study conducted in the USA also revealed that older adults over 75 had a "Western-like" dietary pattern, which included a high consumption of bread, eggs, fried vegetables and fats, and a low consumption of dairy products. In Brazil, South America, the pattern of "coffee with milk and bread and butter" was linked to elderly individuals who were 80 years of age or older and had difficulty chewing their food. Capurso and Capurso recommended whole wheat sourdough bread, particularly for the elderly in Southern Europe, because the fermentation process lowers the glycemic index of bread and because it contains potassium, zinc, iron, magnesium, selenium, calcium, phosphorus, manganese, thiamin, niacin, folate, and riboflavin. When comparing elderly people in Europe to those in other nations, Polish seniors had the highest percentage (38%) and Finnish seniors (26%) of fruit-eating seniors, including fresh and processed berries. Older Italians preferred eating traditional Italian foods like lasagna, pasta and bean soup, and risotto with saffron because of factors like ease of preparation, tradition, and sensory aspects. In Botswana, Africa, grandparents with children and those residing in towns or urban areas were more likely to follow the bread and vegetable pattern. In Asia, South Korea, Shin et al. found that the elderly in Seoul had a strong preference for the following foods: vegetables, fruits, kimchi (traditional fermented cabbage from Korea), fish, tofu, rice cakes, dairy products/eggs, meat, noodles, seafood, poultry, and bread/cookies. Shin and others additionally revealed that the preference of the elderly is for "ready-to-eat" food (53.9%), "ready-to-cook" food (21.2%), "ready-to-heat" food (15.8%), and "fresh-cut" goods (9.1%). Since 2009, Australian New Zealand Food Standards have required folic acid fortification in wheat flour used to make bread (2–3 mg of folic acid per kilogram of wheat flour, or approximately 0.13 mg of folic acid per 100 g of bread). This has had positive effects on the population of Australia. Since bread is one of the main foods consumed by elderly Australians, it has been reported that folic acid fortification is linked to significantly lower plasma homocysteine levels in these individuals. In conclusion, older people from all over the world have a general preference for bread products. Furthermore, according to the food pyramid, bread products should be consumed in large quantities and have little effect on the environment (Arnold et al., 2021).

9.5. FUNCTIONAL FOODS FOR THE ELDERLY

When included in an elderly person's diet, functional foods can lower their risk of illness or postpone the onset of serious conditions, all of which can enhance their quality of life. Functional foods have been researched over the past few decades to support the health properties of the elderly. A dietary recommendation for an elderly person is generally the same as one for a younger adult, but there are some extra considerations because of the physiological changes associated with aging. These physiological alterations, which include a reduction in the secretion of saliva, pancreatic and stomach juices, insulin, and bile, can impact one's capacity to consume and process food. These issues could cause dietary deficits in the elderly. In most cases, increasing meal frequency or serving sizes won't work well for older adults because of physiological issues related to eating and decreased appetite. As a result, numerous studies have created and suggested a wide range of functional foods that can assist senior citizens in enhancing their nutritional status and avoiding deficiencies.

The consumption of functional foods enriched with dietary fiber, omega-3 polyunsaturated fatty acids, phytoestrogens, polyphenols, carotenoids (including α-carotene and β-carotene), lutein, zeaxanthin, probiotics, synbiotics, prebiotics, and plant sterols and stanols, calcium, and vitamin D is advised for the elderly by Taylor and Jędrusek-Golińska et al.

Meydani claims that functional foods enhanced with vitamin E, an antioxidant nutrient, in amounts above recommended may improve immunological function, lower the risk of CVD, and modulate significant aging-related degenerative conditions. Given that older adults have higher populations of fungi and enterobacteria and fewer lactic acid bacteria, particularly bifidobacteria, than younger adults, some studies have suggested that elderly people take probiotics to balance the microflora in their guts and lower their risk of colon cancer, hypertension, urinary tract infections, and enteric infections. Probiotics have been shown by Jędrusek-Golińska et al. to be able to modulate the immune system, as evidenced by a slight rise in prostaglandin E2. On the other hand, no alterations in IgA levels were observed in the elderly (n = 51, 65 years of age and older) who took *Lactobacillus acidophilus*. Additionally, McCabe and colleagues investigated the potential effects of *Lactobacillus reuteri* (ATCC PTA 6475) on intestinal TNF-α levels and bone density in male mice aged 14 weeks following a 4-week treatment period. Increased protein intake has been linked to positive health effects for the elderly, such as quicker recovery from hip fractures, increased lean body mass and strength, and a decreased risk of frailty, according to a study that recommended protein-enriched functional foods. For older adults, Taylor suggested vitamin D and functional foods high in calcium to support bone health and fend off osteoporosis (Arnold et al., 2021).

REFERENCES

1. Abdollahi, M., Larijani, B., Rahimi, R., & Salari, P. (2005). Role of oxidative stress in osteoporosis. *Therapy*, *2*(5), 787–796. https://doi.org/10.2217/14750708.2.5.787.
2. Abiad, M. G., & Meho, L. I. (2018). Food loss and food waste research in the Arab world: A systematic review. *Food Security, 10*, 311–322. https://doi.org/10.1007/s12571-018-0782-7.
3. Afzal, F., Mueen-ud-Din, G., Nadeem, M., Murtaza, M. A., & Mahmood, S. (2020). Effect of eggshell powder fortification on the physicochemical and organoleptic characteristics of muffins. *Pure and Applied Biology, 9*(3), 1488–1496. https://doi.org/10.19045/bspab.2020.90154.
4. Agarwal, S., & Rao, A. V. (2000). Tomato lycopene and its role in human health and chronic diseases. *CMAJ, 163*(6), 739–744.
5. Al Mijan, M., Choi, K. H., & Kwak, H. S. (2014). Physicochemical, microbial, and sensory properties of nanopowdered eggshell-supplemented yogurt during storage. *Journal of Dairy Science, 97*(6), 3273–3280. https://doi.org/10.3168/jds.2013-7367.
6. Al-awwal, N. Y., & Ali, U. L. (2015). Proximate analyses of different samples of egg shells obtained from Sokoto market in Nigeria. *International Journal of Scientific Research, 4*(6), 564–566.
7. Ali, M., & Badawy, W. (2017). Utilization of eggshells by-product as a mineral source for fortification of bread strips. *Journal of Food and Dairy Science, 8*, 455–459. https://doi.org/10.21608/jfds.2017.38960.
8. Ali, R. S., Hasan, S. T., & Alssiraj, M. A. (2019). Effect of chicken eggshell powder fortification on the chemical, physical, and rheological characteristics of bread. *Biochem. Cell. Arch, 19*, 543–547.
9. Alsuhaibani, A. M. A. (2018). Rheological and nutritional properties and sensory evaluation of bread fortified with natural sources of calcium. *Journal of Food Quality, 2018*, 1–8. https://doi.org/10.1155/2018/8308361.
10. Amalraj, A., & Pius, A. (2017). Relative contribution of oxalic acid, phytate, and tannic acid on the bioavailability of calcium from various calcium salts—An *in vitro* study. *International Food Research Journal, 24*(4), 1278–1285.
11. Amarasekara, D. S., Yu, J., & Rho, J. (2015). Bone loss triggered by the cytokine network in inflammatory autoimmune diseases. *Journal of Immunology Research, 2015*, 1–12. https://doi.org/10.1155/2015/832127.
12. Ardawi, M. S. M., Badawoud, M. H., Hassan, S. M., Rouzi, A. A., Ardawi, J. M. S., AlNosani, N. M., Qari, M. H., & Mousa, S. A. (2016). Lycopene treatment against loss of bone mass, microarchitecture, and strength in

relation to regulatory mechanisms in a postmenopausal osteoporosis model. *Bone, 83*, 127–140. https://doi.org/10.1016/j.bone.2015.10.017.

13. Arnold, M., Rajagukguk, Y. V., & Gramza-Michałowska, A. (2021). Functional food for elderly high in antioxidant and chicken eggshell calcium to reduce the risk of osteoporosis—A narrative review. *Foods, 10*(3), 656. https://doi.org/10.3390/foods10030656.
14. Arslan, A., Orkun, S., Aydin, G., Keles, I., Tosun, A., Arslan, M., & Caglayan, O. (2011). Effects of ovariectomy and ascorbic acid supplement on oxidative stress parameters and bone mineral density in rats. *Libyan Journal of Medicine*, *6*, 1–9. https://doi.org/10.3402/ljm.v6i0.5965.
15. Bartter, J., Diffey, H., Yeung, Y. H., O'Leary, F., Häsler, B., Maulaga, W., & Alders, R. (2018). Use of chicken eggshell to improve dietary calcium intake in rural sub-Saharan Africa. *Matern Child Nutr, 14*(1), e12649. https://doi.org/10.1111/mcn.12649.
16. Basri, H., & Sulastri, M. (2021). Physical quality of the first egg of Japanese quail (*Coturnix japonica* L.) after given liquid herbal concoction. *Mangifera Edu, 5*(2), 121–130. https://doi.org/10.31943/mangiferaedu.v5i2.98.
17. Benjakul, S., & Karnjanapratum, S. (2018). Characteristics and nutritional value of whole wheat cracker fortified with tuna bone bio-calcium powder. *Food Chemistry, 259*, 181–187. https://doi.org/10.1016/j.foodchem.2018.03.124.
18. Böhm, V. (2018). Lycopene, tomatoes, and cardiovascular diseases. In A. V. Rao, G. L. Young, & L. G. Rao (Eds.), *Lycopene and Tomatoes in Human Nutrition and Health* (pp. 51–68). CRC Press.
19. Borowska, S., & Brzóska, M. M. (2016). Chokeberries (*Aronia melanocarpa*) and their products as a possible means for the prevention and treatment of noncommunicable diseases and unfavorable health effects due to exposure to xenobiotics. *Comprehensive Reviews in Food Science and Food Safety*, *15*(6), 982–1017. https://doi.org/10.1111/1541-4337.12221.
20. Brun, L. R., Lupo, M., Delorenzi, D. A., Di Loreto, V. E., & Rigalli, A. (2013). Chicken eggshell as a suitable calcium source at home. *International Journal of Food Science & Nutrition, 64*(6), 740–743. https://doi.org/10.3109/09637486.2013.787399.
21. Brunetti, G., Storlino, G., Oranger, A., Colaianni, G., Faienza, M. F., Ingravallo, G., Di Comite, M., Reseland, J. E., Celi, M., Tarantino, U., et al. (2020). LIGHT/TNFSF14 regulates estrogen deficiency-induced bone loss. *Journal of Pathology*, *250*, 440–451. https://doi.org/10.1002/path.5385.

22. Brzezińska, O., Łukasik, Z., Makowska, J., & Walczak, K. (2020). Role of vitamin C in osteoporosis development and treatment: A literature review. *Nutrients*, *12*(8), 2394. https://doi.org/10.3390/nu12082394.
23. Brzóska, M. M., Roszczenko, A., Rogalska, J., Gałażyn-Sidorczuk, M., Mężyńska, M. (2017). Protective effect of chokeberry (*Aronia melanocarpa* L.) extract against cadmium impact on the biomechanical properties of the femur: A study in a rat model of low and moderate lifetime women exposure to this heavy metal. *Nutrients*, *9*(8), 8–10.
24. Cano, A., Chedraui, P., Goulis, D. G., Lopes, P., Mishra, G., Mueck, A., Senturk, L. M., Simoncini, T., Stevenson, J. C., Stute, P., et al. (2018). Calcium in the prevention of postmenopausal osteoporosis: EMAS clinical guide. *Maturitas, 107*, 7–12. https://doi.org/10.1016/j.maturitas.2017.10.004.
25. Cao, J. J. (2011). Effects of obesity on bone metabolism. *Journal of Orthopaedic Surgery and Research*, *6*, 30. https://doi.org/10.1186/1749-799X-6-30.
26. Chakraborty, S., & De, S. D. (2019). Eggshell: An alternative, cheap, bioavailable source of calcium in human diet. *Research Reviews: Journal of Dairy Science and Technology, 8*(1), 25–33.
27. Chavan, S. N., More, U., Mulgund, S., Saxena, V., & Sontakke, A. N. (2007). Effect of supplementation of vitamin C and E on oxidative stress in osteoporosis. *Indian Journal of Clinical Biochemistry*, *22*(2), 101–105. https://doi.org/10.1007/BF02913324.
28. Chen, J. R., Lazarenko, O. P., Wu, X., Kang, J., Blackburn, M. L., Shankar, K., Badger, T. M., & Ronis, M. J. (2010). Dietary-induced serum phenolic acids promote bone growth via P38 MAPK/β-catenin canonical Wnt signaling. *Journal of Bone and Mineral Research*, *25*(11), 2399–2411. https://doi.org/10.1002/jbmr.137.
29. Chilek, T. Z. T., Kairuaman, N. A., Ahmad, F., Wahab, R. A., Zamri, A. I., & Mahmood, A. (2018). Development of white bread fortified with calcium derived from eggshell powder. *Malaysian Applied Biology, 47*(1), 29–39.
30. Corbo, F., Brunetti, G., Crupi, P., Bortolotti, S., Storlino, G., Piacente, L., Carocci, A., Catalano, A., Milani, G., Colaianni, G., et al. (2019). Effects of sweet cherry polyphenols on enhanced osteoclastogenesis associated with childhood obesity. *Frontiers in Immunology*, *10*, 1001. https://doi.org/10.3389/fimmu.2019.01001.
31. Curtis, J. R., & Safford, M. M. (2012). Management of osteoporosis among the elderly with other chronic medical conditions. *Drugs & Aging*, *29*(7), 549–564. https://doi.org/10.2165/11599620-000000000-00000.

32. Daengprok, W., Garnjanagoonchorn, W., & Mine, Y. (2002). Fermented pork sausage fortified with commercial or hen eggshell calcium lactate. *Meat Science, 62*(2), 199–204. https://doi.org/10.1016/S0309-1740(01)00247-9.
33. Devareddy, L., Hooshmand, S., Collins, J. K., Lucas, E. A., Chai, S. C., & Arjmandi, B. H. (2008). Blueberry prevents bone loss in ovariectomized rat model of postmenopausal osteoporosis. *Journal of Nutritional Biochemistry*, *19*(11), 694–699. https://doi.org/10.1016/j.jnutbio.2007.09.004.
34. Domazetovic, V., Marcucci, G., Iantomasi, T., Brandi, M. L., & Vincenzini, M. T. (2017). Oxidative stress in bone remodeling: Role of antioxidants. *Clinical Cases in Mineral and Bone Metabolism, 14*, 209–216. https://doi.org/10.11138/ccmbm/2017.14.1.209.
35. El-Shibiny, S., Abd El-Gawad, M. A. E. K. M., Assem, F. M., & El-Sayed, S. M. (2018). The use of nano-sized eggshell powder for calcium fortification of cow's and buffalo's milk yogurts. *Acta Scientiarum Polonorum Technologia Alimentaria, 17*(1), 37–49.
36. Faienza, M. F., Ventura, A., Marzano, F., & Cavallo, L. (2013). Postmenopausal osteoporosis: The role of immune system cells. *Clinical Development*, 575936. https://doi.org/10.1155/2013/575936.
37. Faienza, M., Corbo, F., Carocci, A., Catalano, A., Clodoveo, M., Grano, M., Wang, D., D'Amato, G., Muraglia, M., Franchini, C., et al. (2020). Novel insights in health-promoting properties of sweet cherries. *Journal of Functional Foods*, *69*, 103945. https://doi.org/10.1016/j.jff.2020.103945.
38. FAO. (2020). Livestock primary. *FAO*. Retrieved from: http://www.fao.org/faostat/en/?#data/QL (accessed on 6 December 2024).
39. Feng, X., & McDonald, J. M. (2011). Disorders of bone remodeling. *Annual Review of Pathology: Mechanisms of Disease, 6*, 121–145. https://doi.org/10.1146/annurev-pathol-011110-130203.
40. Finck, H., Hart, A. R., Jennings, A., & Welch, A. A. (2014). Is there a role for vitamin C in preventing osteoporosis and fractures? A review of the potential underlying mechanisms and current epidemiological evidence. *Nutrition Research Reviews*, *27*(2), 268–283. https://doi.org/10.1017/S0954422414000195.
41. Gabbay, K. H., Bohren, K. M., Morello, R., Bertin, T., Liu, J., & Vogel, P. (2010). Ascorbate synthesis pathway: Dual role of ascorbate in bone homeostasis. *Journal of Biological Chemistry*, *285*, 19510–19520. https://doi.org/10.1074/jbc.M110.110247.
42. Gramza-Michałowska, A. (2014). Caffeine in tea *Camellia sinensis*—

Content, absorption, benefits and risks of consumption. *Journal of Nutrition, Health & Aging, 18*(2), 143–149. https://doi.org/10.1007/s12603-013-0404-1.

43. Gupta, S., Lakshmi, A. J., & Prakash, J. (2006). *In vitro* bioavailability of calcium and iron from selected green leafy vegetables. *Journal of the Science of Food and Agriculture, 86*(14), 2147–2152. https://doi.org/10.1002/jsfa.2589.
44. Hassan, N. M. M. (2015). Chicken eggshell powder as dietary calcium source in biscuits. *World Journal of Dairy & Food Sciences, 10*(2), 199–206.
45. Heaney, R. P. (2002). Effects of caffeine on bone and the calcium economy. *Food and Chemical Toxicology, 40*(9), 1263–1270. https://doi.org/10.1016/S0278-6915(02)00094-7.
46. Heidari, B., Muhammadi, A., Javadian, Y., Bijani, A., Hosseini, R., & Babaei, M. (2017). Associated factors of bone mineral density and osteoporosis in elderly males. *International Journal of Endocrinology and Metabolism*, *15*(1), 1–10. https://doi.org/10.5812/ijem.39662.
47. Huang, C., & Tang, R. (2016). Tea drinking habits and osteoporotic hip/femur fractures: A case-control study. *Pakistan Journal of Medical Sciences*, *32*(2), 408–412. https://doi.org/10.12669/pjms.322.9092.
48. Hubert, P. A., Lee, S. G., Lee, S. K., & Chun, O. K. (2014). Dietary polyphenols, berries, and age-related bone loss: A review based on human, animal, and cell studies. *Antioxidants*, *3*(1), 144–158. https://doi.org/10.3390/antiox3010144.
49. Ishimi, Y., Ohmura, M., Yamaguchi, X. M., & Ikegami, S. (1999). Inhibition by carotenoids and retinoic acid of osteoclast-like cell formation induced by bone resorbing agents *in vitro*. *Journal of Clinical Biochemistry and Nutrition, 27*(2), 113–122. https://doi.org/10.3164/jcbn.27.113.
50. Jakubczyk, K., Kochman, J., Kwiatkowska, A., Kałdunska, J., Dec, K., Kawczuga, D., & Janda, K. (2020). Antioxidant properties and nutritional composition of matcha green tea. *Foods, 9*(4), 483. https://doi.org/10.3390/foods9040483.
51. Jędrusek-Golińska, A., Górecka, D., Buchowski, M., Wieczorowska-Tobis, K., Gramza-Michałowska, A., & Szymandera-Buszka, K. (2020). Recent progress in the use of functional foods for older adults: A narrative review. *Comprehensive Reviews in Food Science and Food Safety, 19*, 835–856. https://doi.org/10.1111/1541-4337.12530.
52. Jelena, C. H., Giorgio, R., Justyna, G., Neda, M. D., Natasa, S., Artur, B., & Giuseppe, G. (2018). Beneficial effects of polyphenols on chronic

diseases and ageing. In C. M. Galanakis (Ed.), *Polyphenols: Properties, Recovery and Applications* (pp. 69–102). Woodhead Publishing.

53. Jin, M., Huang, Q., Zhao, K., & Shang, P. (2013). Biological activities and potential health benefit effects of polysaccharides isolated from *Lycium barbarum* L. *International Journal of Biological Macromolecules, 54*, 16–23. https://doi.org/10.1016/j.ijbiomac.2012.11.023.
54. Kim, M. H., & Lee, H. J. (2016). Osteoporosis, vitamin C intake, and physical activity in Korean adults aged 50 years and over. *Journal of Physical Therapy Science, 28*, 725–730. https://doi.org/10.1589/jpts.28.725.
55. Kobus-Cisowska, J., Szymanowska-Powałowska, D., Szymandera-Buszka, K., Rezler, R., Jarzębski, M., Szczepaniak, O., Marciniak, G., Jędrusek-Golińska, A., & Kobus-Moryson, M. (2020). Effect of fortification with calcium from eggshells on bioavailability, quality, and rheological characteristics of traditional Polish bread spread. *Journal of Dairy Science, 103*(8), 6918–6929. https://doi.org/10.3168/jds.2019-18027.
56. Kowalska, E., Kucharska-Gaca, J., Kuźniacka, J., Lewko, L., Gornowicz, E., Biesek, J., & Adamski, M. (2021). Egg quality depending on the diet with different sources of protein and age of the hens. *Scientific Reports, 11*, 2638. https://doi.org/10.1038/s41598-021-82313-1.
57. Lachowicz, S., Oszmiański, J., Wilczyńska, M., Zaguła, G., Saletnik, B., & Puchalski, C. (2020). Impact mineralization of chokeberry and cranberry fruit juices using a new functional additive on the protection of bioactive compounds and antioxidative properties. *Molecules, 25*(3), 659. https://doi.org/10.3390/molecules25030659.
58. Le Nihouannen, D., Barralet, J. E., Fong, J. E., & Komarova, S. V. (2010). Ascorbic acid accelerates osteoclast formation and death. *Bone, 46*(5), 1336–1343. https://doi.org/10.1016/j.bone.2009.11.021.
59. Lesnierowski, G., & Stangierski, J. (2018). What's new in chicken egg research and technology for human health promotion? A review. *Trends in Food Science & Technology, 71*, 46–51. https://doi.org/10.1016/j.tifs.2017.10.022.
60. Lipinski, B., Hanson, C., Lomax, J., Kitinoja, L., & Waite, R. (2013). Reducing food loss and waste. *World Resources Institute Working Paper*, 1–40.
61. Liu, Z., Ren, Z., Zhang, J., Chuang, C. C., Kandaswamy, E., Zhou, T., & Zuo, L. (2018). Role of ROS and nutritional antioxidants in human diseases. *Frontiers in Physiology, 9*, 477. https://doi.org/10.3389/fphys.2018.00477.

62. Lorenc, R., Głuszko, P., Franek, E., Jabłoński, M., Jaworski, M., Kalinka-Warzocha, E., Karczmarewicz, E., Kostka, T., Księżopolska-Orłowska, K., Marcinowska-Suchowierska, E., et al. (2017). Guidelines for the diagnosis and management of osteoporosis in Poland. Update 2017. *Endokrynologia Polska*, *68*(6), 604–609. https://doi.org/10.5603/EP.2017.0062.

63. Magdelaine, P. (2011). Egg and egg product production and consumption in Europe and the rest of the world. In Y. Nys, M. Bain, & F. V. Immerseel (Eds.), *Improving the Safety and Quality of Eggs and Egg Products: Egg Chemistry, Production and Consumption* (pp. 3–16). Woodhead Publishing.

64. Malmir, H., Shab-Bidar, S., & Djafarian, K. (2018). Vitamin C intake in relation to bone mineral density and risk of hip fracture and osteoporosis: A systematic review and meta-analysis of observational studies. *British Journal of Nutrition*, *119*(8), 847–858. https://doi.org/10.1017/S0007114518000430.

65. Malu, S. P., Abara, A. E., Obochi, G. O., Ita, B. I., & Edem, C. A. (2009). Analysis of *Egeria radiata* and *Thais coronata* shells as alternative sources of calcium for the food industry in Nigeria. *Pakistan Journal of Nutrition*, *8*, 965–969. https://doi.org/10.3923/pjn.2009.965.969.

66. Marchev, S., Temelkova, K., Todorova, M., Eftimov, M., Georgieva, A., Kuzmanova, V., Kuzmanov, A., Bankova, V., Surcheva, S., Vlaskovska, M., et al. (2018). Effects of antioxidants from *Aronia melanocarpa* and *Apium graveolens* on experimental model of osteoporosis. In J. A. Kanis & F. Cosman (Eds.), *Osteoporosis International* (Vol. 29, pp. 297–310). Springer.

67. Morcos, S. R., El-Shobaki, F. A., El-Hawary, Z., & Saleh, N. (1976). Effect of vitamin C and carotene on the absorption of calcium from the intestine. *Zeitschrift für Ernährungswissenschaft*, *15*(4), 387–390. https://doi.org/10.1007/BF02020506.

68. Muraki, S., Yamamoto, S., Ishibashi, H., Oka, H., Yoshimura, N., Kawaguchi, H., & Nakamura, K. (2007). Diet and lifestyle associated with increased bone mineral density: Cross-sectional study of Japanese elderly women at an osteoporosis outpatient clinic. *Journal of Orthopaedic Science, 12*(4), 317–320. https://doi.org/10.1007/s00776-007-1143-0.

69. Namkung, W., Thiagarajah, J. R., Phuan, P., & Verkman, A. S. (2010). Inhibition of Ca2+-activated Cl− channels by gallotannins as a possible molecular basis for health benefits of red wine and green tea. *FASEB Journal, 24*(11), 4178–4186. https://doi.org/10.1096/fj.10-160648.

70. Oliveira, G. R., Vargas-Sanchez, P. K., Fernandes, R. R., Ricoldi, M. S. T., Semeghini, M. S., Pitol, D. L., de Sousa, L. G., Siessere, S., & Bombonato-

Prado, K. F. (2019). Lycopene influences osteoblast functional activity and prevents femur bone loss in female rats submitted to an experimental model of osteoporosis. *Journal of Bone and Mineral Metabolism, 37*(5), 658–667. https://doi.org/10.1007/s00774-018-0970-8.

71. Östman, B., Michaëlsson, K., Helmersson, J., Byberg, L., Gedeborg, R., Melhus, H., & Basu, S. (2009). Oxidative stress and bone mineral density in elderly men: Antioxidant activity of alpha-tocopherol. *Free Radical Biology and Medicine*, *47*(5), 668–673. https://doi.org/10.1016/j.freeradbiomed.2009.05.031.

72. Parfitt, J., Barthel, M., & MacNaughton, S. (2010). Food waste within food supply chains: Quantification and potential for change to 2050. *Philosophical Transactions of the Royal Society B: Biological Sciences, 365*, 3065–3081. https://doi.org/10.1098/rstb.2010.0126.

73. Passali, C., Patsaki, A., Lelovas, P., Aligiannis, N., Makropoulou, M., Kourkoulis, S., Papaioannou, N., Mitakou, S., Skaltsounis, A. L., & Dontas, I. (2019). Red wine polyphenols modulate bone loss in the ovariectomized rat model of postmenopausal osteoporosis. *Journal of the Hellenic Veterinary Medical Society*, *70*, 1541–1550. https://doi.org/10.12681/jhvms.20860.

74. Piskorska-Pliszczynska, J., Mikolajczyk, S., Warenik-Bany, M., Maszewski, S., & Strucinski, P. (2014). Soil as a source of dioxin contamination in eggs from free-range hens on a Polish farm. *Science of the Total Environment, 466, 467*, 447–454. https://doi.org/10.1016/j.scitotenv.2013.07.061.

75. Platon, N., Arus, V. A., Georgescu, A. M., Nistor, I. D., & Barsan, N. (2020). White bread fortified with calcium from eggshell powder. *Revue Chimique, 71*(7), 299–306. https://doi.org/10.37358/RC.20.7.8248.

76. Podmore, I. D., Griffiths, H. R., Herbert, K. E., Mistry, N., Mistry, P., & Lunec, J. (1998). Vitamin C exhibits pro-oxidant properties. *Nature, 392*, 559. https://doi.org/10.1038/33308.

77. Proulx, W. R., Weaver, C. M., & Bock, M. A. (1993). Trypsin inhibitor activity and tannin content do not affect calcium bioavailability of three commonly consumed legumes. *Journal of Food Science, 58*(3), 382–384. https://doi.org/10.1111/j.1365-2621.1993.tb04280.x.

78. Quina, M. J., Soares, M. A. R., & Quinta-Ferreira, R. (2017). Applications of industrial eggshell as a valuable anthropogenic resource. *Resources, Conservation and Recycling, 123*, 176–186. https://doi.org/10.1016/j.resconrec.2016.09.027.

79. Rao, L. G., & Rao, A. V. (2018). Lycopene, tomatoes, and bone health. In A. V. Rao, G. L. Young, & L. G. Rao (Eds.), *Lycopene and Tomatoes in*

Human Nutrition and Health (pp. 89–101). CRC Press.

80. Rao, L. G., Kang, N. N., & Rao, A. V. (2014). Lycopene and other antioxidants in the prevention and treatment of osteoporosis in postmenopausal women. In V. R. Preedy (Ed.), *Aging: Oxidative Stress and Dietary Antioxidants* (1st ed., pp. 247–258). Academic Press.
81. Rao, L. G., Krishnadev, N., Banasikowska, K., & Rao, A. V. (2003). Lycopene I-Effect on osteoclasts: Lycopene inhibits basal and parathyroid hormone-stimulated osteoclast formation and mineral resorption mediated by reactive oxygen species in rat bone marrow cultures. *Journal of Medicinal Food, 6*(1), 69–78. https://doi.org/10.1089/109662003322233459.
82. Rao, L. G., Mackinnon, E. S., Josse, R. G., Murray, T. M., Strauss, A., & Rao, A. V. (2007). Lycopene consumption decreases oxidative stress and bone resorption markers in postmenopausal women. *Osteoporosis International, 18*(1), 109–115. https://doi.org/10.1007/s00198-006-0205-z.
83. Rapuri, P. B., Gallagher, J. C., Kinyamu, H. K., & Ryschon, K. L. (2001). Caffeine intake increases the rate of bone loss in elderly women and interacts with vitamin D receptor genotypes. *American Journal of Clinical Nutrition, 74*(5), 694–700. https://doi.org/10.1093/ajcn/74.5.694.
84. Ray, S., Barman, A. K., Roy, P. K., & Singh, B. K. (2017). Chicken eggshell powder as dietary calcium source in chocolate cakes. *Pharma Innovation Journal, 6*(6), 1–4.
85. Roberts, J. R. (2004). Factors affecting egg internal quality and eggshell quality in laying hens. *Journal of Poultry Science, 41*(3), 161–177. https://doi.org/10.2141/jpsa.41.161.
86. Rowe, S., & Carr, A. C. (2020). Global vitamin C status and prevalence of deficiency: A cause for concern? *Nutrients, 12*(7), 2008. https://doi.org/10.3390/nu12072008.
87. Russo, C., Ferro, Y., Maurotti, S., Salvati, M. A., Mazza, E., Pujia, R., Terracciano, R., Maggisano, G., Mare, R., Giannini, S., et al. (2020). Lycopene and bone: An *in vitro* investigation and a pilot prospective clinical study. *Journal of Translational Medicine, 18*(1), 43. https://doi.org/10.1186/s12967-020-02238-7.
88. Sahni, S., Hannan, M. T., Blumberg, J., Cupples, L. A., Kiel, D. P., & Tucker, K. L. (2009). Protective effect of total carotenoid and lycopene intake on the risk of hip fracture: A 17-year follow-up from the Framingham Osteoporosis Study. *Journal of Bone and Mineral Research, 24*(6), 1086–1094. https://doi.org/10.1359/jbmr.090102.
89. Sahni, S., Kiel, D. P., & Hannan, M. T. (2016). Vitamin C and bone health.

In C. M. Weaver, R. M. Daly, H. A. Bischoff-Ferrari, & editors (Eds.), *Nutritional Influences on Bone Health* (pp. 87–98). Springer.

90. Schaafsma, A., Pakan, I., Hofstede, G. J. H., Muskiet, F. A. J., Van Der Veer, E., & De Vries, P. J. F. (2000). Mineral, amino acid, and hormonal composition of chicken eggshell powder and the evaluation of its use in human nutrition. *Poultry Science, 79*(12), 1833–1838. https://doi.org/10.1093/ps/79.12.1833.

91. Shen, C. L., von Bergen, V., Chyu, M. C., Jenkins, M. R., Mo, H., Chen, C. H., & Kwun, I. S. (2012). Fruits and dietary phytochemicals in bone protection. *Nutrition Research, 32*(12), 897–910. https://doi.org/10.1016/j.nutres.2012.09.018.

92. Shen, C. L., Wang, P., Guerrieri, J., Yeh, J. K., & Wang, J. S. (2008). Protective effect of green tea polyphenols on bone loss in middle-aged female rats. *Osteoporosis International, 19*(7), 979–990. https://doi.org/10.1007/s00198-007-0527-5.

93. Shi, W., Liu, J., Cao, Y., Zhu, Y., Guan, K., & Chen, Y. (2016). Association of dietary and serum vitamin E with bone mineral density in middle-aged and elderly Chinese adults: A cross-sectional study. *British Journal of Nutrition, 115*(1), 113–120. https://doi.org/10.1017/S0007114515004134.

94. Śluszarczyk, B., & Machowska, E. (2019). Food waste in the world and in Poland. *АКАДЕМІЧНИЙ ОГЛЯД, 2019*. https://doi.org/10.32342/2074-5354-2019-1-50-9.

95. Sözen, T., Özışık, L., & Başaran, N. Ç. (2017). An overview and management of osteoporosis. *European Journal of Rheumatology, 4*, 46–56. https://doi.org/10.5152/eurjrheum.2016.048.

96. Sun, K., Wang, L., Ma, Q., Cui, Q., Lv, Q., Zhang, W., & Li, X. (2017). Association between tea consumption and osteoporosis: A meta-analysis. *Medicine (Baltimore), 96*(7), e9034. https://doi.org/10.1097/MD.0000000000009034.

97. Susanna, L., Grilli, G., Ferrari, L., Battelli, G., Pozzo, S., Galasso, I., Russo, R., Brasca, M., Reggiani, R., & Ferrante, V. (2020). Effect of different percentage of Camelina sativa cake in laying hens diet: Performance, welfare, and eggshell quality. *Animals, 10*(8), 1396. https://doi.org/10.3390/ani10081396.

98. Tanabe, S., Santos, J., La, V. D., Howell, A. B., & Grenier, D. (2011). A-type cranberry proanthocyanidins inhibit the RANKL-dependent differentiation and function of human osteoclasts. *Molecules, 16*(3), 2365–2374. https://doi.org/10.3390/molecules16032365.

99. Urban, K., Höhling, H. J., Lüttenberg, B., Szuwart, T., & Plate, U. (2012). An *in vitro* study of osteoblast vitality influenced by the vitamins C and E. *Head & Face Medicine*, *8*, 25. https://doi.org/10.1186/1746-160X-8-25.

100. Vielma, J. R., Picon, D., Gutiérrez, L. V., & Lara, N. D. (2018). Pathophysiology of osteoporosis: Genes, oxidative stress, and immunopathogeny. *Avances en Biomedicina*, *7*, 100–111.

101. Villarreal, A., Stoecker, B. J., Garcia, C., Garcia, K., Rios, R., Gonzales, C., Mandadi, K., Faraji, B., Patil, B. S., & Deyhim, F. (2007). Cranberry juice improved antioxidant status without affecting bone quality in orchidectomized male rats. *Phytomedicine*, *14*(11), 815–820. https://doi.org/10.1016/j.phymed.2007.04.004.

102. Wade, S. W., Strader, C., Fitzpatrick, L. A., Anthony, M. S., & O'Malley, C. D. (2014). Estimating prevalence of osteoporosis: Examples from industrialized countries. *Archives of Osteoporosis*, *9*, 182. https://doi.org/10.1007/s11657-014-0182-3.

103. Waheed, M., Butt, M. S., Shehzad, A., Adzahan, N. M., Shabbir, M. A., Rasul Suleria, H. A., & Aadil, R. M. (2019). Eggshell calcium: A cheap alternative to expensive supplements. *Trends in Food Science & Technology, 91*, 219–230. https://doi.org/10.1016/j.tifs.2019.07.021.

104. Waheed, M., Yousaf, M., Shehzad, A., Inam-Ur-Raheem, M., Khan, M. K. I., Khan, M. R., Ahmad, N., Abdullah, & Aadil, R. M. (2020). Channelling eggshell waste to valuable and utilizable products: A comprehensive review. *Trends in Food Science & Technology, 106*, 78–90. https://doi.org/10.1016/j.tifs.2020.10.009.

105. Walsh, M. C., Kim, N., Kadono, Y., Rho, J., Lee, S. Y., Lorenzo, J., & Choi, Y. (2006). Osteoimmunology: Interplay between the immune system and bone metabolism. *Annual Review of Immunology, 24*, 33–63. https://doi.org/10.1146/annurev.immunol.24.021605.090646.

106. Wilson, D. W., Nash, P., Singh, H., Griffiths, K., Singh, R., De Meester, F., Horiuchi, R., & Takahashi, T. (2017). The role of food antioxidants, benefits of functional foods, and influence of feeding habits on the health of the older person: An overview. *Antioxidants*, *6*(4), 81. https://doi.org/10.3390/antiox6040081.

107. Xhelili, A., Strube, R., Grossi, F., Zvěřinová, I., Taylor, T., Martinez-Juarez, P., Quiroga, S., Suárez, C., & Gjorgjev, D. (2020). A technological scenario for a healthier, more equitable and sustainable Europe in 2040: Citizen perceptions and policy implications. *International Journal of Environmental Research and Public Health, 17*, 231. https://doi.org/10.3390/ijerph17010231.

108. Yang, T. C., & Macdonald, H. M. (2016). Vitamin E homologues: Current

evidence. In C. M. Weaver, R. M. Daly, & H. A. Bischoff-Ferrari (Eds.), *Nutritional Influences on Bone Health* (pp. 107–120). Springer.

109. Zhang, H., & Tsao, R. (2016). Dietary polyphenols, oxidative stress, and antioxidant and anti-inflammatory effects. *Current Opinion in Food Science*, *8*, 33–42. https://doi.org/10.1016/j.cofs.2016.02.002.

110. Zhang, J., Hu, X., & Zhang, J. (2017). Associations between serum vitamin E concentration and bone mineral density in the US elderly population. *Osteoporosis International*, *28*(4), 1245–1253. https://doi.org/10.1007/s00198-016-3855-5.

111. Zhang, J., Lazarenko, O. P., Blackburn, M. L., Shankar, K., Badger, T. M., Ronis, M. J. J., & Chen, J. R. (2011). Feeding blueberry diets in early life prevent senescence of osteoblasts and bone loss in ovariectomized adult female rats. *PLoS ONE*, *6*(9), e24486. https://doi.org/10.1371/journal.pone.0024486.

112. Zhang, Y., Liu, Z., Duan, L., Ji, Y., Yang, S., Zhang, Y., Li, H., Wang, Y., Wang, P., Chen, J., et al. (2020). Effect of low-dose vitamin K_2 supplementation on bone mineral density in middle-aged and elderly Chinese: A randomized controlled study. *Calcified Tissue International*, *106*(5), 476–485. https://doi.org/10.1007/s00223-020-00669-4.

113. Zhu, M., Jinggang, M., ChangSheng, H., Haiping, X., Ning, M., & Caijiao, W. (2010). Extraction, characterization of polysaccharides from *Lycium barbarum* and its effect on bone gene expression in rats. *Carbohydrate Polymers*, *80*(2), 672–676. https://doi.org/10.1016/j.carbpol.2009.11.038.

114. Zhu, Y. F., Wang, J. P., Ding, X. M., Bai, S. P., Qi, S. R. N., Zeng, Q. F., Xuan, Y., Su, Z. W., & Zhang, K. Y. (2020). Effect of different tea polyphenol products on egg production performance, egg quality, and antioxidative status of laying hens. *Animal Feed Science and Technology*, *267*, 114544. https://doi.org/10.1016/j.anifeedsci.2020.114544.

CHAPTER 10

POTENTIAL APPLICATION OF PLANT-BASED FUNCTIONAL FOODS IN THE DEVELOPMENT OF IMMUNE BOOSTERS

Contents

INTRODUCTION

One of the body's most intricate biological systems is the immune system. It is a complex and multifaceted network of specialized cells, organs, proteins, and chemicals that is vital to providing defense against cancer cells and a range of pathogens, including bacteria, viruses, and fungi. It is commonly understood that both innate (non-specific) and adaptive (specific) immunity make up host immunity. A compromised immune system makes a person more vulnerable to infections, which can result in the onset of conditions like inflammatory bowel disorders, rheumatoid arthritis, and allergy diseases. There has been a surge in research efforts aimed at enhancing immune function because immunity is essential to the host's health and helps prevent diseases. Currently, immune adjuvants like aluminum hydroxide, Freund's adjuvant, and albumen adjuvants are primarily used in the clinical application of immunomodulators. Nevertheless, there have been reports of local stimulation, tissue damage, and carcinogenesis caused by these adjuvants. Furthermore, neither Freund's adjuvant nor aluminum hydroxide can significantly boost the cellular immune response. Although albumen adjuvants have recently been the main treatment approach for enhancing an individual's immune function, their use is restricted due to their expensive cost and adverse effects. Therefore, research into safer, more natural adjuvants that can boost immune system function by modulating immune molecules and activating immune cells is required (Figure 10.1).

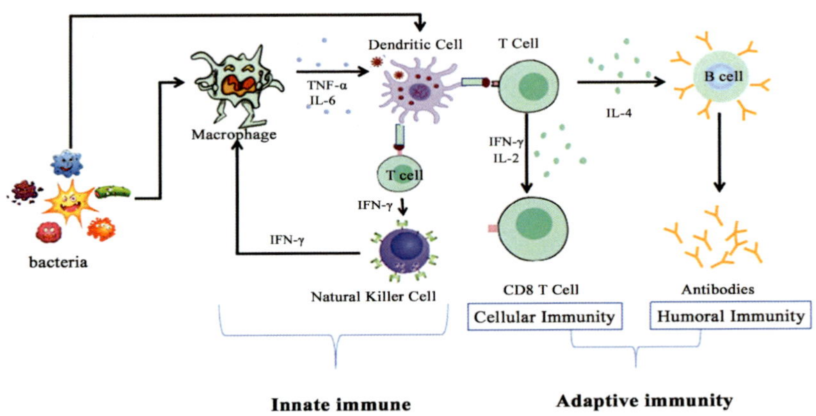

Figure 10.1. *The immune system of innate and adaptive immunity.*

Source: Jiang L, Zhang G, Li Y, Shi G, Li M. (2021). Potential application of plant-based functional foods in the development of immune boosters. Front Pharmacol.;12:637782. doi: 10.3389/fphar.2021.637782. PMID: 33959009; PMCID: PMC8096308.

Plant-based functional foods are made from unprocessed or naturally occurring plant foods that may or may not contain bioactive ingredients. Their high bioactive content and health benefits have led to their widespread consumption in recent years. Notably, an increasing body of pre-clinical research data indicates that plant-based functional foods can lower the risk of developing many illnesses, including hyperlipidemia, hyperuricemia, diabetes, cancer, and cardiovascular disease (CVD). A growing number of people are choosing plant-based functional foods to enhance immune system functions as a result of growing awareness of the potential of these foods to combat diseases. Studies on immune boosters based on plant-based functional foods have drawn a lot of attention. Previous research, for instance, has shown that active polysaccharides derived from Ganoderma can modulate lymphocytes and myeloid cells to trigger an immune defense response against tumor growth, viruses, bacteria, and fungal pathogens. This suggests a possible use for host immune system modulation. Additionally, it has been reported that astragalus, ginseng, and Op hiocordyceps sinensis have significant immune-enhancing properties. Plant functional foods that boost immunity typically contain polysaccharides, saponins, flavonoids, and alkaloids as their active ingredients. This chapter focuses on the biological elements of plant-based functional foods that have immune-boosting properties and their usefulness in immune-enhancing applications, such as the creation of novel disease treatment plans (Jiang et al., 2021).

10.1. BIOACTIVE COMPONENTS OF PLANT-BASED FUNCTIONAL FOODS AND THEIR IMMUNE-BOOSTING EFFECTS

10.1.1. Polysaccharides

Polysaccharides are a class of naturally occurring macromolecules made up of monomers of carbohydrates joined by a glycosidic bond. It is well known that polysaccharides derived from natural sources possess a range of biological properties, such as immune regulation, anti-tumor, and anti-inflammatory properties. In fact, due to their low toxicity and minimal adverse effects, polysaccharides have gained a lot of attention lately for their immune-boosting properties. Polysaccharides have recently been demonstrated to have immune-stimulating properties both *in vivo* and *in vitro*, as demonstrated by the induction of immune organ development and the release of immune-related molecules. Additionally, polysaccharides can boost the body's innate immune system in addition to stimulating the antigen-specific immune system, making them a perfect candidate for adjuvant therapy. Table 10.1 provides an overview of the immune-boosting properties and underlying mechanisms of the polysaccharides found in plant-based functional foods.

Table 10.1. Experiment and Mechanism of Polysaccharides Bioactive Components from Plant-Based Functional Foods on Immune Enhancement

Source	**Bioactive Compound**	**Model Type**	**Effective Dose**	**Effects**	**Mechanisms**	**References**
Dioscorea oppositifolia L./Yam	Yam polysaccharide	Lewis cells of lung cancer (B)	150 mg/kg	↑ IL-2, TNF-α, T lymphocyte proliferation, and the activity of NK cells.	They play an immune-enhancing role by promoting the activity of immune cells and related factors.	Xu et al. (2020a)
Dioscorea oppositifolia L./Yam	Yam polysaccharide	Cyclophosphamide-induced immunosuppressive mice (A)	250 mg/kg	The liver index and white blood cell count significantly improved.	They achieve immune enhancement by regulating immune organs and cells.	Song et al. (2019)
Codonopsis pilosula (Franch.) Nannf./Codonopsis pilosula	*Codonopsis pilosula* polysaccharide	Cyclophosphamide-induced immunosuppressive mice (A)	200 mg/kg	It promotes lymphocyte proliferation *in vitro*. The thymus index and spleen index could be improved and the levels of IFN-γ, IL-2, IL-10, as well as serum IgG could be restored.	The thymus index, spleen index and immune related factors were improved, and the immunity function was enhanced.	Fu et al. (2018); Zou et al. (2019)
Lycium chinense Mill./Lycium chinense	*Lycium chinense* polysaccharide	Cyclophosphamide-induced immunosuppressive mice (A)	200 mg/kg	It enhances immune organ indexes and alleviating immune organ damage, enhance the production of immune-related cytokines (IL-2, IL-6, IL-1β, TNF-α and IFN-γ).	They play an immune-enhancing role by promoting the activity of immune cells and related factors.	Ding et al. (2019)

Gynostemma pentaphyllum (Thunb.) Makino/ Gynostemma pentaphyllum	*Gynostemma pentaphyllum* polysaccharide	Cyclophosphamide-induced immunosuppressive mice (A)	400 mg/kg	↑ IL-4, IgG, IgM, the spleen and thymus indexes, macrophage and splenic lymphocyte activity.	The thymus index, spleen index and immune related factors were enhanced.	Bao et al. (2018); Li et al. (2015)
Ginkgo biloba L./ Ginkgo	*Ginkgo biloba* leaves polysaccharide	SPF chickens (A)	40 g/L	↑ IL-2 and IFN-γ, $CD4^+$ and $CD8^+$ T lymphocyte number	Promote humoral immunity and cellular immunity, and have a good role in immune promotion	Meng et al. (2018b)
Platycodon grandiflorus (Jacq.) A. DC./Platycodon	*Platycodon grandiflorus* polysaccharide	SPF mice cells (B)	100 µg/ml	It can selectively activates B cells and macrophages.	They play an immune-enhancing role by promoting the activity of immune cells.	Lv et al. (2016)
Acanthopanax senticosus (Rupr. Maxim.) Harms/Acanthopanax	*Acanthopanax senticosus* polysaccharide	Solid tumor mouse (A)	200 mg/kg	↑ immune organ index, TNF-α, IL-1β and IL-6	TLR4 signaling pathway may be involved in the immunomodulatory effects.	Zhou et al. (2017)
Panax ginseng C. A. Meyer/ Ginseng	*Panax ginseng* Polysaccharide	chickens (A)	20 g/kg	↑ the immune organs index, IL-2, IL-12 TNF-α, the activity of NK cells.	It can promote the development of immune organs and secretion of related cytokines, so as to improve immune performance.	Hwang et al. (2018);Wei et al. (2019)

Pseudostellaria heterophylla (Miq.) Pax Pseudostellariae Radix	*Pseudostellaria heterophylla* polysaccharide	Cyclophosphamide-induced immunosuppressive mice (A)	50 mg/kg	↑ immune organ index, IgA, IgG, IgM, IL-2, IL-4, IL-6 and IFN-γ.	It can promote the development of immune organs and secretion of related cytokines.	Feng et al. (2020)
Litchi chinensis Sonn./ Litchi	Litchi pulp polysaccharides	Cyclophosphamide-induced immunosuppressive mice (A)	200 mg/ kg	↑ IL-6, TNF-α, IgG, IgA, IgM, immune organ index.	It can stimulate the proliferation of splenocytes, balancing the ratio of spleen lymphocyte subsets, up-regulating the thymus and spleen indices.	Huang et al. (2016)
Atractylodes macrocephala Koidz./ Atractylodes	*Atractylodes macrocephala* Koidz. polysaccharides	Mouse lymphocytes (B)	200 μg/ ml	↑ IL-2, IL-6, IL-10 and the proliferation of T lymphocytes.↓ TNF-α and IgG	It can promote lymphocyte proliferative response	Xu et al. (2020b)
Ziziphus jujuba Mill./ Jujube	Jujube Polysaccharide	Mouse lymphocytes (B)	160 μg/ ml	↑ the mRNA of IL-2, IL-6, IL-10, IL-12 and lymphocyte multiplication	It can improve the immune function by inducing lymphocyte proliferation, lymphocyte cytokine secretion and mRNA expression.	Li et al. (2021)

Source: Jiang L, Zhang G, Li Y, Shi G, Li M. (2021). Potential application of plant-based functional foods in the development of immune boosters. Front Pharmacol.;12:637782. doi: 10.3389/fphar.2021.637782. PMID: 33959009; PMCID: PMC8096308.

Mulberry (*Morus alba* L.) was extracted. leaf polysaccharides (MLP) have a significant chance of enhancing immunity. More precisely, MLP can significantly raise the thymus index and improve the transformation rate of splenic lymphocytes and cytokines. Lately, Chen and colleagues investigated the effects of MLP on immune enhancement *in vitro* and found that spleen B and T cell proliferation was enhanced at concentrations of 125 and 250 μg/ml MLP in comparison with control treatments ($p < 0.05$). MLP (4 and 8 g/kg) was given orally to chickens immunized with the Newcastle disease vaccine for the *in vivo* experiments. MLP significantly increased the serum IgA concentrations in the tracheal and jejunal wash fluids, as well as the serum antibody titer associated with Newcastle disease ($p < 0.05$). Another study assessed how MLP affected the proliferation of lymphocytes in weanling pigs as well as other immune functions such as serum immunoglobulins and cytokines. According to these reports, there was a significant difference ($p < 0.05$) between the thymus and spleen indices (0.6 and 1.2 g/kg) between the MLP groups and the control group. In addition, MLP supplementation boosted the rate of lymphocyte transformation and raised the levels of the serum cytokines interleukin (IL)-1, IL-2, IL-6, IL-8, and interferon (IFN)-γ. These findings imply that MLP has the potential to significantly improve immunity, making it a promising therapeutic candidate.

Long-term clinical trials have demonstrated the non-toxicity of astragalus polysaccharide (APS). More significantly, APS demonstrates a broad spectrum of pharmacological actions associated with immune system regulation, including raising the immune organ index, encouraging immune cell proliferation, inducing cytokine release, and impacting immunoglobulin secretion and immune signal transmission. Chen et al. investigated the effects of APS (0.1, 0.5, and 1.0 mg/ml) on the morphology and immunological function of mouse macrophages (RAW264.7 cells) following lipopolysaccharide (LPS) stimulation. Treatment with 1.0 mg/ml APS significantly inhibited changes in macrophage morphology and the capacity for proliferation caused by LPS stimulation when compared to the control group; additionally, there was a significant increase ($p < 0.05$) in the activity of macrophage acid phosphatase. Additionally, APS significantly decreased levels of the LPS-induced pro-inflammatory cytokines tumor necrosis factor (TNF-α; $p < 0.05$) and alkaline phosphatase activity ($p < 0.05$) at different concentrations. Following administration of 1 mg/ml APS, the macrophages exhibited a significant decrease in the expression of toll-like receptor (TLR) 4, myeloid differentiation primary response gene 88 (MyD88), and nuclear factor-κB (NF-κB) mRNA ($p < 0.05$). As a result, by controlling the expression of genes linked to the TLR4 signaling pathway, APS can be used to enhance macrophage morphology, restore cell proliferation, lower the secretion of pro-inflammatory factors, and lessen the immune stress response. Furthermore, a clinical trial examined the impact of APS injection on 196 patients with bronchial asthma's inflammatory cell count and related factor levels. It was discovered that the inflammatory cell count could be successfully decreased by combining an APS injection with traditional therapy. Following two weeks of therapy, the levels of TNF-α, IL-6, IL-8, IL-13, IL-17, and inflammatory cell counts in control groups were significantly lower; in addition, the treatment group's levels of $CD3^+$, $CD4^+$, and $CD4^+/CD8^+$ were significantly higher ($p < 0.05$). Compared to the control group, there were fewer adverse reactions—two patients with cases of sore throat—with an incidence of 2.04% (Qiu et al., 2018). Therefore, immune cells can be activated by polysaccharides to exert their immune-promoting effects. The intricate structures of polysaccharides make them less soluble in water, which has an impact on their biological activity. Polysaccharide structural modification has the potential to improve their biological activities and shed light on the connection between immune function and polysaccharide structure.

10.1.2. Saponins

One kind of aglycone that contains steroids or triterpenoids is called a saponin. They are frequently discovered to have health-promoting qualities in functional meals. In numerous biological investigations, extracted plant saponins have performed well, exhibiting immune-stimulating and anti-tumor regulatory capabilities. Numerous saponin compounds have been the subject of pharmacological investigations, which have demonstrated their ability to enhance immunity. As an example, the immunological enhancement of gypenosides, which are mainly distributed in *Gynostemma pentaphyllum* (Thunb.) Makino, was investigated for immunosuppression in mice.

Previous research has demonstrated that gypenosides can enhance both specific and non-specific immune functions in rats and stimulate the development of their immune organs. A recent investigation by Dan et al. created a mouse model of immunodeficiency by cyclophosphamide (80 mg/kg) induction. Gypenosides (60, 120, and 180 mg/kg) were then given orally for a duration of 30 days. The peripheral blood white blood cell count, organ index, and $CD4^+/CD8^+$ levels were all significantly higher ($p < 0.01$) than in the model group. In the group receiving high-dose gypenoside, the effect was more noticeable. Furthermore, it was discovered that gypenosides significantly raised the mRNA expression of TNF-α, IFN-γ, IL-10, IL-6, and IL-2 in spleen lymphocytes as well as their expression levels in serum. Therefore, in mice undergoing cyclophosphamide-induced immunosuppression, gypenosides can improve immune function. The species *Codonopsis pilosula* is known to be homologous in medicine and food; additionally, the total saponin content of *C. pilosula* (TSCP) has been investigated for their ability to boost the immune system in mice whose immunosuppression was brought on by hydrocortisone therapy. It is interesting to note that oral TSCP administration improved immune function in a dose-dependent manner (50, 100, and 200 mg/kg) and markedly raised serum IL-2 and IFN-γ levels, spleen T cell proliferation, and the natural killer (NK) cell killing rate ($p < 0.05$). This suggests that TSCP can boost immune function in immunosuppressed mice by counteracting the inhibitory effect of hydrocortisone on T cells and NK cells. Additionally, in the immunosuppressed mice, TSCP markedly improved the phagocytosis index and the half hemolysis value for serum hemolysin when compared to the model group ($p < 0.05$). According to the findings, TSCP can raise mononuclear macrophages' phagocytic activity and stimulate the synthesis of antibodies to bring immunosuppressed mice's antibody levels back to normal. The findings suggest that TSCP has the ability to improve humoral, cellular, and non-specific immune modulatory effects.

In conclusion, saponins can significantly improve immune function in a number of ways. However, saponins have a hemolytic effect, so the dosage for application needs to be carefully considered. Furthermore, more information is needed regarding the way saponins function concerning the immune regulatory system and signal pathways. Further investigation could elucidate the role saponins play in food and drug development, potentially offering an adjuvant for use in the future. Figure 10.2 depicts the bioactive components of saponins and their structures, which have the effect of enhancing immunity; Table 10.2 provides a summary of the experimental studies (Jiang et al., 2021).

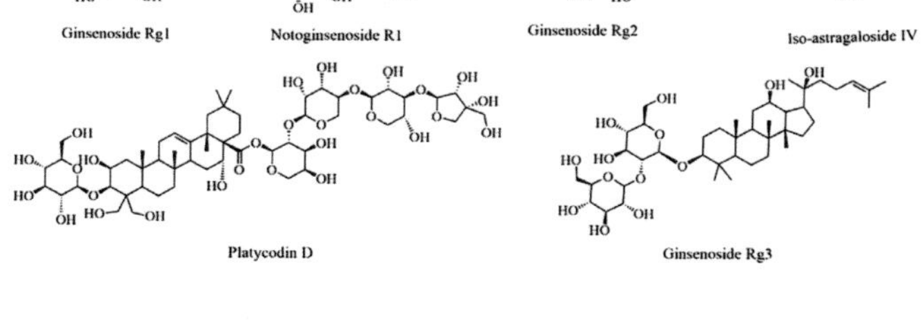

Figure 10.2. *The immune enhancement effects of saponins in plant-based functional foods.*

Source: Jiang L, Zhang G, Li Y, Shi G, Li M. (2021). Potential application of plant-based functional foods in the development of immune boosters. Front Pharmacol.;12:637782. doi: 10.3389/fphar.2021.637782. PMID: 33959009; PMCID: PMC8096308.

Table 10.2. Experiment of Saponins in Plant-Based Functional Foods on Immune Enhancement

Source	**Compound**	**Model Type**	**Effective Dose**	**Effects**	**Mechanisms**	**References**
Polygala tenuifolia Willd/Polygala	Polygala saponins	Cyclophosphamide-induced immunosuppressive mice (A)	400 mg/kg	↑ spleen index, thymus index, IL-2 ↓ IL-6	The mechanism is related to the regulation of cytokines.	Chai et al. (2018)
Panax ginseng C. A. Meyer/Ginseng	Ginsenoside Rg1	Cyclophosphamide-induced immunosuppressive mice (A)	400 mg/kg	It could significantly increase the positive rate of antibody, spleen and bursa index and intestinal total sIgA and specific sIgA content compared with the C group, as well as up-regulated the mRNA expression of TLR4, p65, TGF-β, pIgR and CCR9 genes in the chicken duodenum.	It acts on TLR4 receptors to exert immunomodulatory effects.	Bi et al. (2019)

Panax notoginseng (Burk) F. H. Chen/Notoginseng	Notoginsenoside R1	Traumatic shock rats (A)	50 mg/kg	↓ TNF-α, IL-1β, iNOS, p-p65, ERK1/2/p-ERK1/2 ↑ IL-10	It can enhance the immune response by blocking ERK1/2 and NF-κB signaling pathways, and regulate expression levels of inflammatory factors.	Wang et al. (2019b)
Camellia oleifera Abel./Camellia seed	Tea saponin	Cyclophosphamide-induced chickens (A)	5 mg/kg	The lymphocyte proliferation and serum virus-specific antibodies were increased.	It could increase the specific antibody response to enhance immune function.	Chi et al. (2017)
Panax ginseng C. A. Meyer/Ginseng	Ginseng stem-leaf saponins	Vaccinated mice (A)	50 μg	↑ IgG1, IgG2a, IgG2b, IFN-γ and IL-4 Increased splenocyte proliferative.	It could enhance cellular and humoral immune response to enhance immune function.	Xu et al. (2020c)
Panax quinquefolium L./American ginseng	American ginseng saponins	Rapamycin-induced zebrafish (A)	25 μg/ml	↑ IFN-γ Increased the number of neutrophils and macrophages.	It can increase the function of immune cells and promote the secretion of cytokines.	Lv et al. (2020)

Panax ginseng C. A. Meyer/Ginseng	Ginsenoside Rg1	Cyclophosphamide-induced chickens (A)	1 mg/kg	↑ SIgA, TLR4, p65, TGF-β, pIgR, and spleen and bursa of Fabricius index	It can restore the function of suppressed immune organs to improve the immune response.	Bi et al. (2019)
Panax ginseng C. A. Meyer/Ginseng	Ginsenoside Rg1	Type Ⅲ prostatitis rats (A)	40 mg/kg	↓ IL?8, IL4, TNF-α, TGFβ	It can regulate immune balance by inducing the secretion of cytokines.	Sun et al. (2017)
Panax ginseng C. A. Meyer/Ginseng	Ginsenoside Rg2	Patients with lung cancer (C)	0.5 g/d	↑ $CD3^+$, $CD4^+$, $CD4^+/CD8^+$ ↓ $CD8^+$	It can regulate lymphocytes to protect immune function.	Ma and Bai (2019)
Panax ginseng C. A. Meyer/Ginseng	Ginsenoside Rg3	Patients with gastric cancer (C)	40 mg/d	↑ $CD3^+$, $CD4^+$, $CD4^+/CD8^+$, T lymphocyte transformation rate ↓ CD^8+	It can regulate lymphocytes to protect immune function.	Chen et al. (2017a)
Panax ginseng C. A. Meyer/Ginseng	Ginsenoside Rg3	Nasopharyngeal carcinoma in patients (C)	100 mg/L	↑ $CD4^+$, $CD4^+/CD8^+$, IgG, CD 80/86, IgM, IL-2 ↓ IL-6, $CD8^+$ Promoted the proliferation of lymphocytes	It can significantly enhance the immune function by promoting the proliferation of lymphocytes, and regulating immune factors.	Chen et al. (2017a)

Momordica charantia L./Bitter gourd	Total momordicoside	Nephropathy of rats (A)	20 mg/kg	↓ IL-2, IL-6, TGF-β1	It can inhibit the expression of TGF-β1, thereby reducing the serum IL-2 and IL-6.	Chen et al. (2017b)
Panax notoginseng (Burk.) F. H. Chen/Notoginseng	NotoginsenosideS-6	Con A-Splenic lymphocyte of mice (B)	1 μg/ml	Increased T, and B lymphocyte proliferative	It can significantly promote the proliferation of T and B lymphocytes and enhance immune function.	Qin et al. (2017)
Panax notoginseng (Burk.) F. H. Chen/Pseudo-ginseng	NotoginsenosideS-6	Con A-Splenic lymphocyte of mice (B)	25 μg/ml	↑ IL-2	As an activator, IL-2 induces T lymphocytes to express IL-2R, which in turn induces the production of IL-2 to play an immune role.	Qin et al. (2017)
Platycodon grandiflorus (Jacq.) A. DC./Platycodon	Platycodin D	Lymphocyte and macrophage of mice (B)	50 μg/ml	↑ IL-2, IL-4, TNF-α, and IL-12 Enhanced lymphocyte proliferation and macrophage phagocytosis	It can enhance the activity of lymphocytes and macrophages to regulate the immune function.	Li et al. (2019)

Astragalus mongholicus Bunge/ Radix astragali	Iso-astragaloside VI	Lymphocyte and macrophage of mice (B)	50 µg/ml	↑ IL-1, IL-2, IL-6, and IL-12 p70, the phagocytosis of macrophages	It is related to the enhancement of the secretion capacity of dendritic cells, and the promotion of non-specific immune phagocytosis and lymphocyte proliferation.	Ruan et al. (2021)

Source: Jiang L, Zhang G, Li Y, Shi G, Li M. (2021). Potential application of plant-based functional foods in the development of immune boosters. Front Pharmacol.; 12:637782. doi: 10.3389/fphar.2021.637782. PMID: 33959009; PMCID: PMC8096308.

10.1.3. Flavonoids

Plants contain a significant group of bioactive secondary metabolites called flavonoids. They are significant naturally occurring bioactive components of many functional foods made from plants. The immune-boosting properties of flavonoids have garnered a lot of interest lately. For instance, quercetin, soybean flavone, and sea buckthorn flavone are well-known to have positive immune regulatory effects. They are also very powerful natural immune enhancers that can boost NK cell and cytotoxic T cell killing, encourage cytokine release, raise the immune organ index, improve immune function, and strengthen the body's immune system. Furthermore, flavonoids can control NF-κB expression to prevent the synthesis of pro-inflammatory cytokines.

The beneficial bioactive component hesperidin is primarily present in citrus fruits. In a recent study, the effects of oral hesperidin on rat systemic immune systems following rigorous exercise programs were examined. Hesperidin supplementation (200 mg/kg) increased the cytotoxicity of NK cells, decreased the amount of leukocytosis caused by intense exercise, and increased the percentage of T helper cells and phagocytic monocytes in the spleen, blood, and thymus. Furthermore, the secretion of cytokines (IL-6, IFN-γ) from peritoneal macrophages dropped ($p < 0.01$). Furthermore, *in vitro* research has demonstrated that hesperidin can mediate immune-related pathways, such as the NF-κB pathway, to have an anti-inflammatory effect (Birsu Cincin et al., 2015).

It is known that Lonicera japonica is both a medicinal and edible species (Zhang et al., 2019). In this regard, the immunomodulatory effects of *Lonicera japonica* flavone in mice with immunosuppressive disorders induced by dexamethasone have been investigated. The organ indices were significantly improved ($p < 0.01$) by Lonicera japonica flavone (400 mg/kg) when compared to the model group. This suggests that it has exceptional protective effects on the thymus and spleen. Additionally, the flavone markedly increased serum levels of non-specific immune factors (acid and alkaline phosphatases). In immunosuppressed mice, licorera japonica flavone was shown to significantly reduce the activities of malondialdehyde and monoamine oxidase in the thymus and spleen, as well as to increase the content of superoxide dismutase ($p < 0.01$). A clinical trial was recently carried out to examine the impact of benzathine penicillin and honeysuckle soup on syphilis patients' serum levels of TNF-α and IL-8. Serum levels of TNF-α and IL-8 were significantly lower ($p < 0.05$) than in patients treated with benzathine penicillin. This implies that the extract from Lonicera japonica has positive immunomodulatory effects.

Table 10.3. Flavonoid Components in Plant-Based Functional Foods on Immune Enhancement

Source	Compound	Model Type	Effective Dose	Main Results	Mechanisms	References
Apium graveolens L./Celery	Apigenin	LPS-induced spleen cells (B)	20 μM	↓ IL-1β, IL-6 and TNF-α. The expression of CD80, CD86 and MHCII of DCs were inhibited.	The mechanism is related to inhibiting DCs activation and function.	Liu et al. (2017c)
Glycine max (Linn.) Merr./ Soybean	Soy isoflavones	Rats (A)	50 mg/kg	↓ IL-4, TNF-α, and INF-γ.	It can regulate the expression levels of cytokines to enhance immunity.	Li et al. (2017)
Glycyrrhiza uralensis Fisch./ Licorice	Isoliquiritigenin	LPS-induced bone marrow-derived macrophages (B)	40 μM	↓ the protein levels of IL-1β, IL-6.	It may inhibit Mincle/Syk/ NF-kappa B signaling pathway in macrophage to exert anti-inflammatory effects.	Liao et al. (2020)
Ginkgo biloba L./Gingko	Proanthocyanidins	Cyclophosphamide-immunosuppression mice (A)	100 mg/kg	↑ Thymus and spleen index, TNF-α, and TNF-α mRNA. ↓ IL-10, and IL-10 mRNA.	It can regulate the secretion of cytokines and enhance immunity.	Kong et al. (2018)

Ginkgo biloba L./Gingko	Proanthocyanidins	B16F10 tumor cells (B)	80 μg/ml	↑ IL-12, and TNF-α T cell-mediated immune responses increased.	It enhances the immune response mediated by T cells and plays an immunological enhancement role.	Zhang et al. (2017), Zhang et al. (2019b)
Hippophae rhamnoides L./ Sea-buckthorn	Seabuckthorn flavones	Dendritic cell (B)	200 μg/ml	↑ antigen presenting molecules HLA-DR, CD80, CD83 and CD86.	It can improve the antigen presentation ability of dendritic cell and promote the maturation of dendritic cell phenotype.	Liu et al. (2017c)
Lonicera japonica Thunb./ Flos Lonicera japonica	Luteolin	LPS-induced RAW264.7 cells (B)	1 μM	↓ IL-6, TNF-α, NF-κB and phosphorylation of IκB.	It regulates NF-κB and the secretion of cytokines to enhance the immunoregulative effect.	Cheng and Yeh (2019)
Lonicera japonica Thunb./ Flos Lonicera japonica	Luteolin-7-O-glucoside	LPS-induced mice splenocytes, and macrophages (B)	4.48 μg/ml	NK cell activity and the proliferation of T lymphocytes increased, as well as macrophage lysosomal activity decreased.	It exhibits important immune enhancement activity by regulating NK cell activity and the proliferation of T lymphocytes.	Maatouk et al. (2017)
Sophora japonica L./Fructus Sophorae	Genistein	LPS-induced broiler chicks (A)	5 mg/kg	The immune spleen, thymus and bursa indices were increased.	It can promote the development of immune organs to play immune function.	Kamboh et al. (2016)

Sophora japonica L./Fructus Sophorae	Rutin	Cyclophosphamide-induced rats (A)	50 mg/kg	The phagocytic index, total leukocyte count, and serum immunoglobulin levels were in increased. ↓ TNF-α, and IL-6 ↑ p38-MAPK, NFκB, i-NOS and COX-2 in liver.	It can stimulate humoral and cellular responses to enhance the immunoregulative effect.	Manzoni et al. (2020); Nafees et al. (2015)
Allium cepa L./Onions	Quercetin	Chickens (A)	4 g/kg	The indexes of spleen, bursa of Fabricius and thymus, and the phagocytic function of macrophages were improved. ↑ IL-4, and IL-12	It can promote the development of spleen, bursa and thymus, improve the phagocytosis of macrophages, and induce the secretion of cytokine to improve the immune function of the body.	Li. et al. (2020)
Eucommia ulmoides Oliv./Eucommia ulmoides	Quercetin	ConA or LPS-stimulated lymphocytes (B)	20 μM	↑ IL-2 and IFN-γ in lymphocytes.	It can promote lymphocyte proliferation and cytokine secretion.	Wang et al. (2016)
Eucommia ulmoides Oliv./Eucommia ulmoides	Flavone from *Eucommia ulmoides Oliv.*	ConA or LPS-stimulated lymphocytes (B)	20 μM	↑ IL-2 and IFN-γ in lymphocytes.	It can regulate the secretion of cytokines and enhance immunity.	Wang et al. (2016).

Kaempferia galanga L./Rhizoma kaempferiae	Kaempferol	Cold-stress mice (A)	25 mg/kg	↓ IL-9, IL-13 and $CD8^+$ T cells ↑ $CD4^+$, $CD4^+/CD8^+$ T cells, and IFN-γ	It boosts immunity by inhibiting the activation of pro-inflammatory factors.	Jia et al. (2019)
Pueraria lobata (Willd.) Ohwi/ Pueraria	Puerarin	Gouty arthritis mice (A)	100 mg/ kg	↓ IL-1, and TNF-α ↑ IL-10, and NO The number of neutrophils and lymphocytes was significantly reduced.	It may play an immune-enhancing role by influencing the expression level of inflammatory cytokines.	Zhang et al. (2020b)
Pueraria lobata (Willd.) Ohwi/ Pueraria	Puerarin	Viral myocarditis in children (C)	40 mg/ml	↑ $CD3^+$, $CD4^+$, and $CD4^+/CD8^+$ ↓ IL-6, IL-8, and TNF-α	It can reduce the level of inflammatory factors to improve its immune function.	Wang et al. (2020a)

Source: Jiang L, Zhang G, Li Y, Shi G, Li M. (2021). Potential application of plant-based functional foods in the development of immune boosters. Front Pharmacol.;12:637782. doi: 10.3389/fphar.2021.637782. PMID: 33959009; PMCID: PMC8096308.

Although their immune-boosting properties and mode of action have been thoroughly investigated recently, their poor pharmacokinetic profile limits their clinical use as immunoregulatory medications. Furthermore, the clinical use of flavonoids has been restricted due to their numerous active sites, comparatively slow rates of efficacy, and lack of selectivity or specificity for particular diseases. The initial investigation of clinical immune enhancement applications is anticipated to be facilitated by a gradual elucidation of the immune enhancement effect and mechanism of action of flavonoids. Table 10.3 provides an overview of the immune-boosting properties of flavonoids found in plant-based functional foods, and Figure 10.3 shows the structures of these compounds.

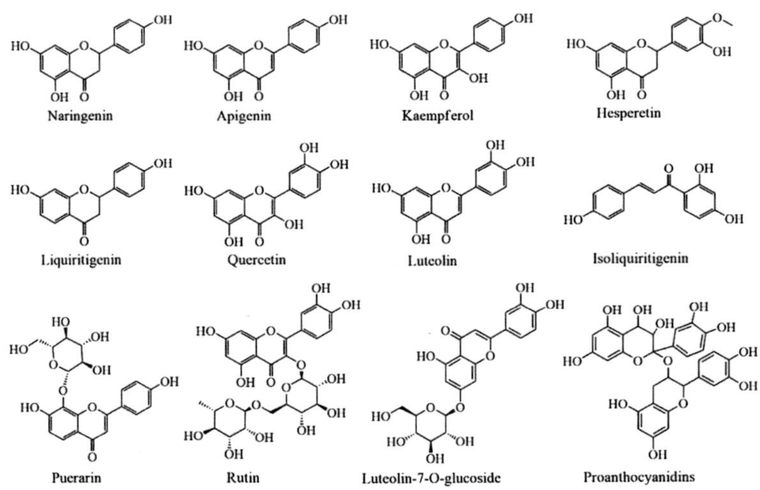

Figure 10.3. *The immune enhancement effects of flavonoids in plant-based functional foods.*

Source: Jiang L, Zhang G, Li Y, Shi G, Li M. (2021). Potential application of plant-based functional foods in the development of immune boosters. Front Pharmacol.;12:637782. doi: 10.3389/fphar.2021.637782. PMID: 33959009; PMCID: PMC8096308.

10.1.4. Alkaloids

Apart from proteins, peptides, amino acids, and vitamin B, alkaloids are nitrogenous substances that are mostly found in plants. Their roles in boosting the immune system should not be disregarded due to their intricate structures and exceptional biological activities. In fact, it is well known that alkaloids improve immunity by controlling cytokine secretion and the growth of splenic and thymic lymphocytes. Figure 10.4 shows how alkaloid structures boost the immune system.

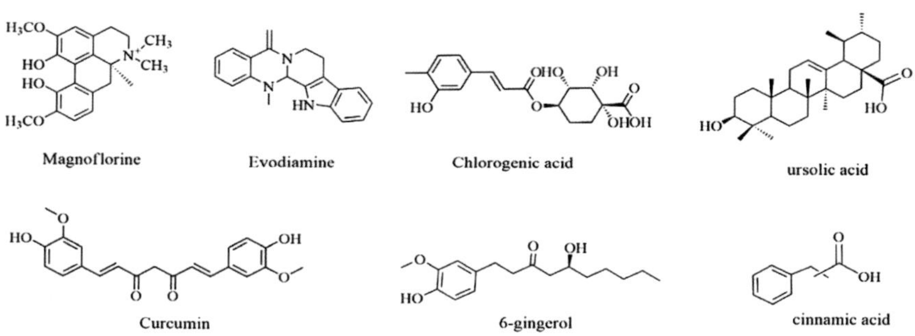

Figure 10.4. *The immune enhancement effects of alkaloids and other components in plant-based functional food.*

Source: Jiang L, Zhang G, Li Y, Shi G, Li M. (2021). Potential application of plant-based functional foods in the development of immune boosters. Front Pharmacol.;12:637782. doi: 10.3389/fphar.2021.637782. PMID: 33959009; PMCID: PMC8096308.

Magnoflorine, which is extracted from *Magnolia officinalis*, has been demonstrated to stimulate the expression of MyD88 and TLR4, which in turn improves the immune function of macrophages stimulated by LPS by activating the NF-κB and phosphatidylinositol 3-kinase-protein kinase B (PI3K-Akt) signaling pathways. The findings demonstrated that 50 μg/ml of magnoflorine can increase the upregulation of IL-1β and TNF-α. Furthermore, treatments with magnoflorine increased the phosphorylation of p38 MAPKs, c-Jun N-terminal kinase (JNK), and extracellular signal-regulated kinase (ERK). Additionally, magnoflorine can increase the number of $CD4^+$ T cells, boost TH1 and TH2 cytokine secretion, and encourage the proliferation of spleen cells induced by LPS in rats.

The primary active ingredient in *Euodia rutaecarpa* (Juss) Benth, evodiamine, has been demonstrated to control mouse immunity (Song et al., 2015). Evodiamine prevents concanavalin A (ConA)-induced thymic and splenic lymphocyte proliferation in mice. To be more precise, ELISA test results revealed that mice treated with evodiamine had significantly less IL-2 and IL-12 released from their spleen and thymus cells than the control group ($p < 0.05$). Reverse transcription polymerase chain reaction analysis further revealed that cells treated with 0.75 μmol/L evodiamine had mRNA levels of Bcl-12 and

CDK2 that were significantly higher than those in the control group. Treatment with 0.75 μmol/L evodiamine caused apoptosis in the thymic and splenic cells of mice, as compared to the control group ($p < 0.05$).

While it has been demonstrated that alkaloids regulate cytokine expression and either stimulate or inhibit immune cell activation or differentiation, more research on the topic of toxic doses and clinical trials is warranted due to the paucity of clinical data.

10.1.5. Others

Other components (such as terpenoids, essential oils, and organic acids) have also been found to exhibit immune enhancement effects, in addition to the bioactive ingredients mentioned above. For instance, the primary ingredient in ginger (*Zingiber officinale* Rosc.) is 6-gingerol. After receiving 6-gingerol treatment, tumor-bearing mice showed an increase in the infiltration of $CD4^+$/$CD8^+$ T cells and B cells as well as a decrease in the number of $CD4^+$ T cells. Furthermore, it is well known that the cinnamic acid in cinnamon raises white blood cell counts. Table 10.4 summarizes the effects of other immune-stimulating agents, and Figure 10.4 shows their structures.

Table 10.4. Other Components in Plant-Based Functional Foods on Immune Enhancement

Source	Compound	Model Type	Effective Dose	Main Results	Mechanisms	References
Curcuma longa L./Turmeric	Curcumin	Colitis mice (A)	200 mg/kg	↓ IL-2, IL-12P40, IL-21, GM-CSF ↑ IL-4, IL-15, IL-23	It can enhance immunity by inhibiting the expression of pro-inflammatory factors and increasing the secretion of anti-inflammatory factors.	Wang et al. (2020c)
Hippophae rhamnoides L./Sea-buckthorn	Ursolic acid	Mouse lymphocytes (A)	200 mg/kg	↓ IL-2, ↑ IL-4, IL-15, IL-23	It can enhance immunity by inhibiting the expression of pro-inflammatory factors and increasing the secretion of anti-inflammatory factors.	Wang et al. (2020c)

Lonicera japonica Thunb./ Flos Lonicera japonica	Chlorogenic acid	LPS-induced RAW264.7 cells (B)	1 μM	↓ IL-6, TNF-α, NF-κB	It regulates NF-κB and the secretion of cytokines to enhance the immunoregulative effect.	Cheng and Yeh (2019)

(A): in vivo.

(B): in vitro.

(C): in human.

↑*: enhanced effects.*

↓*: inhibited effects.*

Source: Jiang L, Zhang G, Li Y, Shi G, Li M. (2021). Potential application of plant-based functional foods in the development of immune boosters. Front Pharmacol.; 12:637782. doi: 10.3389/fphar.2021.637782. PMID: 33959009; PMCID: PMC8096308.

10.2. MECHANISMS OF THE IMMUNE-ENHANCING EFFECTS OF PLANT-BASED FUNCTIONAL FOODS

10.2.1. Influence on Immune Signal Transduction

Transcription of cell genes is regulated by a number of important factors, one of which is the TLR/NF-κB Signaling Pathway. To initiate interaction with NF-κB and catalyze the phosphorylation of IκB, the IKK complex is activated. After being activated, NF-κB is carried to the nucleus, where it directly starts and controls the transcription of immune response-related genes as well as the expression of adhesion molecules and cytokines. Inflammatory cytokine expression depends on NF-κB signaling. When inflammatory stimuli are presented to cells, they work by controlling inflammation via the inflammatory cytokine pathway of IKK-IκB-NF-κB (Figure 10.5). Thus, NF-κB has the ability to control cytokines, adhesion molecules, and inflammatory mediators, which impacts the body's innate or acquired immune response, inflammation, tumor growth, and other biological processes. After inflammation, proper NF-κB signaling intervention may be necessary to prevent further inflammation-related damage and the development of other diseases. Significantly, recent research has demonstrated that the bioactive components of plant-based functional foods can suppress NF-κB activation and consequently migration, having an immunomodulatory effect.

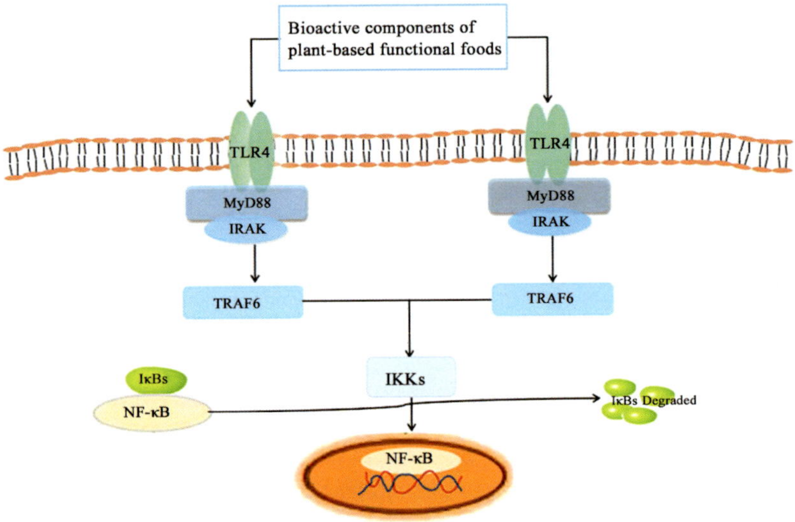

Figure 10.5. *The receptors and signaling pathways triggered by plant-based functional foods.*

Source: Jiang L, Zhang G, Li Y, Shi G, Li M. (2021). Potential Application of Plant-Based Functional Foods in the Development of Immune Boosters. Front Pharmacol.;12:637782. doi: 10.3389/fphar.2021.637782. PMID: 33959009; PMCID: PMC8096308.

TLRs belong to the class of pattern recognition receptors (PRRs) related to the innate immune. TLRs play a crucial defensive role against a range of microbes by identifying the pathogen and triggering acquired immunity through signal transduction, which can start innate immunity as soon as an infection occurs. The signal transduction process of TLRs involves various receptor proteins, which allow the pathways to be classified as either MyD88-dependent or non-MyD88-dependent. The TLR-MyD88/IRAK-mitogen-activated protein kinase (MAPK) pathway and the TLR-MyD88/IRAK-NF-κB pathway are the two primary categories of MyD88-dependent pathways (Fitzgerald and Kagan, 2020). More precisely, MyD88 binds to IRAK to draw in downstream signaling molecules that trigger activator protein-1 and NF-κB activation, as well as release pro-inflammatory cytokines that stimulate innate immune responses. The key components of the non-MyD88-dependent strategy are TLR3 and TLR4, which are crucial for the immune response. Although it also contributes to the recognition of LPS and some endogenous heat shock proteins, TLR4 is primarily expressed in macrophages (Yang et al., 2017).

Research has also demonstrated that elements found in plant-based functional foods can activate TLRs; this activates NF-κB in the nucleus, which in turn promotes cytokine secretion and up-regulates the expression of co-stimulating molecules, ultimately strengthening the immune system. For example, by blocking the expression of the TLR4/NF-κB pathway, APS can provide protection against cardiac dysfunction brought on by sepsis.

MAPK Signaling Pathways: One of the most significant signal transduction systems is made up of MAPKs, a class of threonine-serine protein kinases that are sensitive to a variety of extracellular stimuli. By double site phosphorylation in response to external stimulation, MAPK is activated, promoting the phosphorylation of transcription factors that enter the nucleus and controlling the transcription of genes that are related. ERK, JNK, and p38 MAPK are the three primary members of the MAPK family that are known to be involved in the control of apoptosis, inflammation, and cell proliferation, among other signal transduction pathways. Dong and colleagues demonstrated how hesperidin could control the MAPK pathway in a specific way to affect the immune response within cells. To be more precise, hesperidin increases p38MAPK and JNK expression and activation levels, which strengthens cellular autonomic immunity (Dong et al., 2014). Furthermore, the phosphorylation of p38 and the MAPK pathway of stress-activated protein kinase/JNK can be inhibited by sea buckthorn flavonoids, which lessens the inflammatory effect of the immune system. In mice with myocardial fibrosis, licorice was demonstrated to reduce the immune inflammatory response and to inhibit the expression of p38MAPK, ERK1/2, and JNK. These findings imply that active components may have immune-boosting effects by lowering p38MAPK, JNK, and ERK phosphorylation.

10.2.2. Promotion of Immune Cytokine Production

Cytokines are low-molecular-weight, minuscule proteins that function both autocrinely and paracrinely in immune cells. They play a part in certain inflammatory reactions and regulate the body's immune system. Immune cytokines have a significant impact on the neurological, endocrine, and cardiovascular systems in addition to the immune and hematopoietic systems. The various functions of cytokines—ILs, chemokines, lymphokines, IFNs, and the TNF family—are categorized. Being the cytokine that is secreted the most, interleukin (IL) is crucial for controlling inflammation and intercellular immunity. Recently, there have been significant advancements made in the therapeutic use of IL-2 and IL-10 against cancer. Among the cytokine subsets are IFNs, which have antiviral, anti-tumor, and immune-regulating properties. TNF-α and TNF-β, two members of the TNF family that can contribute to inflammation and immune regulation, are separated into two groups. TNF-α is a class of dual action cytokines that is primarily involved in the regulation of inflammation among these. Cytokines are normally involved in the development, differentiation, and function of immune cells, which helps to maintain the body's homeostasis by modulating the immune response at a low level. Conversely, bodily fluids quickly and significantly produce a wide range of pro-inflammatory cytokines that impact diseases, intracellular signaling pathways, and cell function. Currently, it is understood that plant-based functional foods strengthen immunity by encouraging the synthesis of cytokines. But some elements of functional foods can prevent the body's cells from being damaged by the inflammatory response's rise in cytokines (Zheng et al., 2020).

RAW264.7 macrophages were employed by Wang et al. (2020) to investigate the immunomodulatory and anti-inflammatory properties of TFA. Following a 12-hour treatment period, 10 μg/ml TFA markedly elevated the mRNA levels of IL-1β, IL-6, and TNF-α in normal RAW264.7 cells as well as the secretion of IL-1β, IL-6, and TNF-α. However, in LPS-stimulated RAW2.7 cells, TFA (10, 25, and 100 μg/ml) dose-dependently inhibited the overexpression of IL-1β, IL-6, and TNF-α as well as their increased transcription levels. TFA exerts its anti-inflammatory and immunological bimodulation effect by upregulating cytokine and mRNA levels in the normal state and preventing the excessive release of proinflammatory cytokines and mRNAs when stimulated by LPS (Wang et al., 2020).

10.2.3. Promotion of Innate Immunity Function

A number of defense mechanisms known as innate immunity are essential to the development and function of specific immunity. Thus far, a number of studies have discovered that innate immune cells—such as DCs, macrophages, and NK cells—are regulated in a way that enhances immunity.

10.2.3.1. Effects on NK Cells

The immune system's first line of defense, NK cells are linked to responses during tumor formation, viral infections, and immune regulation. To communicate with other immune cells, NK cells release chemokines and cytokines and express a number of activation and inhibitory receptors. Furthermore, through their capacity for cytotoxic effect and interaction with other immune cells, NK cells contribute significantly to tumor immune surveillance by orchestrating a coordinated anti-tumor immune response. Controlling NK cell activity helps fortify the immune system's defenses against illness. The capacity of naringenin (5, 10, and 21 μg/ml) and heated naringenin (4, 6, and 8 μg/ml) to increase NK activity against K562 myelogenous leukemia cells was examined by Maatouk et al. their results showed that naringenin and heated naringenin, at concentrations of 5 and 6 μg/ml, respectively, enhanced the NK cell lysis activity; the enhancement for heated naringenin was dose-dependent. Notably, NK cytotoxic activity against these target cells was increased upon naringenin treatment. Additionally, Valentová et al. revealed that exposure to rutin, which is frequently present in the typical human diet and is being used more and more in food supplements, also enhanced the killing activity of NK cells.

10.2.3.2. Effects on DCs

The innate immune system's most potent antigen-presenting cells, DCs are essential for immune homeostasis and antitumor activity. These cells serve as important immune sentinels, integrating and relaying a significant amount of afferent signals to lymphocytes to initiate and regulate adaptive immune responses. Furthermore, a wide range of bioactive substances stimulate the growth and maturation of DCs, which has immunomodulatory effects. Consequently, it might be useful to modify DC activity to treat autoimmune diseases.

Food ingredients contain proanthocyanidins. Williams et al. indicated that proanthocyanidins might cause human DCs to develop an anti-inflammatory phenotype, which would cause naive T cells to selectively downregulate the TH1 response. Additionally, Zhang et al. demonstrated that chrysin could enhance the inflammatory response in experimental autoimmune encephalomyelitis by blocking the functional differentiation and maturation of DCs.

10.2.3.3. Effects on the Mononuclear Phagocyte System (MPS)

Tissues, lymph, and blood all contain immune effector cells that make up the MPS. Macrophages are essential components of the host's defense against pathogens and are involved in antigen presentation, pathogen phagocytosis, cytokine secretion, and immune response activation. In addition to performing immune functions by secreting cytokines like TNF-α and IL-6 and releasing cytotoxic molecules like nitric oxide (NO), macrophages can also directly phagocytize pathogens. Functional foods made from plants contain biologically active components that can stimulate macrophages, improve phagocytosis and antigen presentation, and encourage the release of pertinent active molecules. For this reason, macrophages are prime targets for many immunomodulatory and anti-inflammatory medications.

The root of Panax ginseng C is used to extract ginseng polysaccharide (GPS). A. Meyer. The effects of GPS on LPS-induced mouse macrophage morphology and immune function (RAW264.7) were investigated in one study. The outcomes demonstrated that GPS dramatically enhanced the morphology and proliferation of macrophages. Treatments with 0.5 and 1 mg/ml GPS significantly decreased alkaline phosphatase activity brought on by LPS stimulation, while treatments with 1 mg/ml GPS significantly increased macrophage acid phosphatase activity when compared to the LPS group. The primary saponin constituent of *Platycodon grandiflorus* is platycodin D, which makes up the majority of the plant's active ingredients. Li et al. investigated, using an MTT assay, how platycodin D affected the immune function of mouse macrophage phagocytosis. Plasmid D treatment at 25, 50, 75, and 100 μg/ml markedly increased lymphocyte proliferation and improved macrophage phagocytosis when compared to the control group ($p < 0.01$). Compared to the other groups, the group treated with 50 μg/ml platycodin D showed a more significant effect ($p < 0.01$). Moreover, platycodin D treatment increased TNF-α and IL-12 secretion in macrophages; the ideal concentration was 50 μg/ml. Overall, it was discovered that mouse macrophage activity was increased by platycodin D.

10.2.4. Enhanced Adaptive Immunity

The cells of the adaptive immune system display a highly specific immune response to a particular antigen, the initial immune cells show broad specificity against pathogens.

10.2.4.1. Enhanced Cellular Immunity

The immune response that T lymphocytes mediate is known as cellular immunity. Antigen-presenting cells in cellular immunity break down antigens into peptides that attach to the major histocompatibility complex (MHC) and trigger an activated T-cell receptor signal. T cells multiply and differentiate quickly after the antigen attaches to appropriate receptors on their surface, and some of them develop into sensitized lymphocytes. Among these are T helper (Th) cells, which release cytokines like ILs, which trigger Tc and other phagocytes to assemble around foreign cells and ultimately eliminate them, and cytotoxic T cells (Tc), which can rupture and kill exogenous cells. As the immune response comes to an end, T suppressor (Ts) cells start to operate and stop the immune response by stopping other lymphocytes from acting. Numerous investigations have demonstrated the ability of plant-based functional foods to enhance cellular immune function and encourage lymphocyte transformation (Yang et al., 2017).

For instance, it has been demonstrated that lentinan increases the toxicity of macrophages to metastatic tumors by controlling the activity of immune cells on several levels, including T lymphocytes and macrophages. To be more precise, lentinan can improve the host body balance by stimulating T lymphocyte activity and proliferation. Furthermore, it has been demonstrated that lentinan activates T cells and increases cancer patients' chances of survival.

10.2.4.2. Enhanced Humoral Immunity

B cells, or plasma cells, are the primary mediators of humoral immunity; they create antibodies to safeguard the immune system. The affinity and immunological response of the antibody to the antigen are reflected in the antibody titer. Hemolysin concentration in serum serves as a crucial gauge of humoral immune activity. In this regard, it has been discovered that the active ingredients in plant-based functional meals provide immunity by generating antibodies that quicken lymphocytic phagocytosis and phagocyte clearance.

After cyclophosphamide induction, the impact of polysaccharide extracted from the herb *Gastrodia elata* on humoral immunity in immunodeficient mice was investigated. Serum IgA, IgG, and hemolysin levels were significantly elevated ($p < 0.01$) in the *Gastrodia elata* polysaccharide medium-dose (200 mg/kg) and high-dose (400 mg/kg) groups following a 10-day course of treatment. Additionally, the high-dose polysaccharide group showed increases in the spleen and thymus indices, while the medium-dose polysaccharide group showed a significant ($p < 0.05$) increase in serum IgM levels. *G. elata* polysaccharide has the ability to mitigate the cyclophosphamide-induced suppression of humoral immune function in mice.

10.2.5. Promotion Effect on Immune Organs Function

The body's entire defense network is the immune system. The production of a significant number of immune cells by the immune organ is necessary for immune function to control the immune response and stop the spread of infection. The precursor hematopoietic stem cells from which all blood cells develop and reside are different, and the cells are primarily classified into two categories: the central and peripheral immune organs.

The spleen is a peripheral immune organ that is involved in humoral immunity, while the thymus, as the primary immune organ, is primarily involved in the cellular immune response. As of right now, several studies have demonstrated how different immune organs can be affected by the components of plant-based functional foods (Figure 10.6). These effects include boosting the development of partial visceral organs, improving the organ index, which indicates changes in the immune system, and increasing the weight of the organs (McComb et al., 2019).

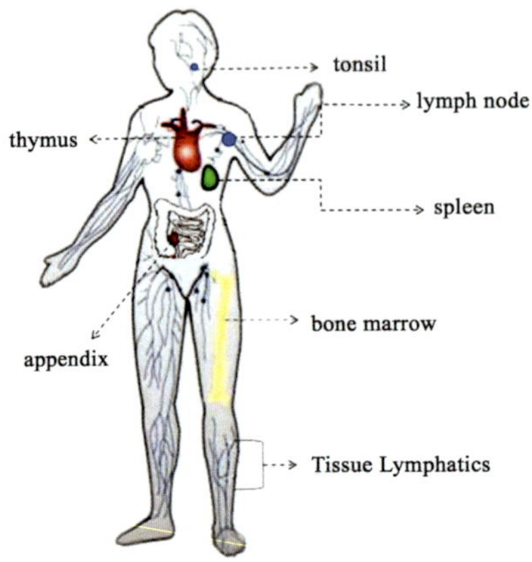

Figure 10.6. *The immune system of immune organs.*

Source: Jiang L, Zhang G, Li Y, Shi G, Li M. (2021). Potential application of plant-based functional foods in the development of immune boosters. Front Pharmacol.;12:637782. doi: 10.3389/fphar.2021.637782. PMID: 33959009; PMCID: PMC8096308.

For instance, several studies have confirmed that *Lycium barbarum* polysaccharides (LBP) have an enhancing effect on the immune organ index. Additionally, Tang and He assessed the impact of LBP using a D-galactose aging mouse model and discovered that the LBP (3 g/kg) groups had significantly higher spleen and thymus indices than the control group ($p < 0.01$). As a result, LBP can strengthen the body's immune system and efficiently safeguard the immune organs. Additionally, Kamboh et al. examined the immunomodulatory effects of hesperidin on broilers induced with lipopolysaccharide (LPS) and found that treatment with 20 mg/kg of hesperidin significantly increased the spleen and bursa index at 21 and 42 days in comparison to the control group ($p < 0.05$). These findings suggest that the bioactive ingredients in plant-based functional foods support immune system development and growth as well as have immune-boosting properties (Jiang et al., 2021).

REFERENCES

1. Ahmad, W., Jantan, I., Kumolosasi, E., & Bukhari, S. N. (2015). Immunostimulatory effects of the standardized extract of *Tinospora crispa* on innate immune responses in Wistar Kyoto rats. *Drug Design, Development and Therapy, 9*, 2961–2973. https://doi.org/10.2147/DDDT.S85405.
2. Ahn, H., Jeon, E., Kim, J. C., Kang, S. G., Yoon, S. I., Ko, H. J., et al. (2017). Lentinan from shiitake selectively attenuates AIM2 and non-canonical inflammasome activation while inducing pro-inflammatory cytokine production. *Scientific Reports, 7*, 314. https://doi.org/10.1038/s41598-017-01462-4.
3. Andrea, K., Patrik, S., Peter, K., Pavol, Z., Sona, U., Martin, K., et al. (2017). Are plant-based functional foods a better choice against cancer than single phytochemicals? A critical review of current breast cancer research. *Biomedicine and Pharmacotherapy, 96*, 1465–1477. https://doi.org/10.1016/j.biopha.2017.11.134.
4. Bao, F. X., Tao, L. X., & Zhang, H. Y. (2018). Research progress on pharmacological effects of *Gynostemma pentaphyllum* active ingredients. *Chinese Journal of New Drugs and Clinical Remedies, 37*, 11–17. https://doi.org/10.14109/j.cnki.xyylc.2018.01.003.
5. Bi, S. C., Ma, X. D., Wu, Y., Cui, X. M., & Hu, S. H. (2019). Oral administration of ginsenoside Rg1 enhances gut mucosal immunity in chickens. *Chinese Journal of Veterinary Science, 39*, 2215–2221. https://doi.org/10.16303/j.cnki.1005-4545.2019.11.21.
6. Birsu Cincin, Z., Unlu, M., Kiran, B., Bireller, S. E., Baran, Y., & Cakmakoglu, B. (2015). Anti-proliferative, apoptotic, and signal transduction effects of hesperidin in non-small cell lung cancer cells. *Cancer Cell International, 38*(3), 195–204. https://doi.org/10.1007/s13402-015-0222-z.
7. Cai, G., Sun, K., Xia, S., Feng, Z., Zou, H., Gu, J., et al. (2020). Decrease in immune function and the role of mitogen-activated protein kinase (MAPK) overactivation in apoptosis during T lymphocyte activation induced by zearalenone, deoxynivalenol, and their combinations. *Chemosphere, 255*, 126999. https://doi.org/10.1016/j.chemosphere.2020.126999.
8. Chai, Z., Zhang, J. J., Sun, S. J., Wei, K. Z., Yan, J. L., & Wei, J. Z. (2018). Study on effects of anti-aging and immune regulation of *Polygala* saponins. *China Journal of Traditional Chinese Medicine and Pharmacy, 33*, 704–707.

9. Chen, G. Y., Han, G. J., Zhang, L. L., Li, H., & Yang, C. M. (2020). Immune regulation of ginseng polysaccharide on lipopolysaccharide-stimulated mouse macrophages. *Chinese Journal of Animal Science*, 1–11. https://doi.org/10.19556/j.0258-7033.20200323-10.

10. Chen, H. Z., Luo, H. N., Quan, B. Y., & Yi, C. X. (2017). Effect of Rg3 on the expression of CD80/86 and cellular immune function in patients with nasopharyngeal carcinoma after radiotherapy. *Jilin Journal of Traditional Chinese Medicine, 37*(12), 1211–1214. https://doi.org/10.13463/j.cnki.jlzyy.2017.12.007.

11. Chen, M. Z., Li, Y., Chen, W., & He, W. (2017). Regulating effect of total momordicoside on immune function in IgA nephropathy of rats. *Journal of Liaoning University of Traditional Chinese Medicine, 19*(07), 46–49. https://doi.org/10.13194/j.issn.1673-842x.2017.07.011.

12. Chen, Q. Y., & Zhan, J. H. (2019). The immunomodulatory effect of traditional Chinese medicine and its research progress. *Jiangxi Medical Journal, 54*, 181–184. https://doi.org/10.3969/j.issn.1006-2238.2019.2.032.

13. Chen, X. L., Sheng, Z. C., Qiu, S. L., Yang, H. F., Jia, J. P., Wang, J., et al. (2019). Purification, characterization, and *in vitro* and *in vivo* immune enhancement of polysaccharides from mulberry leaves. *PLoS ONE, 14*, e0208611. https://doi.org/10.1371/journal.pone.0208611.

14. Chen, Z., Liu, L., Gao, C., Chen, W., Vong, C. T., Yao, P., et al. (2020). *Astragali radix* (Huangqi): A promising edible immunomodulatory herbal medicine. *Journal of Ethnopharmacology, 258*, 112895. https://doi.org/10.1016/j.jep.2020.112895.

15. Cheng, C. Y., & Yeh, C. C. (2019). Adaptive immunoregulation of luteolin and chlorogenic acid in lipopolysaccharide-induced interleukin-10 expression. *Chinese Journal of Integrative Medicine, 32*, 186–192. https://doi.org/10.4103/tcmj.tcmj_23_19.

16. Cheng, Q., Guo, S. S., Ding, B. Y., Li, Y. H., & Xia, Y. (2017). Effects of dietary coated-cinnamon on non-specific immune function of broiler chickens. *Chinese Poultry Science, 39*(18), 28–33. https://doi.org/10.16372/j.issn.1004-6364.2017.18.006.

17. Chi, X., Bi, S., Xu, W., Zhang, Y., Liang, S., & Hu, S. (2017). Oral administration of tea saponins to relieve oxidative stress and immune suppression in chickens. *Poultry Science, 96*, 3058–3067. https://doi.org/10.3382/ps/pex127.

18. Conlon, K. C., Miljkovic, M. D., & Waldmann, T. A. (2019). Cytokines in the treatment of cancer. *Journal of Interferon and Cytokine Research, 39*, 6–21. https://doi.org/10.1089/jir.2018.0019.

19. Cronkite, D. A., & Strutt, T. M. (2018). The regulation of inflammation by innate and adaptive lymphocytes. *Journal of Immunology Research, 2018*, 1467538. https://doi.org/10.1155/2018/1467538.
20. Dan, B., Li, L. Y., He, J., & Chen, X. H. (2020). Immunoregulatory effects of gypenosides on cyclophosphamide-induced immunosuppression in mice. *Northwest Pharm. J., 35*, 680–684. https://doi.org/10.3969/j.issn.1004-2407.2020.05.011.
21. Davoodvandi, A., Sahebnasagh, R., Mardanshah, O., Asemi, Z., Nejati, M., Shahrzad, M. K., et al. (2019). Medicinal plants as natural polarizers of macrophages: Phytochemicals and pharmacological effects. *Curr. Pharm. Des., 25*(30), 3225–3238. https://doi.org/10.2174/1381612825666190829154934.
22. De Ferrars, R. M., Czank, C., Zhang, Q., Botting, N. P., Kroon, P. A., Cassidy, A., et al. (2014). The pharmacokinetics of anthocyanins and their metabolites in humans. *Br. J. Pharmacol., 171*(13), 3268–3282. https://doi.org/10.1111/bph.12676.
23. Ding, S. J., Jiang, H. M., & Fang, J. (2018). Regulation of immune function by polyphenols. *J. Immunol. Res., 2018*, 1264074. https://doi.org/10.1155/2018/1264074.
24. Ding, Y., Yan, Y., Chen, D., Ran, L., Mi, J., Lu, L., et al. (2019). Modulating effects of polysaccharides from the fruits of *Lycium barbarum* on the immune response and gut microbiota in cyclophosphamide-treated mice. *Food Funct., 10*, 3671–3683. https://doi.org/10.1039/c9fo00638a.
25. Dong, W., Wei, X., Zhang, F., Hao, J., Huang, F., Zhang, C., et al. (2014). A dual character of flavonoids in influenza A virus replication and spread through modulating cell-autonomous immunity by MAPK signaling pathways. *Sci. Rep., 4*, 7237. https://doi.org/10.1038/srep07237.
26. Dong, Z., Zhang, M. M., Li, H. X., Zhan, Q. P., Lai, F. R., & Wu, H. (2019). Structural characterization and immunomodulatory activity of a novel polysaccharide from *Pueraria lobata* (Willd.) Ohwi root. *Int. J. Biol. Macromol., 154*, 1556–1564. https://doi.org/10.1016/j.ijbiomac.2019.11.040.
27. Du, F. Y., Xue, G. J., Liu, Z. B., & Chen, G. L. (2021). Research advances of cytokine storm and therapeutics. *Chin. J. Med. Chem., 31*(01), 39–54. https://doi.org/10.14142/j.cnki.cn21-1313/r.2021.01.005.
28. Fan, Q. L., Li, H., Jiang, S. Q., Li, L., & Ye, J. L. (2021). Effect of capsaicin, gingerol, allicin, and essential oil on growth performance, carcass performance, antioxidant and immune function of broiler chickens. *Feed Industry, 42*(02), 7–12. https://doi.org/10.13302/j.cnki.fi.2021.02.002.

29. Feng, D., Hao, S. Y., Fu, Y., Wang, X. H., Du, X. H., & Liu, H. R. (2020). Effect of Radix Pseudostellariae polysaccharide injection on immune function of immunosuppressive mice. *J. Trad. Chin. Vet. Med.*, *39*, 74–77. https://doi.org/10.13823/j.cnki.jtcvm.2020.03.019.

30. Ferreira, S. S., Passos, C. P., Madureira, P., Vilanova, M., & Coimbra, M. A. (2015). Structure-function relationships of immunostimulatory polysaccharides: A review. *Carbohydr. Polym.*, *132*, 378–396. https://doi.org/10.1016/j.carbpol.2015.05.079.

31. Fitzgerald, K. A., & Kagan, J. C. (2020). Toll-like receptors and the control of immunity. *Cell*, *180*, 1044–1066. https://doi.org/10.1016/j.cell.2020.02.041.

32. Fu, Y. P., Feng, B., Zhu, Z. K., Feng, X., Chen, S. F., Li, L. X., et al. (2018). The polysaccharides from *Codonopsis pilosula* modulates the immunity and intestinal microbiota of cyclophosphamide-treated immunosuppressed mice. *Molecules*, *23*, 1801. https://doi.org/10.3390/molecules23071801.

33. García, M. J., Pascual, M., Del Pozo, C., Díaz-González, A., Castro, B., Rasines, L., et al. (2020). Impact of immune-mediated diseases in inflammatory bowel disease and implications in therapeutic approach. *Sci. Rep.*, *10*(1), 10731. https://doi.org/10.1038/s41598-020-67710-2.

34. Gong, X. W., & Jiang, Y. (2003). The structural basis of biological function of mitogen-activated protein kinases. *Chin. J. Biochem. Mol. Biol.*, *19*(1), 5–11.

35. Gong, X., Ji, M. Y., Xu, J. P., Zhang, C. H., & Li, M. H. (2019). Hypoglycemic effects of bioactive ingredients from medicine food homology and medicinal health food species used in China. *Crit. Rev. Food Sci. Nutr.*, *60*, 2303–2326. https://doi.org/10.1080/10408398.2019.1634517.

36. Gong, X., Li, X., Xia, Y., Xu, J. F., Li, Q. Y., Zhang, C. H., et al. (2020). Effects of phytochemicals from plant-based functional foods on hyperlipidemia and their underpinning mechanisms. *Trends Food Sci. Technol.*, *103*, 304–320. https://doi.org/10.1016/j.tifs.2020.07.026.

37. Hao, W. L., Chen, Z. B., Zhao, R., & Bo, L. (2015). Effect of *Lycium barbarum* polysaccharide on immune function and anti-fatigue of sub-healthy mice. *Chin. J. Biologicals*, *28*, 693–697. https://doi.org/10.13200/j.cnki.cjb.000954.

38. Haque, M. A., Jantan, I., Harikrishnan, H., & Abdul Wahab, S. M. (2018). Magnoflorine enhances LPS-activated pro-inflammatory responses via MyD88-dependent pathways in U937 macrophages. *Planta Med.*, *84*(17), 1255–1264. https://doi.org/10.1055/a-0637-9936.

39. He, Y., Hu, Z., Li, A., Zhu, Z., Yang, N., Ying, Z., et al. (2019). Recent advances in biotransformation of saponins. *Molecules*, *24*, 2365. https://doi.org/10.3390/molecules24132365.

40. Hou, D. R., Liu, Z., Cui, S. T., & Ma, J. (2021). Tanshinone -A inhibited LPS-induced cell inflammation by regulating the TLR4/IκBα/NFκB signaling pathway. *Chin. Pharmacol. Bull.*, *37*(2), 210–214. https://doi.org/10.3969/j.issn.1001-1978.2021.02.012.

41. Huang, F., Zhang, R., Liu, Y., Xiao, J., Liu, L., Wei, Z., et al. (2016). Dietary litchi pulp polysaccharides could enhance immunomodulatory and antioxidant effects in mice. *Int. J. Biol. Macromol.*, *92*, 1067–1073. https://doi.org/10.1016/j.ijbiomac.2016.08.021.

42. Huang, H., Luo, S. H., Huang, D. C., Cheng, S. J., Cao, C. J., & Chen, G. T. (2019). Immunomodulatory activities of proteins from *Astragalus membranaceus* waste. *J. Sci. Food Agr.*, *99*, 4174–4181. https://doi.org/10.1002/jsfa.9650.

43. Huang, Q., Li, L. Y., Liu, Q. Q., & Wang, Z. (2020). Advances in immunoregulation effects of *Ganoderma lucidum* polysaccharide and/or *Polyporus umbellatus* polysaccharide. *Food Sci.*, *41*, 275–282. https://doi.org/10.7506/spkx1002-6630-20190813-149.

44. Huang, W., Li, M. L., Xia, M. Y., & Shao, J. Y. (2018). Fisetin-treatment alleviates airway inflammation through inhibition of MyD88/NF-κB signaling pathway. *Int. J. Mol. Med.*, *42*(1), 208–218. https://doi.org/10.3892/ijmm.2018.3582.

45. Hwang, S. H., Shin, M. S., Yoon, T. J., & Shin, K. S. (2018). Immunoadjuvant activity in mice of polysaccharides isolated from the leaves of *Panax ginseng* C. *Int. J. Biol. Macromolecules*, *107*, 2695–2700. https://doi.org/10.1016/j.ijbiomac.2017.10.160.

46. Ji, M. Y., Bo, A., Yang, M., Xu, J. F., Jiang, L. L., Zhou, B. C., et al. (2020). The pharmacological effects and health benefits of *Platycodon grandiflorus*—a medicine food homology species. *Foods*, *9*, 142. https://doi.org/10.3390/foods9020142.

47. Jia, Z., Chen, A., Wang, C., He, M., Xu, J., Fu, H., et al. (2019). Amelioration effects of kaempferol on immune response following chronic intermittent cold-stress. *Res. Vet. Sci.*, *125*, 390–396. https://doi.org/10.1016/j.rvsc.2019.08.012.

48. Jiang, F., Guan, H., Liu, D., Wu, X., Fan, M., & Han, J. (2017). Flavonoids from sea buckthorn inhibit the lipopolysaccharide-induced inflammatory response in RAW264.7 macrophages through the MAPK and NF-κB pathways. *Food Funct.*, *8*(3), 1313–1322. https://doi.org/10.1039/c6fo01873d.

49. Jiang, L. L., Gong, X., Ji, M. Y., Wang, C. C., Wang, J. H., & Li, M. H. (2020). Bioactive compounds from plant-based functional foods: A promising choice for the prevention and management of hyperuricemia. *Foods*, *9*(8), 973. https://doi.org/10.3390/foods9080973.

50. Jiang, L., Zhang, G., Li, Y., Shi, G., & Li, M. (2021). Potential application of plant-based functional foods in the development of immune boosters. *Frontiers in Pharmacology, 12*, 637782. https://doi.org/10.3389/fphar.2021.637782.

51. Jiang, Y. P., Liu, Y. F., Guo, Q. L., Xu, C. B., Zhu, C. G., & Shi, J. G. (2016). Sesquiterpene glycosides from the roots of *Codonopsis pilosula*. *Acta Pharm. Sin. B.*, *6*, 46–54. https://doi.org/10.1016/j.apsb.2015.09.007.

52. Kamboh, A. A., Hang, S. Q., Khan, M. A., & Zhu, W. Y. (2016). *In vivo* immunomodulatory effects of plant flavonoids in lipopolysaccharide-challenged broilers. *Animal, 10*, 1619–1625. https://doi.org/10.1017/S1751731116000562.

53. Kim, J. H., Doo, E. H., Jeong, M., Kim, S., Lee, Y. Y., Yang, J., et al. (2019). Enhancing immunomodulatory function of Red Ginseng through fermentation using *Bifidobacterium animalis* subsp. *lactis* LT 19-2. *Nutrients*, *11*(7), 1481. https://doi.org/10.3390/nu11071481.

54. Kong, N., Liu, C., Yang, F. H., Zhang, T. Y., Wang, B. K., & Gao, H. J. (2018). Effects of procyanidins on immune function in cyclophosphamide-induced immunosuppression mice. *Tianjin Med. J.*, *46*, 1291–1294. https://doi.org/10.11958/20181203.

55. Kumar, A., Mosa, K. A., Ji, L., Kage, U., Dhokane, D., Karre, S., et al. (2018). Metabolomics-assisted biotechnological interventions for developing plant-based functional foods and nutraceuticals. *Crit. Rev. Food Sci. Nutr.*, *58*(11), 1791–1807. https://doi.org/10.1080/10408398.2017.1285752.

56. Li, F. J., Li, J. S., Wang, Y. L., Jin, X., Li, X., & Yu, Y. (2021). Research on the immunoregulatory effect of jujube polysaccharide on lymphocyte in mice. *Sci. Technol. Cereals, Oils Foods, 29*, 141–147. https://doi.org/10.16210/j.cnki.1007-7561.2021.01.019.

57. Li, J. S., Feng, H. H., Wang, M., & Yu, Y. (2019). Effect of platycodin D on lymphocyte and macrophage immune function of mice. *J. Northwest A&F Univ. (Nat. Sci. Ed).*, *47*, 39–44. https://doi.org/10.13207/j.cnki.jnwafu.2019.01.005.

58. Li, L. K., Luo, Q. H., Huang, C., Chen, X. L., Chen, P., Li, Y. F., et al. (2017). Effects of soy isoflavones on expression of IL-2, IL-4, TNF-α, and INF-γ in male rats' spleen. *Acta Agri. Zhejiangensis*, *29*(9), 1458–1464. https://doi.org/10.3969/j.issn.1004-1524.2017.09.06.

59. Li, Q., & Verma, I. M. (2002). NF-kappaB regulation in the immune system. *Nat. Rev. Immunol.*, *2*(10), 725–734. https://doi.org/10.1038/nri910.

60. Li, X. B., Wang, C., Chen, Y. L., Zhan, J. P., Xie, Z. L., & Zhang, Y. T. (2015). Effect on spleen immune function of *Gynostemma pentaphyllum* polysaccharide to immunosuppressed mice induced by cyclophosphamide. *Lishizhen Med. Materia Med. Res.*, *26*, 2308–2310. https://doi.org/10.3969/j.issn.1008-0805.2015.10.002.

61. Li, X. B., Zhan, J. P., Zhang, Y. T., Xie, Z. L., Zhu, Y. Q., Chen, Y., et al. (2016). Effect of polysaccharide from *Gastrodia elata* B1 on humoral immune function in immunosuppressed mice induced by cyclophosphamide. *Chin. J. Gerontol.*, *36*(5), 1027–1028. https://doi.org/10.3969/j.issn.1005-9202.2016.05.002.

62. Li, X. Q., Yang, W. W., & Li, S. Z. (2020). Research progress on the immune function of active ingredients in food. *J. Shenyang Med. Coll., 22*, 277–288.

63. Li, X., Yang, K., Yu, Y., Qiu, Z. Y., & Li, J. S. (2020). Effect of onion quercetin on immune organ index, phagocytic function of macrophages and secretion cytokine in chickens. *Feed Research, 43*(03), 33–38. https://doi.org/10.13557/j.cnki.issn1002-2813.2020.03.009.

64. Liao, Y., Tan, R. Z., Li, J. C., Liu, T. T., Zhong, X., Yan, Y., et al. (2020). Isoliquiritigenin attenuates UUO-induced renal inflammation and fibrosis by inhibiting mincle/Syk/NF-kappa B signaling pathway. *Drug Design, Development and Therapy, 14*, 1455–1468. https://doi.org/10.2147/DDDT.S243420.

65. Lin, W. Q., Wang, W. T., Wang, D. L., & Ling, W. H. (2017). Quercetin protects against atherosclerosis by inhibiting dendritic cell activation. *Molecular Nutrition & Food Research, 61*, 1700031. https://doi.org/10.1002/mnfr.201700031.

66. Lin, Y. T., Chen, L. L., & Hu, X. F. (2020). A review of the types, functions and related diseases of immune cells. *Biological Teaching, 45*, 77–80. CNKI:SUN:SWJX.0.2020-04-037.

67. Liu, B. C., Qiu, Y., Zhao, R. D., Han, X., Yun, F. Y., Tui, X., et al. (2017). Digital gene expression profiling of dendritic cells treated with seabuckthorn flavones. *Chinese Journal of Microbiology and Immunology, 37*, 840–848. https://doi.org/10.3760/cma.j.issn.0254-5101.2017.11.007.

68. Liu, C. H., Liu, H., & Ge, B. (2017). Innate immunity in tuberculosis: Host defense vs. pathogen evasion. *Cell & Molecular Immunology, 14*, 963–975. https://doi.org/10.1038/cmi.2017.88.

69. Liu, J., Wang, X., Yong, H., Kan, J., & Jin, C. (2018). Recent advances in flavonoid-grafted polysaccharides: Synthesis, structural characterization, bioactivities and potential applications. *International Journal of Biological Macromolecules, 116*, 1011–1025. https://doi.org/10.1016/j.ijbiomac.2018.05.149.

70. Liu, M., Li, S. S., Wang, X. X., Zhu, Y. F., Zhang, J. J., Liu, H., et al. (2018). Characterization, anti-oxidation and anti-inflammation of polysaccharides by *Hypsizygus marmoreus* against LPS-induced toxicity on lung. *International Journal of Biological Macromolecules, 111*, 121–128. https://doi.org/10.1016/j.ijbiomac.2018.01.010.

71. Liu, Y. F., Xue, X. X., Li, Z. Y., Wang, J. P., & Zhang, Y. J. (2017). Effect of apigenin on dendritic cell maturation and function in murine splenocytes. *Acta Pharmaceutica Sinica, 52*(3), 397–402. https://doi.org/10.16438/j.0513-4870.2016-0949.

72. Lv, J., Gao, Y., Li, C., Yang, L. F., & Zhao, B. N. (2020). Effect of American ginseng saponins on enhancing immunity based on zebrafish model organisms. *Chinese Traditional Herbal Drugs, 51*(14), 3728–3733. https://doi.org/10.7501/j.issn.0253-2670.2020.14.016.

73. Lv, X. C., Zhang, L. S., & Wang, F. J. (2016). Recent advances on immune regulation of herbal polysaccharides. *Acta Universitatis Traditionis Medicinae Sinensis Pharmacologiaeque Shanghai, 30*, 97–101. https://doi.org/10.16306/j.1008-861x.2016.03.022.

74. Ma, H. X., & Bai, W. M. (2019). Efficacy of ginsenoside Rh2 combined with chemotherapy for lung cancer and its effect on tumor markers and immune function. *Medical Journal of the Air Force, 35*(05), 406–409. https://doi.org/10.3969/j.issn.2095-3402.2019.05.012.

75. Maatouk, M., Elgueder, D., Mustapha, N., Chaaban, H., Bzéouich, I. M., Loannou, I., et al. (2016). Effect of heated naringenin on immunomodulatory properties and cellular antioxidant activity. *Cell Stress & Chaperones, 21*, 1101–1109. https://doi.org/10.1007/s12192-016-0734-0.

76. Maatouk, M., Mustapha, N., Mokdad-Bzeouich, I., Chaaban, H., Abed, B., Iaonnou, I., et al. (2017). Thermal treatment of luteolin-7-O-β-glucoside improves its immunomodulatory and antioxidant potencies. *Cell Stress & Chaperones, 22*, 775–785. https://doi.org/10.1007/s12192-017-0808-7.

77. Manzoni, A. G., Passos, D. F., Leitemperger, J. W., Storck, T. R., Doleski, P. H., Jantsch, M. H., et al. (2020). Hyperlipidemia-induced lipotoxicity and immune activation in rats are prevented by curcumin and rutin. *International Immunopharmacology, 81*, 106217. https://doi.org/10.1016/j.intimp.2020.106217.

78. McComb, S., Thiriot, A., Akache, B., Krishnan, L., & Stark, F. (2019).

Introduction to the immune system. In *Methods in Molecular Biology, 2024*, 1–24. https://doi.org/10.1007/978-1-4939-9597-4_110.1007/978-1-62703-589-7_1.

79. Mehmood, A., Zhao, L., Wang, C. T., Nadeem, M., Raza, A., Ali, N., et al. (2019). Management of hyperuricemia through dietary polyphenols as a natural medicament: A comprehensive review. *Critical Reviews in Food Science and Nutrition, 59*(9), 1433–1455. https://doi.org/10.1080/10408398.2017.1412939.
80. Meng, F. C., Li, Q., Qi, Y. M., He, C. W., Wang, C. M., & Zhang, Q. W. (2018). Characterization and immunoregulatory activity of two polysaccharides from the root of *Ilex asprella*. *Carbohydrate Polymers, 197*, 9–16. https://doi.org/10.1016/j.carbpol.2018.05.066.
81. Meng, X. Y., Chu, Z. F., Zang, J., Zhang, H., Wang, S. J., Wei, K., et al. (2018). Immunomodulatory functions of Ginkgobiloba leaves polysaccharide on vvIBDV vaccine. *Chin. J. Vet. Sci.*, *39*, 640–645. https://doi.org/10.16303/j.cnki.1005-4545.2019.04.07.
82. Mohamad, N. E., Romli, M. F., Alitheen, N. B., Abu, N., Yeap, S. K., Lim, K. L., et al. (2020). Apoptosis and metastasis inhibitory potential of pineapple vinegar against mouse mammary gland cells *in vitro* and *in vivo*. *Nutr. Metab. (Lond)*, *16*, 49. https://doi.org/10.1186/s12986-019-0380-5.
83. Morsink, M. A. J., Willemen, N. G. A., Leijten, J., Bansal, R., & Shin, S. R. (2020). Immune organs and immune cells on a chip: An overview of biomedical applications. *Micromachines*, *11*, 849. https://doi.org/10.3390/mi11090849.
84. Nafees, S., Rashid, S., Ali, N., Hasan, S. K., & Sultana, S. (2015). Rutin ameliorates cyclophosphamide-induced oxidative stress and inflammation in Wistar rats: Role of NFκB/MAPK pathway. *Chem. Biol. Interact.*, *231*, 98–107. https://doi.org/10.1016/j.cbi.2015.02.021.
85. Ni, C. N., & Zhu, J. T. (2020). Clinical efficacy of honeysuckle soup combined with benzathine penicillin for syphilis patients and the effects on the serum IL-8 and TNF-α levels. *Chin. J. Hum. Sex.*, *29*(6), 105–107. https://doi.org/10.3969/j.issn.1672-1993.2020.06.030.
86. Ning, Y. C., Qiao, H. X., Pan, C. M., & Zhang, X. J. (2016). Effects of fermented gypenosides on immune function in mice. *Hubei Agr. Sci.*, *55*, 2304–2307. https://doi.org/10.14088/j.cnki.issn0439-8114.2016.09.038.
87. Orlowsky, E. W., & Kraus, V. B. (2015). The role of innate immunity in osteoarthritis: When our first line of defense goes on the offensive. *J. Rheumatol.*, *42*(3), 363–371. https://doi.org/10.3899/jrheum.140382.
88. Pang, M. X., Fang, Y. Y., Chen, S. H., Zhu, X. X., Shan, C. W., Su, J., et

al. (2017). Gypenosides inhibits xanthine oxidoreductase and ameliorates urate excretion in hyperuricemic rats induced by high cholesterol and high fat food (Lipid Emulsion). *Med. Sci. Monit.*, *23*, 1129–1140. https://doi.org/10.12659/msm.903217.

89. Pi, J. H., Tan, J., Hu, C. T., & Xiang, D. B. (2015). Effects of Lonicera Japonica flavone on immunomodulation in mice. *Chin. J. Appl. Physiol.*, *31*, 89–92. https://doi.org/10.13459/j.cnki.cjap.2015.01.026.

90. Qiao, J., & Fu, Y. X. (2020). Cytokines that target immune killer cells against tumors. *Cell Mol. Immunol.*, *17*(7), 722–727. https://doi.org/10.1038/s41423-020-0481-0.

91. Qin, F., Zhu, S. Y., Cheng, D. R., Chen, Y. Y., Zuo, W. Y., Wang, A. P., et al. (2017). Study on the hemolytic activity and immune activity of S-6 *in vitro*. *Chin. J. Vet. Parasitol.*, *25*(04), 45–49.

92. Qiu, Z. W., Gao, W., Wu, J., Gou, S. D., & Lu, M. Z. (2018). Effect of Astragalus polysaccharide injection on inflammatory cell count and levels of related factors in bronchial lavage fluid or sputum of patients with asthma. *J. Prev. Med. Chin. PLA.*, *36*, 746–749. https://doi.org/10.13704/j.cnki.jyyx.2018.06.015.

93. Rajput, Z. I., Hu, S. H., Xiao, C. W., & Arijo, A. G. (2007). Adjuvant effects of saponins on animal immune responses. *J. Zhejiang Univ. Sci. B.*, *8*(3), 153–161. https://doi.org/10.1631/jzus.2007.B0153.

94. Rasouli, E., & Jahanian, R. (2015). Improved performance and immunological responses as the result of dietary genistein supplementation of broiler chicks. *Animal*, *9*, 1473–1480. https://doi.org/10.1017/S1751731115000853.

95. Ren, L., Zhang, J., & Zhang, T. H. (2020). Immunomodulatory activities of polysaccharides from Ganoderma on immune effector cells. *Food Chem.*, *340*, 127933. https://doi.org/10.1016/j.foodchem.2020.127933.

96. Ruan, M., Yu, B., & Zhou, F. (2021). Effect of iso-astragaloside on immunologic function of lymphocytes *in vitro* and dendritic cells. *Chin. Trad. Herbal Med.*, *52*, 196–202. https://doi.org/10.7501/j.issn.0253-2670.2021.01.023.

97. Ruiz-Iglesias, P., Estruel-Amades, S., Camps-Bossacoma, M., Massot-Cladera, M., Franch, À., Pérez-Cano, F. J., et al. (2020). Influence of hesperidin on systemic immunity of rats following intensive training and exhausting exercise. *Nutrients*, *12*(5), 1291. https://doi.org/10.3390/nu12051291.

98. Russick, J., Joubert, P. E., Gillard-Bocquet, M., Torset, C., Meylan, M., Petitprez, F., et al. (2020). Natural killer cells in the human lung tumor

microenvironment display immune inhibitory functions. *J. Immunother. Cancer*, *8*, e001054. https://doi.org/10.1136/jitc-2020-001054.

99. Shafabakhsh, R., Pourhanifeh, M. H., Mirzaei, H. R., Sahebkar, A., Asemi, Z., & Mirzaei, H. (2019). Targeting regulatory T cells by curcumin: A potential for cancer immunotherapy. *Pharmacol. Res.*, *147*, 104353. https://doi.org/10.1016/j.phrs.2019.104353.
100. Song, C. Y., Deng, L., Hu, H. Y., & Zhang, X. M. (2008). Effects of evodiamine on immune function in mice. *Journal of Chinese Materia Medica, 31*, 885–888. https://doi.org/10.3321/j.issn:1001-4454.2008.06.030.
101. Song, D. X., & Jiang, J. G. (2017). Hypolipidemic components from medicine food homology species used in China: Pharmacological and health effects. *Arch. Med. Res.*, *48*, 569–581. https://doi.org/10.1016/j.arcmed.2018.01.004.
102. Song, J. J., Fan, J. C., Wang, Y., & Zheng, X. Z. (2019). The protective effect of Chinese yam polysaccharide on spleen injury in mice induced by cyclophosphamide. *Asia-Pacific Traditional Medicine, 15*, 15–17. https://doi.org/10.11954/ytctyy.201902003.
103. Song, Y., Li, J., Cheng, Y., Lin, Z., He, B. Y., & Wang, C. Y. (2015). Lowering effect of evodiamine dispersible tablets on uric acid in chickens. *Chinese Journal of New Drugs, 24*, 1057–1060.
104. Sun, B. N., Yu, S., Zhao, D. Y., Guo, S. H., Wang, X. H., & Zhao, K. (2018). Polysaccharides as vaccine adjuvants. *Vaccine, 36*, 5226–5234. https://doi.org/10.1016/j.vaccine.2018.07.040.
105. Sun, R., Zhang, J. W., Liang, H., Wang, H. L., Zhang, S. P., & Zhi, F. (2017). Effect of ginseng saponin for serum immune factors in rats with type III prostatitis. *Journal of Sichuan Traditional Chinese Medicine, 35*(2), 54–56. https://doi.org/10.3969/j.issn:1672-1993.2020.06.030.
106. Sun, T. T., Gao, Y. H., Sun, Z., Lou, Y. J., & Zhou, H. Z. (2018). Research advances of Lycium barbarum polysaccharides. *Chinese Journal of Veterinary Drug, 52*, 75–80. https://doi.org/10.11751/ISSN.1002-1280.2018.12.12.
107. Tang, T., & He, B. X. (2013). Treatment of D-galactose induced mouse aging with Lycium barbarum polysaccharides and its mechanism study. *African Journal of Traditional, Complementary, and Alternative Medicines, 10*, 12–17. https://doi.org/10.4314/ajtcam.v10i4.3.
108. Tao, S., Zhao, Z., Zhang, X., Guan, X., Wei, J., Yuan, B., et al. (2020). The role of macrophages during breast cancer development and response to chemotherapy. *Clinical and Translational Oncology, 22*, 1938–1951. https://doi.org/10.1007/s12094-020-02348-0.

109. Tejada, S., Pinya, S., Martorell, M., Capó, X., Tur, J. A., Pons, A., et al. (2018). Potential anti-inflammatory effects of hesperidin from the genus citrus. *Current Medicinal Chemistry, 25*(37), 4929–4945. https://doi.org/10.2174/0929867324666170718104412.

110. Valentová, K., Šíma, P., Rybková, Z., Křížan, J., Malachová, K., & Křen, V. (2016). Antimutagenic and immunomodulatory properties of quercetin glycosides. *Journal of the Science of Food and Agriculture, 96*, 1492–1499. https://doi.org/10.1002/jsfa.7251.

111. Vivier, E., Artis, D., Colonna, M., Diefenbach, A., Di Santo, J. P., Eberl, G., et al. (2018). Innate lymphoid cells: 10 years on. *Cell, 174*, 1054–1066. https://doi.org/10.1016/j.cell.2018.07.017.

112. Wang, H., Chen, X. Y., Jing, J., Liu, X., Tian, C. L., & Liu, M. C. (2019). Progress on regulation effects of flavonoids on immune-related signaling pathways. *Progress in Veterinary Medicine, 40*(12), 102–105. https://doi.org/10.16437/j.cnki.1007-5038.2019.12.021.

113. Wang, J. T., Wang, H. C., & Liu, L. (2019). Progress in research on structure-function relationship of lentinan. *Food Science, 40*, 363–369. https://doi.org/10.7506/spkx10026630-20181016-162.

114. Wang, L., Lu, W., Ding, Y., & Qiao, F. F. (2020). Effect of pueraria injection combined with ribavirin on immune function and inflammatory factors in children with viral myocarditis. *International Medical Health Guidance News, 26*, 697–700. https://doi.org/10.3760/cma.j.issn.1007-1245.2020.05.030.

115. Wang, M., Guo, Z., Zhou, H. Y., Ge, B. J., Wang, Z., Li, H. T., et al. (2020). Study on the bi-directional regulation of anti-inflammatory immune response by total flavonoids of *Astragalus* in macrophages RAW264.7. *Chinese Journal of Preventive Veterinary Medicine, 42*(8), 822–829. https://doi.org/10.3969/j.issn.1008-0589.201912014.

116. Wang, Q., Li, H. R., Li, P. Y., Xiao, J. B., & Wang, B. L. (2019). Experimental study on effects of notoginsenoside R1 on cardiac function, lung tissue injury, and immune imbalance in rats with traumatic shock. *Medical Journal of West China, 21*, 1826–1836. https://doi.org/10.3969/j.issn.1672-3511.2019.12.005.

117. Wang, X., & Lin, Z. (2019). Immunomodulating effect of Ganoderma (lingzhi) and possible mechanism. *Advances in Experimental Medicine and Biology, 1182*, 1–37. https://doi.org/10.1007/978-981-32-9421-9_1.

118. Wang, Y., Fu, X., Zhang, C., & Ji, Y. T. (2020). Regulatory effect of curcumin on expression to relevant cytokines in colonic tissue about colitis mice. *Journal of Tianjin University of Traditional Chinese Medicine, 39*(6), 686–689. https://doi.org/10.11656/j.issn.1673-9043.2020.06.18.

119. Wang, Y., Zhang, Q., Chen, Y., Liang, C. L., Liu, H., Qiu, F., et al. (2020). Antitumor effects of immunity-enhancing traditional Chinese medicine. *Biomedicine & Pharmacotherapy, 121*, 109570. https://doi.org/10.1016/j.biopha.2019.109570.

120. Wang, Z. H., Lei, M. S., Peng, S., Zheng, Y., Shi, L. J., & Peng, M. J. (2016). Effects of flavonoid and flavone from *Eucommia ulmoides* Oliv. on proliferation of splenocytes and production of IL-2 and IFN-γ in mice. *Natural Product Research and Development, 28*(4), 514–518+489. https://doi.org/10.16333/j.1001-6880.2016.4.008.

121. Wei, X. C., Chen, T., Sun, J. J., Luo, J. Y., Ni, Y. C., Shu, G., et al. (2019). Effects of ginseng leaf polysaccharide on immune performance of Yuehuang chickens during 1 to 28 days of age. *Chin. J. Anim. Nutr., 31*, 2330–2339. https://doi.org/10.3969/j.issn.1006-267x.2019.05.040.

122. Williams, A. R., Klaver, E. J., Laan, L. C., Ramsay, A., Fryganas, C., Difborg, R., et al. (2017a). Co-operative suppression of inflammatory responses in human dendritic cells by plant proanthocyanidins and products from the parasitic nematode *Trichuris suis*. *Immunology, 150*, 312–328. https://doi.org/10.1111/imm.12687.

123. Williams, A. R., Krych, L., Fauzan Ahmad, H., Nejsum, P., Skovgaard, K., Nielsen, D. S., et al. (2017b). A polyphenol-enriched diet and *Ascaris suum* infection modulate mucosal immune responses and gut microbiota composition in pigs. *PLOS One, 12*(10), e0186546. https://doi.org/10.1371/journal.pone.0186546.

124. Xing, J., Zhang, Z. Q., Luo, K. K., Tang, X. Q., Sheng, X. Z., & Zhan, W. B. (2020). T and B lymphocytes immune responses in flounder (*Paralichthys olivaceus*) induced by two forms of outer membrane protein K from *Vibrio anguillarum*: Subunit vaccine and DNA vaccine. *Mol. Immunol., 118*, 40–51. https://doi.org/10.1016/j.molimm.2019.12.002.

125. Xu, H., Niu, Y., Hong, W., Liu, W., Zuo, X., Bao, X., et al. (2020c). Development of a water-in-oil-in-water adjuvant for foot-and-mouth disease vaccine based on ginseng stem-leaf saponins as an immune booster. *Comp. Immunol. Microbiol. Infect. Dis., 71*, 101499. https://doi.org/10.1016/j.cimid.2020.101499.

126. Xu, W., Fang, S. J., Guan, R., Zhang, C. R., & Hu, S. H. (2020b). Immunomodulatory effect of *Atractylodis macrocephalae* Koidz. polysaccharides on mouse lymphocytes. *Chin. J. Immunol., 36*, 1573–1577. https://doi.org/10.3969/j.issn.1000-484X.2020.13.007.

127. Xu, X. Y., Rui, S. Z., Chen, C., Zhang, G. C., Li, Z., Wang, J. H., et al. (2019). Protective effects of astragalus polysaccharide nanoparticles on septic cardiac dysfunction through inhibition of TLR4/NF-kB signaling

pathway. *Int. J. Biol. Macromol.*, *153*, 977–985. https://doi.org/10.1016/j.ijbiomac.2019.10.227.

128. Xu, Y. Z., Pang, H. G., Li, H. Y., & Fan, J. C. (2020a). Research on anti-tumor and immunomodulatory effects of yam polysaccharides on tumor mice. *J. Med. Forum*, *41*(8–10), 15. https://doi.org/10.3969/j.issn.1006-267x.2017.12.008.
129. Xue, M., Sun, H., Cao, Y., Wang, G., Meng, Y., Wang, D., et al. (2015). Mulberry leaf polysaccharides modulate murine bone-marrow-derived dendritic cell maturation. *Hum. Vaccin. Immunother.*, *11*, 946–950. https://doi.org/10.1080/21645515.2015.1011977.
130. Yang, J., Sha, J. D., Gao, X., Lin, S. T., Sun, T. T., Tian, C., et al. (2017). Immune regulation function of flavonoids and its mechanisms. *Chin. J. Anim. Nutr.*, *29*, 4295–4300. https://doi.org/10.3969/j.issn.1006-267x.2017.12.008.
131. Yu, X. C., Zhang, L. S., & Wang, F. J. (2016). Recent advances on immune regulation of herbal polysaccharide. *Acta Univ. Trad. Med. Sinensis Pharmacologiaeque Shanghai*, *30*, 97–101. https://doi.org/10.16306/j.1008-861x.2016.03.022.
132. Zhang, B., Nan, T. G., Xin, J., Zhan, Z. L., Kang, L. P., Yuan, Y., et al. (2019). Development of a colloidal gold-based lateral flow dipstick immunoassay for rapid detection of chlorogenic acid and luteoloside in *Flos Lonicerae Japonicae*. *J. Pharm. Biomed. Anal.*, *170*, 83–88. https://doi.org/10.1016/j.jpba.2019.03.035.
133. Zhang, K., Ge, Z. Z., Xue, Z. Q., Huang, W. J., Mei, M., Zhang, Q., et al. (2015). Chrysin suppresses human CD14(+) monocyte-derived dendritic cells and ameliorates experimental autoimmune encephalomyelitis. *J. Neuroimmunol.*, *288*, 13–20. https://doi.org/10.1016/j.jneuroim.2015.08.017.
134. Zhang, L. H., Wu, T. T., Zhao, L. G., Xiao, W., Zhang, S. H., Wang, C. Y., et al. (2019b). Advances in studies on anticancer activity of flavonoids from ginkgo biloba extract. *Chin. Pharm. J.*, *54*(06), 444–449. https://doi.org/10.11669/cpj.2019.06.003.
135. Zhang, L. L., Wang, Y., & Liu, L. (2020). Research progress of cytokines and inflammatory immune diseases. *Pharm. Clin. Res.*, *28*(03), 202–205. https://doi.org/10.13664/j.cnki.pcr.2020.03.011.
136. Zhang, L., Wang, S., Liu, Z., Zhang, L., Wang, S., & Wang, B. (2017). Procyanidin, a kind of biological flavonoid, induces protective anti-tumor immunity and protects mice from lethal B16F10 challenge. *Int. Immunopharmacol.*, *47*, 251–258. https://doi.org/10.1016/j.intimp.2017.04.007.

137. Zhang, X. X., Yao, K. N., Ren, L. H., Chen, T., & Yao, D. G. (2016). Protective effect of Astragalus polysaccharide on endothelial progenitor cells injured by thrombin. *Int. J. Biol. Macromol.*, *82*, 711–718. https://doi.org/10.1016/j.ijbiomac.2015.09.051.

138. Zhang, Y., Zhang, L., Zhang, Y., Xu, J. J., Sun, L. L., & Li, S. Z. (2016). The protective role of liquiritin in high fructose-induced myocardial fibrosis via inhibiting NF-κB and MAPK signaling pathway. *Biomed. Pharmacother.*, *84*, 1337–1349. https://doi.org/10.1016/j.biopha.2016.10.036.

139. Zhang, Y., Zhong, X. M., Yang, J., Dong, J. L., & Tian, L. (2020b). Effect of puerarin on the expression of inflammatory cytokines in gouty arthritis mice. *Immunol. J.*, *36*(10), 903–907+913. https://doi.org/10.13431/j.cnki.immunol.j.20200140.

140. Zhao, B., Lv, C., & Lu, J. (2019). Natural occurring polysaccharides from *Panax ginseng* C. A. meyer: A review of isolation, structures, and bioactivities. *Int. J. Biol. Macromol.*, *133*, 324–336. https://doi.org/10.1016/j.ijbiomac.2019.03.229.

141. Zhao, R., Hao, W. L., Ma, B. L., & Chen, Z. B. (2015). Improvement effect of Lycium barbarum polysaccharide on subhealth mice. *Iran J. Basic Med. Sci.*, *18*, 1245–1252. https://doi.org/10.22038/IJBMS.2015.6281.

142. Zhao, X. J., Yang, R. L., Bi, Y., Bilal, M., Kuang, Z., Iqbal, H. M. N., et al. (2019). Effects of dietary supplementation with mulberry (*Morus alba* L.) leaf polysaccharides on immune parameters of weanling pigs. *Animals (Basel)*, *10*, 35. https://doi.org/10.3390/ani10010035.

143. Zheng, Y. J., Ren, W. Y., Zhang, L. N., Zhang, Y. M., Liu, D. L., & Liu, Y. Q. (2020). A review of the pharmacological action of Astragalus polysaccharide. *Front. Pharmacol.*, *11*, 349. https://doi.org/10.3389/fphar.2020.00349.

144. Zhou, L. H., Yao, T., Guo, A. L., Lin, J. J., Pan, S. Q., & Chang, Y. M. (2020). Progress in the study of two-way immune regulation of traditional Chinese medicine in recent ten years. *J. Basic Chin. Med.*, *26*, 1016–1033.

145. Zhou, L. J., Long, T. T., Zhou, X., & Bao, Y. X. (2017). Immunomodulatory effects of *Acanthopanax Senticosus* polysaccharides on Lewis tumor-bearing mice through TLR4 signaling pathway. *Chin. J. Immunol.*, *33*, 849–853. https://doi.org/10.3969/j.issn.1000-484X.2017.06.009.

146. Zou, Y. F., Zhang, Y. Y., Fu, Y. P., Inngjerdingen, K. T., Paulsen, B. S., Feng, B., et al. (2019). A polysaccharide isolated from *Codonopsis pilosula* with immunomodulation effects both *in vitro* and *in vivo*. *Molecules*, *24*, 3632. https://doi.org/10.3390/molecules24203632.

INDEX

K

L

M

N

O